Springer Undergraduate Mathematics Series

For further volumes:
http://www.springer.com/series/3423

Frazer Jarvis

Algebraic Number Theory

 Springer

Frazer Jarvis
School of Mathematics and Statistics
University of Sheffield
Sheffield
UK

ISSN 1615-2085   ISSN 2197-4144   (electronic)
ISBN 978-3-319-07544-0 ISBN 978-3-319-07545-7   (eBook)
DOI 10.1007/978-3-319-07545-7
Springer Cham Heidelberg New York Dordrecht London

Library of Congress Control Number: 2014941941

Mathematics Subject Classification: 11Rxx, 11R04, 11R09, 11R18, 11R27, 11R29, 11R47, 11-01

Printed on acid-free paper

Springer is part of Springer Science+Business Media (www.springer.com)

Preface

This book, like others in the SUMS series, is designed to be suitable for undergraduate courses. While many institutions may not offer such courses in algebraic number theory, some may instead offer the possibility for students to take reading courses, or to write projects, in more diverse areas of pure mathematics. It is my hope that this book is suitable for any of these options.

Outline of the Book

Let us now summarise the content of the book. The first chapter begins with a review of the Euclidean algorithm and its importance in the proof of the Fundamental Theorem of Arithmetic, which states that every integer can be uniquely factorised into prime numbers. We then consider some questions about the integers which can be addressed by working with the Gaussian integers $\mathbb{Z}[i]$, and for which analogues of unique factorisation are essential to the proof.

Motivated by the need to consider larger sets than $\mathbb{Z}$, the second chapter defines algebraic numbers, and number fields and their rings of integers. These are the sets which are going to generalise the rational numbers and the integers and in which we are going to be studying arithmetic throughout the rest of the book. The third chapter continues this theme, studying natural questions about rings of integers in number fields. Some of these further questions are of a harder nature than those in Chap. 2, and the reader (or instructor) may wish to omit some of this material on a first reading.

Unique factorisation forms the topic of the next two chapters; we would like to know whether the ring of integers in a given number field has unique factorisation or not, in order to deduce interesting results such as those at the end of the first chapter. Although unique factorisation does not always hold (indeed, it's quite rare!), Kummer suggested a way to recover it, by adding "ideal numbers". Later, Dedekind reformulated this to give a definition of ideals in the rings of integers, and Chap. 4 considers properties of ideals. Prime ideals, analogues of prime numbers, are studied in the fifth chapter, whose main result is the uniqueness of factorisation of ideals into prime ideals. The chapter also introduces the class number and class group, which measure the failure of unique factorisation of elements.

Chapter 6 considers the case of imaginary quadratic fields, which is probably the simplest family of worked examples. After considering problems such as finding the imaginary quadratic fields with unique factorisation, the relation with the theory of positive definite quadratic forms is explained, which gives a convenient way to make computations of class numbers of imaginary quadratic fields.

Before considering further examples, we develop some theory in Chap. 7. This chapter studies geometry of numbers and the class group, proving some useful general theoretical results, which then get illustrated in Chap. 8, which focuses on other classes of fields with small degree. To treat the case of real quadratic fields, we introduce Pell's equation, and its solution via continued fractions. Next, Chap. 9 considers number fields generated by roots of unity, and in particular, Kummer's work on Fermat's Last Theorem.

These nine chapters are already more than enough to form a full one-semester course in a typical UK University. But this is just the start of the theory, almost all understood by the mid-nineteenth century. I have also included additional chapters, which could form the basis for further reading, or for student projects. These concern explicit class number formulae, and the number field sieve. The first chapter, on analytic methods, considers the Riemann zeta function, and zeta functions of number fields, culminating in the analytic class number formula, and a brief discussion of p-adic numbers, of great importance for modern number theorists. The advent of the computer has led to increased interest in the problem of factorisation of integers, through the RSA cryptosystem, and the current method of choice, which uses many techniques of algebraic number theory, is the number field sieve. But today the field is extremely active, and has expanded far beyond the contents of this book. I am very conscious that there are topics which might have been treated as part of this list which have been omitted: the relation to Galois Theory, further ramification theory, class field theory, p-adic methods and transcendence theory, amongst others.

A Note for Instructors

There are many possible combinations of topics that could form a suitable course. A basic one-semester course at an undergraduate level at a typical UK University might review Chap. 1 quickly, spend quite a lot of time on Chap. 2, briefly touch on some of the topics in Chap. 3 and cover Chaps. 4 and 5 fairly fully, with examples taken from parts of Chaps. 6 and 8. Any remaining time might be spent on additional topics from some of the later chapters, such as quadratic forms (from Chap. 6), the finiteness of the class number (from Chap. 7), or roots of unity and Fermat's Last Theorem (from Chap. 9). The relationship between class numbers of imaginary quadratic fields and quadratic forms (Chap. 6), analytic methods (Chap. 10) and the number field sieve (Chap. 11) are more likely to be suitable for undergraduate project work. Some of this material is also likely to be suitable for beginning graduate students.

Prerequisites

I have tried to keep the prerequisites to a minimum. Nevertheless, there are a number of topics which might be useful, and which most readers will have met in their previous studies. It is my intention that readers should be able to read most of the text without previous courses in these topics, but it is inevitably true that previous acquaintance with them will be helpful, and readers without this knowledge may occasionally need to consult additional sources for this background. These topics include: elementary number theory (here [7] is an excellent source), a first course in linear algebra (including vector spaces over fields), a first course in group theory (little beyond Lagrange's Theorem is required) and some basic understanding of rings and fields. Even here, it should be possible for a reader without this knowledge to follow the material, with the understanding that a ring is just a set in which one can add, subtract and multiply, in such a way that all the algebraic properties of $\mathbb{Z}$ are satisfied (in particular, all our rings are commutative and have a multiplicative identity). Slightly more advanced concepts such as ideals and quotient rings are an essential part of algebraic number theory (and were, in fact, motivated by the theory, as we will see), and they are introduced from scratch. Similarly, I hope that readers would be able to follow the text with the idea that a field is just a set in which one can add, subtract, multiply and divide by non-zero elements, in such a way that all the algebraic properties satisfied by $\mathbb{Q}$ continue to hold.

A Note on Computer Packages

It was a difficult decision to exclude explicit mention of computer packages within the text. There is considerable scope for using computer packages to study algebraic number theory; many have been written and new ones continue to appear; had I written the book 5 years ago, I might well have written examples with a different package to my package of choice today, and 10 years ago I might have preferred a different choice again.

But it would seem to be a mistake to write an undergraduate text with no mention of resources available for students to study the subject further, or to develop understanding by finding interesting examples with the aid of a computer package. Partly to make amends, here is a brief guide to available resources at the time of writing.

An excellent list of packages and online tables of examples can be found at the *Number Theory Web*; see http://www.numbertheory.org/ntw/N1.html. All the packages listed here are linked to from this page.

Algebraic number theory is a relatively specialist area, and is not so well served by well-known general-purpose computer packages such as Maple or Mathematica, although both have their uses. The MAGMA Computational Algebra System (http://www.magma.maths.usyd.edu.au/magma/) has the best treatment of algebraic number theory of the major packages; however, it is comparatively expensive, and less likely to be installed on University computer systems.

However, there are free alternatives which have been particularly developed for use in algebraic number theory. (Of course, the disadvantage of such programs is that there may be little in the way of documentation or support.)

One such program is PARI-GP (http://www.pari.math.u-bordeaux.fr/), developed originally by Henri Cohen (Bordeaux) and a team of co-workers. Cohen has also written several nice textbooks on computational number theory.

More recently, William Stein has been heading the SAGE project to develop an open-source equivalent to Maple, Mathematica and MAGMA. It can be downloaded from http://www.sagemath.org/; the project incorporates parts of PARI-GP.

Acknowledgments

My own University currently has no undergraduate course in *Algebraic Number Theory*. I have, however, been very fortunate to have been able to supervise a number of undergraduate projects on topics covered by this book, and I would like to thank those students whose projects have helped me reflect on ways to present some of this material: Emma Aitken, Callum Argyle, Fran Barrett, Emma Bentley, Huw Birch, Hannah Cooke, Anna Duckett, Matt Grogan, Christian Jones and Jacqueline Lyons.

I am also grateful to those who read parts of the manuscript and made helpful suggestions, particularly to Tobias Berger and Daniel Fretwell. The anonymous referees for the book sent many helpful suggestions, and I am very grateful to them for their time.

This book has been some time in writing; my thanks to successive editors at Springer for their patience. I would like to thank the University of Sheffield for encouraging me to write this book, and particularly my colleagues and students in the number theory group, Tobias Berger, Neil Dummigan and Jayanta Manoharmayum; Daniel Fretwell, Andrew Jones and Konstantinos Tsaltas.

I would like to thank all those who developed my interest and knowledge in mathematics in general, and algebraic number theory in particular.

Above all, I would like to thank my parents and family for their love and support. Mostly I would like to thank my wife, Nicole, and children, Martin and Laura, the latter two of whom did what they could to stop this book ever being finished.

Contents

1 Unique Factorisation in the Natural Numbers   1
   1.1 The Natural Numbers   2
   1.2 Euclid's Algorithm   3
   1.3 The Fundamental Theorem of Arithmetic   8
   1.4 The Gaussian Integers   9
   1.5 Another Application of the Gaussian Integers 14

2 Number Fields 17
   2.1 Algebraic Numbers 17
   2.2 Minimal Polynomials 20
   2.3 The Field of Algebraic Numbers 21
   2.4 Number Fields 25
   2.5 Integrality 28
   2.6 The Ring of All Algebraic Integers 31
   2.7 Rings of Integers of Number Fields 35

3 Fields, Discriminants and Integral Bases 39
   3.1 Embeddings 40
   3.2 Norms and Traces 44
   3.3 The Discriminant 47
   3.4 Integral Bases 51
   3.5 Further Theory of the Discriminant 53
   3.6 Rings of Integers in Some Cubic and Quartic Fields 57
      3.6.1 $K = \mathbb{Q}(\sqrt{2}, \sqrt{3})$ 57
      3.6.2 $K = \mathbb{Q}(\sqrt{-2}, \sqrt{-5})$ 59
      3.6.3 $K = \mathbb{Q}(\sqrt[3]{2})$ 60
      3.6.4 $K = \mathbb{Q}(\sqrt[3]{175})$ 61

4 Ideals 65
   4.1 Uniqueness of Factorisation Revisited 66
   4.2 Non-unique Factorisation in Quadratic Number Fields 67
   4.3 Kummer's Ideal Numbers 71
   4.4 Ideals 73
   4.5 Generating Sets for Ideals 76

4.6   Ideals in Quadratic Fields   79
4.7   Unique Factorisation Domains and Principal
       Ideal Domains   81
4.8   The Noetherian Property   84

5   Prime Ideals and Unique Factorisation   87
5.1   Some Ring Theory   87
5.2   Maximal Ideals   94
5.3   Prime Ideals.   96
5.4   Unique Factorisation into Prime Ideals   99
5.5   Coprimality   102
5.6   Norms of Ideals   104
5.7   The Class Group   105
5.8   Splitting of Primes   107
5.9   Primes in Quadratic Fields.   112

6   Imaginary Quadratic Fields   113
6.1   Units.   113
       6.1.1 $d \equiv 2, 3 \pmod 4$ 114
       6.1.2 $d \equiv 1 \pmod 4$ 114
       6.1.3   Summary.   115
6.2   Euclidean Imaginary Quadratic Fields.   116
       6.2.1 $d \equiv 2, 3 \pmod 4$ 116
       6.2.2 $d \equiv 1 \pmod 4$ 117
6.3   Quadratic Forms.   120
6.4   Reduction Theory.   123
6.5   Class Numbers and Quadratic Forms   131
       6.5.1 $d \equiv 2, 3 \pmod 4$ 134
       6.5.2 $d \equiv 1 \pmod 4$ 142
6.6   Counting Quadratic Forms.   143

7   Lattices and Geometrical Methods   149
7.1   Lattices   149
7.2   Geometry of Number Fields.   153
7.3   Finiteness of the Class Number   158
7.4   Dirichlet's Unit Theorem.   164

8   Other Fields of Small Degree   169
8.1   Continued Fractions   170
8.2   Continued Fractions of Square Roots   176
8.3   Real Quadratic Fields   180
8.4   Biquadratic Fields.   184
8.5   Cubic Fields   188

9   Cyclotomic Fields and the Fermat Equation   191
       9.1   Definitions.   191
       9.2   Discriminants and Integral Bases   195
       9.3   Gauss Sums and Quadratic Reciprocity   198
       9.4   Remarks on Fermat's Last Theorem   202

10   Analytic Methods   207
       10.1   The Riemann Zeta Function.   207
       10.2   The Functional Equation of the Riemann Zeta Function   211
       10.3   Zeta Functions of Number Fields   214
       10.4   The Analytic Class Number Formula   215
       10.5   Explicit Class Number Formulae   223
       10.6   Other Embeddings   226

11   The Number Field Sieve   231
       11.1   The RSA Cryptosystem and the Problem of Factorisation   231
       11.2   The Quadratic Sieve   233
       11.3   The Number Field Sieve: A First Example   237
       11.4   Index Calculus   238
       11.5   Prime Ideals and the Algebraic Factorbase   241
       11.6   Further Obstructions   244
       11.7   The General Case   248
       11.8   Closing Comments   253

Appendix A: Solutions and Hints to Exercises   257

Index   289

Chapter 1
Unique Factorisation in the Natural Numbers

We are so used to working with the natural numbers from infancy onwards that we take it for granted that natural numbers may be factorised uniquely into prime numbers. For example, $360 = 2^3 3^2 5$ is *the* prime factorisation of 360. However, we should notice that there are already senses in which this factorisation is not really unique; we can write $360 = 2 \times 3 \times 5 \times 2 \times 3 \times 2$, or even $360 = (-2) \times 5 \times 3 \times (-3) \times 2 \times 2$. Nevertheless, we can see that all these factorisations are "essentially the same", in a way which we could make precise, and we will do so later.

It was not until the early nineteenth century that mathematicians became aware that this uniqueness of factorisation is actually a rather special property of the natural numbers, and that it required a proof. It seems that Gauss was aware of this around 1800, but, as with much of modern number theory, the issue was brought to the fore as a consequence of work on Fermat's Last Theorem. Let's recall the statement:

Conjecture 1.1 (Fermat's Last Theorem) *The equation $x^n + y^n = z^n$ has no solutions with x, y and z positive integers when $n \geq 3$.*

Fermat seems to have made this conjecture around 1640; it was not finally proven until work of Andrew Wiles, partly with Richard Taylor, in 1994.

Since Fermat proved the result for $n = 4$, it is easy to see that it suffices to treat the case where $n = p$, an odd prime. Then we can write the equation in the conjecture as

$$\prod_{m=0}^{p-1} (x + e^{\frac{2\pi i m}{p}} y) = z^p.$$

As we shall see in Chap. 9, there is a rather simple "proof" of many special cases of this conjecture if we have some sort of unique factorisation statements in the *cyclotomic fields* $\mathbb{Q}(e^{\frac{2\pi i}{p}})$. Unfortunately, it turns out that these cyclotomic fields do not, in general, have such a unique factorisation property. This failure of unique factorisation led Kummer to develop the theory of ideals, and factorisation of ideals, which is the starting point for algebraic number theory.

F. Jarvis, *Algebraic Number Theory*, Springer Undergraduate
Mathematics Series, DOI: 10.1007/978-3-319-07545-7_1,
© Springer International Publishing Switzerland 2014

At the end of this chapter, we will see further examples where naturally arising problems about the integers $\mathbb{Z}$ can be addressed using arithmetic in a larger set, and where unique factorisation in this set is a crucial requirement.

We will begin, therefore, with a reminder of the main definitions and results concerning the uniqueness of factorisation into primes for the natural numbers. This will also provide useful clues when it comes to generalisations to other settings.

1.1 The Natural Numbers

In this book, the natural numbers will be taken to be $\mathbb{N} = \{1, 2, 3, \ldots\}$. In order to study factorisation in the natural numbers, we need some basic definitions.

Definition 1.2 Let a and b be integers. Then b *divides* a, or b is a *factor* or *divisor* of a, if $a = bc$ for some integer c. Write $b|a$ to mean that b divides a and $b \nmid a$ to mean that b does not divide a.

When b divides a, we also say that a is a *multiple* of b.

Thus $17|323$ as $323 = 17 \times 19$, but $17 \nmid 324$. For all $a \in \mathbb{Z}$, we have $1|a$ and $a|a$ (since $a = a \times 1$) and $a|0$ (since $0 = a \times 0$). Notice that $0 \nmid a$, if $a \neq 0$. If $b|a$, then $-b|a$, so the non-zero divisors of an integer occur naturally in pairs. Clearly if $b \neq 0$, then $b|a$ means that the remainder when a is divided by b is 0.

Definition 1.3 A *prime number* is a natural number $p > 1$ which is not divisible by any natural number other than 1 and p itself.

A *composite number* is a natural number $n \neq 1$ which is divisible by natural numbers other than 1 and itself. Notice that 1 is neither prime nor composite.

For example, 37 is prime, but $39 = 3 \times 13$ is not.

Theorem 1.4 (Euclid) *There are infinitely many prime numbers.*

Proof Suppose that there are only finitely many primes, $p_1, p_2, \ldots, p_n$.
Define
$$N = p_1 p_2 \ldots p_n + 1.$$

Suppose that p is a prime factor of N. None of the primes $p_1, \ldots, p_n$ is a factor of N, as N is 1 more than a multiple of each p_i. So p is not one of the primes $p_1, \ldots, p_n$, and we have found another prime, which contradicts our assumption that $p_1, \ldots, p_n$ are all the primes. So there must be infinitely many primes.   □

The importance of prime numbers is that they are the building blocks multiplicatively for all of the natural numbers; every natural number will be the product of prime numbers.

Exercise 1.1

1. Adapt the proof of Theorem 1.4, using $N = 4p_1 \ldots p_n - 1$, to show that there are infinitely many primes $p \equiv 3 \pmod 4$. (Recall that $a \equiv b \pmod m$ is equivalent to $m \mid a - b$.)
2. Similarly, show that there are infinitely many primes $p \equiv 5 \pmod 6$, using $N = 6p_1 \ldots p_n - 1$.

Exercise 1.2

1. Let x be an even integer. Show that any prime divisor of $x^2 + 1$ is necessarily of the form 1 (mod 4). [*Hint. Explain that the congruence $x^2 + 1 \equiv 0 \pmod p$ shows that x is an element of order 4 in the group of non-zero integers modulo p, and then use Lagrange's Theorem.*]
2. Adapt the proof of Theorem 1.4, using $N = (2p_1 \ldots p_n)^2 + 1$, to show that there are infinitely many primes $p \equiv 1 \pmod 4$.

Exercise 1.3 Using the method of the previous exercise with the polynomial $x^2 + x + 1$, where x is an integer divisible by 6, show that there are infinitely many prime numbers $p \equiv 1 \pmod 6$.

1.2 Euclid's Algorithm

The notion of common divisors, and especially of the greatest common divisor turns out to be surprisingly important. Indeed, the existence of the greatest common divisor, which for $\mathbb{Z}$ is a corollary of Euclid's algorithm, is the key to proving unique factorisation of integers into prime numbers. Although most readers will have seen this as part of a course in elementary number theory, we dwell on it a little here for two reasons: firstly because we want to stress its importance for unique factorisation, and secondly because we will want to generalise the arguments later.

If two integers a and b are both multiples of another integer c, so that $c \mid a$ and $c \mid b$, then c is a *common factor* of a and b. For example, 8 and 36 have common factors ± 1, ± 2 and ± 4. The *highest common factor* will just be the largest of the common factors; for example, the highest common factor of 8 and 36 is 4. Unless both a and b are zero, there will be a highest common factor of a and b, and we write this as (a, b). Here's a more formal definition.

Definition 1.5 The integer $h = (a, b)$ is a *highest common factor* (or *greatest common divisor*) of given integers a, b if

1. $h \mid a$ and $h \mid b$ (so that h really is a common factor of a and b);
2. if $c \mid a$ and $c \mid b$, then $c \le h$ (in words: if c is a common factor of a and b, then c is at most h).

Clearly $(a, b) = (b, a)$ and $(0, b) = (b, b) = |b|$ when b is non-zero. The case where a and b have no common factor except ± 1 will occur particularly often, and we have special terminology:

Definition 1.6 Let a and b be integers. Say that a and b are *coprime* (or *relatively prime*) if $(a, b) = 1$, i.e., a and b have no common factor except $(\pm)1$.

An integer a can always be divided by a positive integer b to give a unique quotient q and a unique remainder r in the range $0 \leq r < b$, i.e.,

$$a = qb + r \qquad (0 \leq r < b).$$

The quotient and remainder are always assumed to be integers. Thus, for example,

$$78 = 8(9) + 6 \quad \text{and} \quad -78 = -9(9) + 3.$$

The simple process of finding a quotient and remainder is known as the *division algorithm*.

The key theoretical result is the following simple lemma:

Lemma 1.7 *Suppose that $a = qb + r$. Then $(a, b) = (b, r)$.*

Proof Suppose that d divides a and b. Then, since $r = a - qb$, we also have $d|r$. Thus every common divisor of a and b also divides r. In particular, (a, b) divides r, and since it also divides b, we see that (a, b) is a common factor of b and r. Therefore $(a, b) \leq (b, r)$, as (b, r) is the *highest* common factor of b and r.

Conversely, any common divisor of b and r also divides $a = qb + r$. In particular, (b, r) divides a, and as in the first paragraph, we conclude that $(b, r) \leq (a, b)$.

Combining these inequalities, we see that $(a, b) = (b, r)$, as required. $\square$

Repeatedly applying the division algorithm gives Euclid's algorithm, which allows us to compute highest common factors extremely efficiently.

Example 1.8 Let's work out the highest common factor of 630 and 132. The division algorithm gives

$$630 = 4 \times 132 + 102.$$

But now Lemma 1.7 gives $(630, 132) = (132, 102)$. It follows that we just have to work out the highest common factor of these two smaller numbers, 132 and 102. But we can repeat the division algorithm:

$$132 = 1 \times 102 + 30,$$

and Lemma 1.7 gives $(132, 102) = (102, 30)$. Repeat again: the division algorithm gives

$$102 = 3 \times 30 + 12,$$

and Lemma 1.7 gives $(102, 30) = (30, 12)$. Next the division algorithm gives

$$30 = 2 \times 12 + 6,$$

and Lemma 1.7 gives $(30, 12) = (12, 6)$. Finally, the division algorithm gives

$$12 = 2 \times 6 + 0,$$

and Lemma 1.7 gives $(12, 6) = (6, 0)$. We've already explained that the highest common factor of 0 and b is just $|b|$, so $(6, 0) = 6$. We conclude that

$$(630, 132) = (132, 102) = (102, 30) = (30, 12) = (12, 6) = (6, 0) = 6,$$

and so we have shown that $(630, 132) = 6$.

It is usual to write these equations in tabular form:

$$
\begin{aligned}
630 &= 4 \times 132 + 102 \\
132 &= 1 \times 102 + 30 \\
102 &= 3 \times 30 + 12 \\
30 &= 2 \times 12 + 6 \\
12 &= 2 \times 6 + 0
\end{aligned}
$$

and the argument shows that the highest common factor is the last non-zero remainder.

We get interesting information by running the algorithm backwards too. We can use all the lines above to write the highest common factor as the sum of a multiple of 630 and a multiple of 132:

$$
\begin{aligned}
6 &= 30 - 2 \times 12 \\
&= 30 - 2 \times (102 - 3 \times 30) = 7 \times 30 - 2 \times 102 \\
&= 7 \times (132 - 1 \times 102) - 2 \times 102 = 7 \times 132 - 9 \times 102 \\
&= 7 \times 132 - 9 \times (630 - 4 \times 132) = 43 \times 132 - 9 \times 630.
\end{aligned}
$$

Thus the highest common factor of 630 and 132 has been written in the form $630s + 132t$ for certain integers s and t. At each stage, we use one of the equations in the forward direction of Euclid's algorithm to express the highest common factor using numbers appearing at earlier steps in the algorithm.

The same argument, applied more generally, gives the general result:

Theorem 1.9 *Let $a, b \in \mathbb{Z}$, with $b \neq 0$. Then there exist $s, t \in \mathbb{Z}$ such that $(a, b) = sa + tb$.*

Exercise 1.4

1. In Euclid's Algorithm, show that if r_n, r_{n+1} and r_{n+2} are three consecutive remainders, then $r_{n+2} < r_n/2$.
2. Deduce that if $a > b$, then $r_{2n} < b/2^n$.
3. Deduce that if $a > b$, Euclid's Algorithm terminates after at most $2 \log_2 b$ steps.

Exercise 1.5 We can modify the algorithm to allow negative remainders, by using a division algorithm of the form

$$a = \tilde{q}b + \tilde{r},$$

where $-b/2 < \tilde{r} \le b/2$.

1. Use this modified algorithm to compute the highest common factor of 630 and 132.
2. Show that if $a > b$, the modified Euclid's Algorithm terminates after at most $\log_2 b$ steps.

Exercise 1.6 Explain that if $f(X)$ and $g(X)$ are in $\mathbb{Q}[X]$, the polynomials with coefficients in $\mathbb{Q}$, then there is a similar division algorithm (coming from long division of polynomials): there are polynomials $q(X)$ and $r(X)$ in $\mathbb{Q}[X]$ with

$$f(X) = q(X)g(X) + r(X)$$

with the degree of the polynomial $r(X)$ strictly less than the degree of $g(X)$, or $r(X) = 0$.

Extend this in the obvious way to give a Euclidean algorithm, and find the highest common factor of $x^5 + 4x^4 + 10x^3 + 15x^2 + 14x + 6$ and $x^4 + 4x^3 + 9x^2 + 12x + 9$.

Recall that integers a and b are said to be *coprime* or *relatively prime* if their highest common factor (a, b) is 1. Thus, if a and b are coprime, there exist integers s and t such that $sa + tb = 1$. Conversely, if there exist s and t such that $sa + tb = 1$, a common factor of a and b will divide $sa + tb$, and so will divide 1. This implies that a and b are coprime. Thus:

Corollary 1.10 *Let $a, b \in \mathbb{Z}$. Then a and b are coprime if and only if there exist integers s and t such that $sa + tb = 1$.*

Exercise 1.7 Using the Euclidean Algorithm, show that 999 and 700 are coprime, and find integers s and t such that $999s + 700t = 1$.

We can use this result to prove several elementary properties of the highest common factor which shouldn't be surprising to you, given your long experience with the natural numbers!

Corollary 1.11 *Let $a, b \in \mathbb{Z}$, not both zero. If $h = (a, b)$, then a/h and b/h are coprime.*

Proof By Euclid's algorithm, there exist integers s and t such that $sa + tb = (a, b)$ $= h$, so that $s(a/h) + t(b/h) = 1$. Then a/h and b/h are coprime. $\square$

Another easy result is the following:

Lemma 1.12 *Suppose that* $(a, bc) = 1$. *Then* $(a, b) = 1$ *and* $(a, c) = 1$.

Proof If a and bc are coprime, then there are integers s and t such that

$$sa + tbc = 1.$$

But then

$$sa + (tc)b = 1.$$

Put $m = tc$, so that there are integers s and m with $sa + mb = 1$. It follows that a and b must be coprime. Similarly, a and c are coprime, using the bracketing $sa + (tb)c = 1$. □

In fact, the converse is also true—if a is coprime to b and to c, then it is coprime to bc.

Lemma 1.13 *Suppose that* $(a, b) = 1$ *and* $(a, c) = 1$. *Then* $(a, bc) = 1$.

Proof If $(a, b) = 1$, then there are integers s and t so that

$$sa + tb = 1.$$

If also $(a, c) = 1$, then there are integers p and q so that

$$pa + qc = 1.$$

Rearrange these:

$$tb = 1 - sa$$
$$qc = 1 - pa$$

and multiply:

$$(tq)bc = 1 - sa - pa + spa^2 = 1 - (s + p - spa)a.$$

Put $m = s + p - spa$ and $n = tq$; then the equation becomes $ma + nbc = 1$, and so a and bc are coprime. □

The last of our simple results is surprisingly important. We'll use it in the next section to deduce the so-called *Fundamental Theorem of Arithmetic*.

Lemma 1.14 *Suppose that* $a|bc$ *and* $(a, b) = 1$. *Then* $a|c$.

Proof As $(a, b) = 1$, there exist integers s and t such that $sa + tb = 1$. Multiply this equation by c:

$$sac + tbc = c.$$

Notice that a clearly divides sac. The hypothesis that $a|bc$ implies that a divides the left-hand side, and since this is equal to the right-hand side c, we get that $a|c$. □

1.3 The Fundamental Theorem of Arithmetic

In this section, we prove that every number can be written *uniquely* as a product of prime numbers. This fact will not seem surprising to you, and it wasn't until about 1800 that Gauss pointed out that it fails in a number of similar situations. We'll explore these issues in some depth in Chap. 4. Indeed, the failure of unique factorisation is the main topic of this book.

The following mild reformulation of Lemma 1.14 is the key result:

Lemma 1.15 *Suppose that $p|ab$, where $a, b \in \mathbb{Z}$, and p is prime. Then either $p|a$ or $p|b$.*

Proof If $p|a$, we are done. If $p \nmid a$, we need to show that $p|b$.

But if $p \nmid a$, then $(a, p) = 1$: a common divisor of a and p must divide p, but the only divisors of p are 1 and p, and we are assuming that $p \nmid a$. So the only possible common divisor is 1.

Then the result follows as in Lemma 1.14: as $(a, p) = 1$, we can write $sa + tp = 1$ for some integers s and t; multiply by b to get $sab + tpb = b$, and as $p|ab$, it divides the left-hand side which equals the right-hand side, so $p|b$ as required.   □

Repeated application of this lemma gives:

Corollary 1.16 *Suppose that $p|a_1a_2 \ldots a_n$. Then $p|a_i$ for some $i = 1, \ldots, n$.*

Let's use this result to prove the Fundamental Theorem of Arithmetic:

Theorem 1.17 (Fundamental Theorem of Arithmetic) *Every integer n greater than 1 can be expressed uniquely (apart from the order of factors) as a product of primes.*

Proof Suppose (for a contradiction) that there is an integer n with two different factorisations. Dividing out any primes occurring in both factorisations, we get an equality of the form

$$p_1 p_2 \ldots p_r = q_1 q_2 \ldots q_s$$

where the factors p_i and q_j are all primes, not necessarily all distinct, but where no prime on the left-hand side also occurs on the right-hand side. But p_1 divides the left-hand side and therefore the right-hand side. So $p_1|q_1 \ldots q_s$. But Corollary 1.16 now shows that p_1 must divide one of the q_j, and therefore must be identical with one of the q_j, since the only divisors of the prime q_j are 1 and q_j itself. This contradicts the hypothesis that no prime occurs on both sides of the equality.   □

Thus each integer $n > 1$ can be written uniquely in the form

$$n = p_1^{n_1} p_2^{n_2} \cdots p_k^{n_k}$$

where $p_1, p_2, \ldots, p_k$ are primes with $p_1 < p_2 < \cdots < p_k$, and $n_1, n_2, \ldots, n_k$ are natural numbers. For example,

$$360 = 2^3 \times 3^2 \times 5, \qquad 4725 = 3^3 \times 5^2 \times 7, \qquad 714420 = 2^2 \times 3^6 \times 5 \times 7^2.$$

It is worth taking a little time to reflect on this argument. The key result is really Lemma 1.15, which gave the Corollary 1.16. The proof of this lemma depends strongly on the elementary properties of the highest common factor. In more general situations, we may or may not have an analogue of the Euclidean algorithm; if we do, then it is likely that we can easily define a notion of highest common factor, and that the argument above can be generalised. On the other hand, we may have no analogue of the Euclidean algorithm, and this will make our task harder; with luck, it may be possible to define a highest common factor anyway, and to recover unique factorisation, but we shall see that this is rare. We shall return to this theme later.

1.4 The Gaussian Integers

Before launching into a general theory, let us consider an example of a question where it is useful to consider more general settings than just the integers $\mathbb{Z}$.

Fermat asked which natural numbers could be written as the sum of two squares. That is, given a natural number n, are there integers a and b so that $n = a^2 + b^2$?

One fruitful way to think of this is to factorise the right-hand side, as a product of the two complex numbers $a + ib$ and $a - ib$. That is, we are working in

$$\mathbb{Z}[i] = \{x + iy \mid x, y \in \mathbb{Z}\},$$

and asking how the number n factorises in the larger set $\mathbb{Z}[i]$. Elements of the set $\mathbb{Z}[i]$ are known as *Gaussian integers*. Write

$$N(x + iy) = |x + iy|^2 = (x + iy)\overline{(x + iy)} = (x + iy)(x - iy) = x^2 + y^2,$$

the *norm* of $x + iy$.

Here is our first elementary observation:

Lemma 1.18 *Suppose that n_1 and n_2 can be written as the sum of two squares. Then their product $n_1 n_2$ is also the sum of two squares.*

Proof Suppose that $n_1 = a^2 + b^2$ and that $n_2 = c^2 + d^2$. Equivalently,

$$n_1 = N(a + ib) = (a + ib)\overline{(a + ib)},$$
$$n_2 = N(c + id) = (c + id)\overline{(c + id)}.$$

As multiplication of complex numbers is commutative,

$$|zw|^2 = zw\overline{z}\overline{w} = z\overline{z}w\overline{w} = |z|^2|w|^2,$$

and we see that $N(zw) = N(z)N(w)$. In particular, we have

$$N\left((a + ib)(c + id)\right) = N(a + ib)N(c + id) = n_1 n_2,$$

but also

$$N\left((a + ib)(c + id)\right) = N\left((ac - bd) + i(ad + bc)\right) = (ac - bd)^2 + (ad + bc)^2.$$

Combining these gives

$$n_1 n_2 = (a^2 + b^2)(c^2 + d^2) = (ac - bd)^2 + (ad + bc)^2,$$

as required.   □

(Of course, we could have proven the lemma simply by writing down the final identity.)

The lemma suggests that we should start by working out the prime numbers p which can be written as the sum of two squares. For example, if every prime number p could be written as the sum of two squares, then every natural number could be written as the sum of two squares.

One doesn't have to try too many examples before realising that $p = 3$ cannot be written as the sum of two squares. Indeed, square numbers give a remainder which is 0 or 1 modulo 4. So the only possible sums of two squares are $0 + 0$, $0 + 1$ or $1 + 1$ modulo 4, and no number which is 3 (mod 4) can be written as the sum of two squares.

The situation for prime numbers which are congruent to 1 (mod 4) is considerably better: it turns out that every such prime number is the sum of two squares, as you might like to investigate numerically.

Let's prove this now. We'll first give a proof which is missing one crucial detail. Try and spot where the gap is.

Theorem 1.19 *Every prime number $p \equiv 1$ (mod 4) can be written as the sum of two squares.*

Proof Since $p \equiv 1$ (mod 4), we can solve the equation $x^2 + 1 \equiv 0$ (mod p). Indeed, one way to do this explicitly is the following. Write $p = 4k + 1$, and set $x = (2k)!$. Then

$$(2k)!(2k)! = 1.2.\ldots.(2k-1)(2k)(2k)(2k-1)\ldots 2.1$$
$$= (-1)^{2k}1.2.\ldots.(2k-1)(2k)(2k)(2k-1)\ldots 2.1$$
$$= (-1)(-2)\ldots.(-2k+1)(-2k)(2k)(2k-1)\ldots 2.1$$
$$\equiv (p-1)(p-2)\ldots.(2k+2)(2k+1)(2k)(2k-1)\ldots 2.1 \pmod{p}$$
$$\equiv (p-1)! \pmod{p}$$
$$\equiv -1 \pmod{p}$$

where the last line is Wilson's Theorem, that if p is prime, then $(p-1)! \equiv -1 \pmod{p}$ (see [7, Corollary 4.5]). Thus $x^2 \equiv -1 \pmod{p}$ as required.

With this value of x, then, $p|x^2+1 = (x+i)(x-i)$ in $\mathbb{Z}[i]$. If p were prime in $\mathbb{Z}[i]$, we would have $p|x+i$ or $p|x-i$. However, $\frac{x\pm i}{p} \notin \mathbb{Z}[i]$, which is a contradiction (neither the real nor imaginary parts are integers). So p is not prime, and it therefore factorises in $\mathbb{Z}[i]$.

Suppose that p factorises as $\alpha\beta$. Then $N(p) = p^2 = N(\alpha)N(\beta)$. We have three possibilities:

$$N(\alpha) = 1, \quad N(\beta) = p^2;$$
$$N(\alpha) = p, \quad N(\beta) = p;$$
$$N(\alpha) = p^2, \quad N(\beta) = 1.$$

Suppose first that $N(\alpha) = 1$, with $\alpha = a + ib$. Since the only solutions to $a^2 + b^2 = 1$ are $a = \pm 1$, $b = 0$ and $a = 0$, $b = \pm 1$, this means that $\alpha = \pm 1$ or $\alpha = \pm i$, and so $\beta = \pm p$ or $\pm ip$. Notice that this doesn't involve factorising p; merely writing it in an equivalent way using *units*. The case $N(\alpha) = p^2$ and $N(\beta) = 1$ is similar.

Thus p must factorise as $\alpha\beta$ with $N(\alpha) = N(\beta) = p$. If we write $\alpha = a + ib$, we deduce that $a^2 + b^2 = p$, and we have found a representation of p as the sum of two squares. $\qquad\square$

Hopefully you spotted the gap in the argument, although it's quite subtle. We used the claim that if $p|(x+i)(x-i)$ in $\mathbb{Z}[i]$ and p is prime, then $p|x+i$ or $p|x-i$. This should seem reasonable; in $\mathbb{Z}$, if a prime number p divides a product ab, then $p|a$ or $p|b$ (Lemma 1.15). This was one of a sequence of lemmas which followed in a reasonably straightforward manner from the existence of the Euclidean algorithm for $\mathbb{Z}$.

However, we have not explained that there is any analogous algorithm for $\mathbb{Z}[i]$, and so we cannot yet deduce the analogous statement over $\mathbb{Z}[i]$. Fortunately there is a version of the Euclidean algorithm for the Gaussian integers:

Lemma 1.20 *Let α, $\beta \in \mathbb{Z}[i]$, with $\beta \neq 0$. Then there exist Gaussian integers κ and ρ such that*

$$\alpha = \kappa\beta + \rho$$

with $N(\rho) < N(\beta)$.

Proof We start by finding $\kappa \in \mathbb{Z}[i]$ with $|\frac{\alpha}{\beta} - \kappa| < 1$.

Simply take the quotient $\alpha/\beta = x + iy \in \mathbb{C}$. Then we choose integers m and n such that $|x - m| \leq \frac{1}{2}$ and $|y - n| \leq \frac{1}{2}$. Write $\kappa = m + in \in \mathbb{Z}[i]$, and write $\rho = \alpha - \kappa\beta$. This makes κ the closest point of $\mathbb{Z}[i]$ to $\frac{\alpha}{\beta}$. Then

$$\left|\frac{\alpha}{\beta} - \kappa\right| = |(x + iy) - (m + in)| = |(x - m) + i(y - n)| \leq \sqrt{\left(\tfrac{1}{2}\right)^2 + \left(\tfrac{1}{2}\right)^2} \doteq \frac{1}{\sqrt{2}}.$$

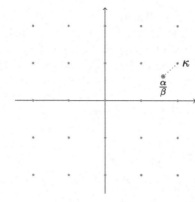

κ is chosen as the closest point of $\mathbb{Z}[i]$ to $\frac{\alpha}{\beta}$

Then

$$N(\rho) = |\rho|^2 = |\alpha - \kappa\beta|^2 = \left|\frac{\alpha}{\beta} - \kappa\right|^2 \cdot |\beta|^2 \leq \frac{1}{2}|\beta|^2 < |\beta|^2 = N(\beta),$$

as required.   □

You might like to check that this leads to the conclusion that if $\pi \,|\, \alpha\beta$ in $\mathbb{Z}[i]$ for a prime π, then $\pi \,|\, \alpha$ or $\pi \,|\, \beta$, exactly as in Lemma 1.15. We shall omit that here; later we will develop a more general theory of *Euclidean rings*, with $\mathbb{Z}[i]$ an example, and will see that the theory given in Sects. 1.2 and 1.3 can be generalised to such rings.

This fills in the gap in the proof of the lemma above, and shows that every prime number $p \equiv 1 \pmod 4$ can be written as the sum of two squares. This is the key ingredient in the following classification of those integers which can be written as the sum of two squares:

Theorem 1.21 *A natural number n can be written as the sum of two squares if and only if n has prime power factorisation $n = \prod_p p^{n_p}$ where n_p is even for all primes $p \equiv 3 \pmod 4$.*

Exercise 1.8 Show that the set $\mathbb{Z}[\sqrt{-2}] = \{a + b\sqrt{-2} \mid a, b \in \mathbb{Z}\}$ also has a Euclidean algorithm, by imitating Lemma 1.20.

Exercise 1.9 Prove that an odd prime p can be written as $x^2 + 2y^2$ with x and $y \in \mathbb{Z}$ if and only if the congruence $x^2 \equiv -2 \pmod p$ is soluble.

Exercise 1.10 Show that the set $\mathbb{Z}[\sqrt{2}] = \{a + b\sqrt{2} \mid a, b \in \mathbb{Z}\}$ also has a Euclidean algorithm, using the norm function $N(a + b\sqrt{2}) = |a^2 - 2b^2|$.

Exercise 1.11 Find $\alpha, \beta \in \mathbb{Z}[\sqrt{-7}] = \{a + b\sqrt{-7} \mid a, b \in \mathbb{Z}\}$ with $|\alpha/\beta - \kappa| > 1$ for each $\kappa \in \mathbb{Z}[\sqrt{-7}]$, and deduce that the natural attempt to generalise the Euclidean algorithm to $\mathbb{Z}[\sqrt{-7}]$ fails.

Remark 1.22 We saw above, in the proof of Lemma 1.19, that $x = (\frac{p-1}{2})!$ gives a solution to $x^2 \equiv -1 \pmod{p}$. However, in practice this is hard to calculate.

Here is a better way, using Legendre symbols (see [7, Chap. 7]). Recall that a is a quadratic residue modulo p if the equation $x^2 \equiv a \pmod{p}$ has two solutions, and is a non-residue if there are no solutions, and that the Legendre symbol $(\frac{a}{p})$ is defined to be $+1$ if a is a quadratic residue, and -1 if not. Legendre symbols have various properties which enable them to be calculated easily, such as multiplicativity, $(\frac{ab}{p}) = (\frac{a}{p})(\frac{b}{p})$, explicit formulae for $(\frac{-1}{p})$ and $(\frac{2}{p})$, and also quadratic reciprocity (see Theorem 9.15 for a statement and proof), which relates $(\frac{p}{q})$ and $(\frac{q}{p})$ for two primes p and q.

Recall that $a^{(p-1)/2} \equiv \pm 1 \pmod{p}$ for all a not divisible by p, the answer being $+1$ when a is a quadratic residue, and -1 if not. That is, $a^{(p-1)/2} \equiv (\frac{a}{p}) \pmod{p}$ (this is due to Euler).

Compute the Legendre symbols $(\frac{a}{p})$ for $a = 2, a = 3$, and so on, until you find one with $(\frac{a}{p}) = -1$. (You should be able to persuade yourself, using the multiplicativity of the Legendre symbol, that the smallest such a will be prime.) Then $a^{(p-1)/2} \equiv -1 \pmod{p}$; we simply put $x = a^{(p-1)/4} \pmod{p}$ (remember that $p \equiv 1 \pmod{4}$); then $x^2 \equiv -1 \pmod{p}$.

As an example, consider $p = 73$. We compute $(\frac{2}{73}) = (\frac{3}{73}) = 1$, but that $(\frac{5}{73}) = -1$. We therefore compute $x = 5^{18} \pmod{73}$; this is easily done by successively squaring modulo 73:

$$5^2 \equiv 25, \quad 5^4 \equiv 25^2 \equiv 41, \quad 5^8 \equiv 41^2 \equiv 2, \quad 5^{16} \equiv 2^2 \equiv 4,$$

and then
$$5^{18} = 5^{16} \times 5^2 \equiv 4 \times 25 \equiv 27 \pmod{73}.$$

So $x = 27$ gives a solution to $x^2 \equiv -1 \pmod{73}$.

Exercise 1.12 Use this method (perhaps with a computer equipped with a suitable computer algebra package) to find a solution to $x^2 \equiv -1 \pmod{1009}$.

We shall discuss generalisations of the Euclidean algorithm to imaginary quadratic fields in Chap. 6.

I can't resist ending the section with two related results published in American Mathematical Monthly 97 (1990). The first is a remarkable proof, due to Don Zagier [18], that every prime number $p \equiv 1 \pmod{4}$ is the sum of two squares. It is based on an earlier argument by Roger Heath-Brown. Sadly the potential for generalisation to other situations seems considerably more limited than our argument above.

Exercise 1.13

1. Recall that an *involution* on a set S is a map $f : S \longrightarrow S$ such that $f \circ f = \text{id}$,
   i.e., for all $s \in S$, $f(f(s)) = s$. If f is an involution on a finite set S, let
   $\text{Fix}_S(f) = \{s \in S \mid f(s) = s\}$ denote the set of *fixed points* of S. Show that
   $|S| \equiv |\text{Fix}_S(f)| \pmod{2}$.
   [*Hint: Consider all of the subsets of S of the form $\{s, f(s)\}$.*]
   Deduce that if $|S|$ is odd, then there is always a fixed point in S.
2. Now let $p = 4k + 1$ be prime, and consider the set

$$S = \{(x, y, z) \in \mathbb{N}^3 \mid x^2 + 4yz = p\}.$$

   Show that the map $f : S \longrightarrow S$ defined by

$$f : (x, y, z) \mapsto \begin{cases} (x + 2z, z, y - x - z), & \text{if } x < y - z, \\ (2y - x, y, x - y + z), & \text{if } y - z < x < 2y, \\ (x - 2y, x - y + z, y), & \text{if } 2y < x, \end{cases}$$

   is an involution on S.
3. Show that f has a unique fixed point, and deduce that $|S|$ is odd.
4. Show that $f' : (x, y, z) \mapsto (x, z, y)$ is another involution on S, and deduce that
   $\text{Fix}_S(f')$ is non-empty.
5. Deduce that p can be written as the sum of two squares.

   The proofs given above that every prime number congruent to 1 (mod 4) are not
constructive; that is, given p, the proof doesn't really help to find integers a and
b such that $p = a^2 + b^2$. However, there is a constructive algorithm which works
quickly, which we will outline without proof in the next exercise. For a proof of why
this works, as well as a historical discussion, see the article by Stan Wagon [15] in
the same issue of the journal mentioned above:

Exercise 1.14 Here is a constructive algorithm to solve $p = a^2 + b^2$ for
$p \equiv 1 \pmod{4}$ which works quickly. There are just two steps:

Step 1:   Find an integer x such that $x^2 \equiv -1 \pmod{p}$ (see Remark 1.22).

Step 2:   Run the Euclidean algorithm on p and x; then a and b can be taken to be
the first two remainders less than $\sqrt{p}$.

Using Exercise 1.12, hence factor 1009 in the Gaussian integers.

1.5 Another Application of the Gaussian Integers

Let's finish the chapter with another application of uniqueness of factorisation in
$\mathbb{Z}[i]$. We will find all integer solutions to $x^3 = y^2 + 1$. (Recall that equations where
only integer solutions are sought are known as *Diophantine equations*, after the Greek
mathematician Diophantus.)

Remark 1.23 This is a special case of *Catalan's conjecture*, which predicts that the only consecutive perfect powers are $8 = 2^3$ and $9 = 3^2$. The conjecture was finally proven, after a long history, by Preda Mihăilescu, in 2002.

Remark 1.24 We have remarked that some care has to be taken in defining uniqueness of factorisation. For example,

$$6 = 2 \times 3 = (-2) \times (-3)$$

should really be counted as equivalent factorisations – we've simply multiplied both by -1, and since $(-1)(-1) = 1$, this shouldn't really matter. The same considerations will apply in $\mathbb{Z}[i]$; given a factorisation $\alpha = \beta\gamma$, and if u and v in $\mathbb{Z}[i]$ satisfy $uv = 1$, then we will consider $\alpha = (u\beta)(v\gamma)$ as an *equivalent* factorisation. In $\mathbb{Z}[i]$, the possible values of such *units* u are ± 1 or $\pm i$, exactly those elements u with $N(u) = 1$.

Suppose that x and y are integers satisfying $x^3 = y^2 + 1$. The first observation is that if x is even, $y^2 + 1 \equiv 0 \pmod 4$, which is not possible. So x is odd, and therefore y is even.

Now we make use of the theory of the Gaussian integers started in the previous section. As in the previous section, we will use the word "prime" rather loosely, and assume that primes in $\mathbb{Z}[i]$ satisfy the same properties as prime numbers in $\mathbb{Z}$ do (we will justify this more carefully in later chapters). In $\mathbb{Z}[i]$, we can write

$$x^3 = (y + i)(y - i).$$

Let's first show that any common factor of $y + i$ and $y - i$ must be a unit ± 1 or $\pm i$ (i.e., $y + i$ and $y - i$ are coprime).

Lemma 1.25 *Suppose that $\alpha | y + i$ and also $\alpha | y - i$. Then α is a unit.*

Proof Suppose that $\alpha | y + i$ and also $\alpha | y - i$, and suppose that α is *not* a unit. Then $\alpha | ((y + i) - (y - i))$, and so is a factor of $2i = (1 + i)^2$. However, any factorisation $1 + i = \beta\gamma$ must satisfy $N(\beta)N(\gamma) = N(1 + i) = 2$, so either $N(\beta) = 1$ or $N(\gamma) = 1$, and then β or γ is ± 1 or $\pm i$, a unit. So $1 + i$ is a prime in $\mathbb{Z}[i]$. As $\alpha | (1 + i)^2$, if α is not a unit, we must have $1 + i | \alpha$, using unique factorisation in $\mathbb{Z}[i]$. Then $1 + i | x^3$, and so $1 + i | x$ (again using unique factorisation). But then $(1 + i)^2 | x^2$, so $2i | x^2$, and so x^2 is even. This contradicts the observation that x is odd. $\quad\square$

We conclude that $y + i$ and $y - i$ are coprime (in the sense that any common divisor must be a unit). If $\pi | x$ and π is a prime (so not a unit), then $\pi^3 | x^3 = (y + i)(y - i)$. Since $y + i$ and $y - i$ have no factor in common, either $\pi^3 | y + i$ and $\pi \nmid y - i$, or vice versa. In particular,

$$y + i = u\beta^3$$
$$y - i = v\gamma^3$$

where u and v are units. In fact, since the units are ± 1 and $\pm i$, they are all already cubes, and we can absorb them into β and γ.

We can therefore suppose that $y + i = (a + bi)^3$ for some integers a and b. Expanding:

$$y + i = (a^3 - 3ab^2) + i(3a^2 b - b^3),$$

and equating imaginary parts gives

$$(3a^2 - b^2)b = 1.$$

The only way that a product of two integers can give 1 is if both are 1, or both are -1. If $b = 1$, there is no possible solution for a (we would need $3a^2 = 2$). However, if $b = -1$, we see that $a = 0$ gives the only solution.

It follows that $y + i = (-i)^3 = i$, so that the only solution in integers to the original equation $x^3 = y^2 + 1$ is when $y = 0$, which implies that $x = 1$.

Exercise 1.15 Find all solutions of $x^3 = y^2 + 2$ by the same method (you may assume that $\mathbb{Z}[\sqrt{-2}]$ has unique factorisation by Exercise 1.8).

Exercise 1.16 Find all solutions of $x^5 = y^2 + 1$ by the same method.

Chapter 2
Number Fields

We've just seen examples where questions about integers were naturally treated by working in the slightly bigger set $\mathbb{Z}[i]$ of Gaussian integers. In this chapter we begin the development of some more general theory.

The aim of algebraic number theory is to study generalisations of the usual arithmetic in the natural numbers in more general settings. While studying these generalisations, it will become clear that there are new phenomena of interest in their own right.

In Chap. 1, we saw that unique factorisation plays a prominent role, and that this was a consequence of the Euclidean algorithm. In order to be able to talk about a Euclidean algorithm, we need to work in sets closed under addition and multiplication, i.e., in rings. As in the examples at the end of Chap. 1, we will also want to work in sets contained in the complex numbers, that is, with subrings of $\mathbb{C}$.

It turns out that there is a good theory of integers inside any field K which is a finite degree extension of $\mathbb{Q}$, and these finite extensions, known as number fields, form the setting for algebraic number theory. Finite degree extensions of $\mathbb{Q}$ are constructed by adjoining complex numbers which are the roots of polynomial equations with rational (or integer) coefficients. Although many complex numbers are roots of polynomial equations with integer coefficients, it turns out that not every complex number can be written in this way.

2.1 Algebraic Numbers

Definition 2.1 A complex number α is said to be *algebraic* if it is the root of a polynomial equation with integer coefficients. If α is not algebraic, it is *transcendental*.

Every rational number $\frac{m}{n}$ is algebraic, as it is a root of $nX - m = 0$; also, $\pm\sqrt{2}$ are roots of $X^2 - 2 = 0$, so $\pm\sqrt{2}$ are both algebraic. Indeed, every polynomial with integer coefficients of degree n will have n algebraic numbers as roots. This gives

F. Jarvis, *Algebraic Number Theory*, Springer Undergraduate
Mathematics Series, DOI: 10.1007/978-3-319-07545-7_2,
© Springer International Publishing Switzerland 2014

such a large collection of algebraic numbers, that one might wonder whether every complex number could be written as a root of a polynomial.

However, this is not true. Liouville (1844) was the first to construct an explicit example of a transcendental number, while Hermite (1873) and Lindemann (1882) proved that e and π respectively are transcendental.

For readers with some knowledge of Cantor's theory of countability, the simplest proof of the existence of transcendental numbers is due to Cantor himself (1874), although it does not give any way to construct such numbers. We will give Liouville's construction of an explicit transcendental number below.

Write $\mathcal{A} \subseteq \mathbb{C}$ for the collection of all algebraic numbers.

Theorem 2.2 (Cantor) *The set $\mathcal{A}$ is countable; that is, there are only countably many algebraic numbers.*

Proof Given a polynomial equation $p(X) = c_0 X^d + c_1 X^{d-1} + \cdots + c_d = 0$ with all $c_i \in \mathbb{Z}$ and $c_0 \neq 0$, define the quantity

$$H(p) = d + |c_0| + \cdots + |c_d| \in \mathbb{Z}.$$

This process associates an integer to every polynomial with integer coefficients. Notice that $\deg(p) = d < H(p)$.

Let H be any natural number. Then it is easy to see that there are only finitely many polynomials $p(X)$ which satisfy $H(p) \leq H$. Say that an algebraic number $\alpha \in \mathcal{A}$ is *of level H* if α is a root of some polynomial p with $H(p) \leq H$. As there are only finitely many polynomials with $H(p) \leq H$, and all have at most H roots (since the degree of such a polynomial is bounded by H), there are only finitely many algebraic numbers of level H, for any given H.

On the other hand, every algebraic number is a root of such a polynomial, and is therefore of level H for some H. The collection of algebraic numbers can therefore be written as a union

$$\mathcal{A} = \bigcup_{H=1}^{\infty} \{\alpha \in \mathcal{A} \mid \alpha \text{ is of level } H\}.$$

We have already remarked that each set on the right-hand side is finite. Furthermore, the union is a countable union, as the indexing set consists of the natural numbers. Therefore $\mathcal{A}$ is a countable union of finite sets, and is therefore countable. ☐

Since $\mathbb{C}$ is uncountable, and its subset $\mathcal{A}$ is countable, we conclude that transcendental numbers exist. Even more, we see that the set of transcendental numbers is actually *uncountable*, so that, in some sense, almost every complex number is transcendental.

Now let us give Liouville's explicit construction of a transcendental number, which avoids use of countability arguments.

We need the following theorem:

Theorem 2.3 (Liouville) *Let α be a real algebraic number which is a root of an irreducible polynomial $f(X)$ over $\mathbb{Z}$ of degree $n > 1$. Then there is a constant c such that for all rational numbers $\frac{p}{q}$,*

$$\left| \alpha - \frac{p}{q} \right| > \frac{c}{q^n}.$$

Proof The result is clear in the case where $|\alpha - p/q| > 1$; choosing $c = 1$ covers these values. We therefore consider the remaining case where $|\alpha - p/q| \leq 1$.

Apply the Mean Value Theorem to $f(X)$ at the points α and p/q to deduce the existence of a number γ strictly between α and p/q such that

$$\frac{f(\alpha) - f(p/q)}{\alpha - p/q} = f'(\gamma).$$

As α is a root of f, we see that $f(\alpha) = 0$. Also, as $f(X)$ is an irreducible polynomial of degree $n > 1$, it has no rational roots, so $f(p/q) \neq 0$. However, as $f(X)$ has integer coefficients, the denominator of $f(p/q)$ must divide q^n, so $q^n f(p/q)$ is a non-zero integer. Therefore $|f(p/q)| \geq 1/q^n$.

Now $|\gamma - \alpha| < 1$ as γ is strictly between α and p/q, and $|\alpha - p/q| \leq 1$. By continuity of f' at α, we see that $|f'(\gamma)| < 1/c_0$ for some constant c_0 for all γ within 1 of α, where the constant c_0 depends only on α. Then

$$\left| \alpha - \frac{p}{q} \right| = \left| \frac{f(p/q)}{f'(\gamma)} \right| > \frac{c_0}{q^n}.$$

Choose $c = \min(c_0, 1)$ to cover both cases, and the result follows. $\square$

In order to find a transcendental number, we simply need to find an α where the inequality of Theorem 2.3 fails for all n. Liouville suggested choosing

$$\alpha = \sum_{k=1}^{\infty} 10^{-k!} = 0.110001000000000000000000100\ldots.$$

Define $p_r/q_r = \sum_{k=1}^{r} 10^{-k!}$. The first three numbers are $p_1/q_1 = 0.1$, $p_2/q_2 = 0.11$ and $p_3/q_3 = 0.110001$. Then $q_r = 10^{r!}$, and

$$\left| \alpha - \frac{p_r}{q_r} \right| = \sum_{k=r+1}^{\infty} 10^{-k!} < \frac{2}{10^{(r+1)!}} = \frac{2}{(10^{r!})^{r+1}} = \frac{2}{q_r^{r+1}}. \tag{2.1}$$

If α were algebraic of some degree n, there would be a constant c such that $|\alpha - p/q| > c/q^n$ for all rationals p/q. However, choosing $p/q = p_r/q_r$ for large enough $r > n$ gives a contradiction, using (2.1).

Exercise 2.1 Show that all numbers $\sum_{k=1}^{\infty} s_k 10^{-k!}$ are transcendental, where $s_k \in \{1, -1\}$.

By considering a variant of Cantor's diagonal argument, prove that the set of numbers of this form is uncountable. This gives another proof of the uncountability of the set of transcendental numbers.

Hermite's proof of the transcendence of e, and Lindemann's proof of the transcendence of π, are (only just) beyond the scope of this book. References for the arguments include [11].

2.2 Minimal Polynomials

As already mentioned, we will be able to do arithmetic in fields which are obtained by adjoining roots of polynomials to $\mathbb{Q}$, i.e., algebraic numbers. In this section, we will look at some properties of polynomials.

Recall that a *monic* polynomial is one whose leading coefficient is 1.

Lemma 2.4 *If α is algebraic, then there is a unique monic polynomial $f(X) \in \mathbb{Q}[X]$ of smallest degree with α as a root.*

Proof If α is a root of a polynomial $f(X) = c_0 X^n + c_1 X^{n-1} + \cdots + c_n$ with $c_0 \neq 0$, it will also be a root of $X^n + \frac{c_1}{c_0} X^{n-1} + \cdots + \frac{c_n}{c_0}$ got by dividing through by the leading coefficient.

Amongst all the monic polynomials with α as a root, let $f(X)$ be one with smallest degree. We claim that $f(X)$ is unique.

Suppose that $g(X)$ is another monic polynomial of the same degree with α as a root. Then α is also a root of $(f - g)(X)$, and since the leading terms of $f(X)$ and $g(X)$ cancel, the degree of $f - g$ is smaller than that of f or g. If $f - g \neq 0$, then we can divide through by its leading coefficient to find a monic polynomial of smaller degree than f with α as a root, contradicting the choice of $f(X)$.   □

Definition 2.5 Let α be an algebraic number. The *minimal polynomial* of α over $\mathbb{Q}$ is the monic polynomial over $\mathbb{Q}$ of smallest degree with α as a root.

Lemma 2.6 *If $m(X)$ is the minimal polynomial of the algebraic number α, then it is irreducible.*

Proof Indeed, if $m(X)$ were to factorise as the product $f(X)g(X)$ of two polynomials over $\mathbb{Q}$ of smaller degree, then since $m(\alpha) = 0$, we would have $f(\alpha)g(\alpha) = 0$, and α would be a root of either f or g, and this contradicts the choice of m as the polynomial *of smallest degree* with α as a root.   □

The minimal polynomial has a particularly useful property: every polynomial with α as a root is necessarily a multiple of the minimal polynomial of α. We'll prove that next, with the aid of Euclid's algorithm for polynomials.

Lemma 2.7 *Suppose that α is a root of some polynomial $f(X) \in \mathbb{Q}[X]$. If $m(X)$ is the minimal polynomial of α, then $m(X) \mid f(X)$.*

Proof Just like $\mathbb{Z}$, the ring of rational polynomials $\mathbb{Q}[X]$ has an obvious division algorithm (see Exercise 1.6), and we can find polynomials $q(X), r(X) \in \mathbb{Q}[X]$ such that

$$f(X) = q(X)m(X) + r(X),$$

where $r(X)$ is the zero polynomial, or has smaller degree than $m(X)$. Substitute $X = \alpha$:

$$f(\alpha) = q(\alpha)m(\alpha) + r(\alpha);$$

as $f(\alpha) = m(\alpha) = 0$, we must have $r(\alpha) = 0$. However, $r(X)$ has smaller degree than $m(X)$, and $m(X)$ was the monic polynomial of smallest degree with α as a root. If $r(X)$ were non-zero, we could scale it to get a monic polynomial of smaller degree than $m(X)$ with α as a root, and this would contradict the definition of $m(X)$. Therefore $r(X)$ must be the zero polynomial. In particular, $f(X) = q(X)m(X)$, and so $f(X)$ is a multiple of $m(X)$. $\qquad\qquad\square$

If K is any field, and if α satisfies some equation over K, then we also have a notion of *minimal polynomial over K*, the monic polynomial with coefficients in K of smallest degree with α as a root; any other polynomial with coefficients in K with α as a root is a multiple of the minimal polynomial. The argument is identical to the one just given.

Exercise 2.2 Find the minimal polynomial of $\sqrt{2} + \sqrt{3}$ (over $\mathbb{Q}$). Would the minimal polynomial over $\mathbb{Q}(\sqrt{2})$ be the same?

Exercise 2.3 Find the minimal polynomial of $\alpha = \frac{1+i}{\sqrt{2}}$ over $\mathbb{Q}$. What is the minimal polynomial of α over each of $\mathbb{Q}(i)$, $\mathbb{Q}(\sqrt{2})$ and $\mathbb{Q}(\sqrt{-2})$?

2.3 The Field of Algebraic Numbers

The main aim of this section is to establish the basic algebraic properties of algebraic numbers; indeed, we prove that the set $\mathcal{A}$ of algebraic numbers is actually a *field*. Recall that a field is a set which satisfies exactly the same algebraic properties as $\mathbb{Q}$, so that we must be able to add, subtract, multiply and divide (by non-zero elements) in $\mathbb{Q}$, and the usual algebraic rules (e.g., addition and multiplication are commutative and associative) are satisfied. Since we are dealing with subsets of the complex numbers $\mathbb{C}$, all these rules are inherited from $\mathbb{C}$, and we just have to check that the collection of algebraic numbers is closed under the usual arithmetic operations. That is, we want to see that if α and β are algebraic numbers, then so are $\alpha + \beta$, $\alpha - \beta$ and $\alpha\beta$, and if $\beta \neq 0$, so is α/β.

Although one could write a proof using no abstract algebra (using the ideas of the end of Sect. 2.6), we will give a more algebraic argument now, since some of the definitions will reappear later in the book.

We are going to give an equivalent algebraic formulation for what it means for the complex number α to be algebraic. Let's recall that for any complex number α, $\mathbb{Q}(\alpha)$ denotes the smallest *field* one can obtain by applying all the usual arithmetic operations (addition, subtraction, multiplication, division) to the rational numbers and α; it consists of all quotients $p(\alpha)/q(\alpha)$ where $p(X)$ and $q(X)$ are polynomials with rational coefficients, and where $q(\alpha) \neq 0$.

On the other hand, we also have the notation $\mathbb{Q}[\alpha]$, which means the *ring* of all *polynomial expressions* in α. It is the smallest ring one can obtain by applying the arithmetic operations of addition, subtraction and multiplication (but not division) to the rational numbers and α.

For example, $\frac{3\alpha^3+\alpha-1}{\alpha^2+2}$ is in $\mathbb{Q}(\alpha)$, but not necessarily in $\mathbb{Q}[\alpha]$. Of course, $\mathbb{Q}[\alpha] \subseteq \mathbb{Q}(\alpha)$.

Let's begin with a simple remark about algebraic numbers:

Proposition 2.8 *If α is algebraic, $\mathbb{Q}[\alpha] = \mathbb{Q}(\alpha)$, and so every element of $\mathbb{Q}(\alpha)$ can be written as a polynomial in α.*

Proof We have to explain that every quotient of polynomials $p(\alpha)/q(\alpha)$ with $q(\alpha) \neq 0$ can be written alternatively as a polynomial in α.

Let $m(X)$ denote the minimal polynomial of α over $\mathbb{Q}$. The highest common factor of the two polynomials $m(X)$ and $q(X)$ must be a factor of $m(X)$; as $m(X)$ is irreducible, its only factors are 1 and $m(X)$ itself. However, $m(X)$ is not a factor of $q(X)$, as $m(\alpha) = 0$, but $q(\alpha) \neq 0$. By a similar argument to the discussion of the Euclidean algorithm in Chap. 1 (see Exercise 1.6), there are polynomials $s(X)$ and $t(X)$ over $\mathbb{Q}$ such that

$$s(X)q(X) + t(X)m(X) = 1.$$

In particular, $s(\alpha)q(\alpha) + t(\alpha)m(\alpha) = 1$, and therefore $s(\alpha)q(\alpha) = 1$, because $m(\alpha) = 0$. We conclude that $1/q(\alpha) = s(\alpha)$, and so $p(\alpha)/q(\alpha) = p(\alpha)s(\alpha)$, a polynomial expression in α, as required. $\qquad\square$

Note that if α is transcendental, there is no way to write $1/\alpha$ as a polynomial in α; otherwise, we could multiply through by α and find a rational polynomial with α as a root. Therefore $1/\alpha$ is in $\mathbb{Q}(\alpha)$, but not in $\mathbb{Q}[\alpha]$. Thus if α is not algebraic, then $\mathbb{Q}(\alpha)$ is strictly bigger than $\mathbb{Q}[\alpha]$, and the property of Proposition 2.8 therefore characterises algebraic numbers.

Recall that the *degree* of the field extension $\mathbb{Q}(\alpha)/\mathbb{Q}$ is the dimension of the set $\mathbb{Q}(\alpha)$ when regarded as a vector space over $\mathbb{Q}$; that is, it is the number of elements in a basis $\{\omega_1, \ldots, \omega_n\}$ so that every element of $\mathbb{Q}(\alpha)$ can be expressed uniquely as a sum $a_1\omega_1 + \cdots + a_n\omega_n$ with $a_i \in \mathbb{Q}$. It is denoted $[\mathbb{Q}(\alpha) : \mathbb{Q}]$. For example, $\mathbb{Q}(\sqrt{2})$ has degree 2 over $\mathbb{Q}$, as each element can be written as $a + b\sqrt{2}$.

Exercise 2.4 Show that $\mathbb{Q}(\sqrt{2}, \sqrt{3})$ has degree 4 over $\mathbb{Q}$ by proving that $1, \sqrt{2}, \sqrt{3}$ and $\sqrt{6}$ are linearly independent.

Now we can give an equivalent formulation of what it means for a complex number to be algebraic:

Proposition 2.9 *Let α be a complex number. Then the following are equivalent:*

1. α *is algebraic;*
2. *the field extension $\mathbb{Q}(\alpha)/\mathbb{Q}$ is of finite degree.*

Proof (1) $\Rightarrow$ (2). Suppose that α is algebraic. Then let $m(X) = X^n + c_1 X^{n-1} + \cdots + c_n$ denote the minimal polynomial for α, so that

$$\alpha^n + c_1 \alpha^{n-1} + \cdots + c_n = 0,$$

or, rearranging:

$$\alpha^n = -(c_1 \alpha^{n-1} + \cdots + c_n). \tag{2.2}$$

As α is algebraic, every element of $\mathbb{Q}(\alpha)$ can be written as a polynomial in α. Further, if this polynomial has degree n or above, we can reduce the degree by replacing all occurrences of α^r for $r \geq n$ using (2.2). It follows that every element of $\mathbb{Q}(\alpha)$ can be written as an expression

$$a_{n-1} \alpha^{n-1} + a_{n-2} \alpha^{n-2} + \cdots + a_0$$

with $a_i \in \mathbb{Q}$.

Furthermore, this expression is unique—if an element can be written in two different ways

$$a_{n-1} \alpha^{n-1} + a_{n-2} \alpha^{n-2} + \cdots + a_0 = b_{n-1} \alpha^{n-1} + b_{n-2} \alpha^{n-2} + \cdots + b_0$$

then subtracting one side from the other gives a polynomial of degree strictly smaller than n with α as a root. However, the minimal polynomial is $m(X)$, of degree n, so there can be no polynomial of degree less than n with α as a root.

So every element of $\mathbb{Q}(\alpha)$ is a unique rational linear combination of the n elements $1, \alpha, \ldots, \alpha^{n-1}$. Thus $\mathbb{Q}(\alpha)$ is n-dimensional as a vector space over $\mathbb{Q}$, and therefore $[\mathbb{Q}(\alpha) : \mathbb{Q}]$ is finite.

(2) $\Rightarrow$ (1). If $[\mathbb{Q}(\alpha) : \mathbb{Q}]$ is some finite number, n, say, then any $n + 1$ elements of the $\mathbb{Q}$-vector space $\mathbb{Q}(\alpha)$ are linearly dependent. In particular, the elements $1, \alpha, \ldots, \alpha^n$ are linearly dependent, so that there exists a linear relationship

$$a_0 + a_1 \alpha + \ldots + a_n \alpha^n = 0,$$

and consequently α satisfies a polynomial equation over $\mathbb{Q}$, and is algebraic. $\square$

The proof actually shows something more:

Corollary 2.10 *Suppose that α is algebraic. Then the degree of the extension $[\mathbb{Q}(\alpha) : \mathbb{Q}]$ is the same as the degree of the minimal polynomial of α over $\mathbb{Q}$. Every element of $\mathbb{Q}(\alpha)$ can be written as a polynomial in α of degree less than $[\mathbb{Q}(\alpha) : \mathbb{Q}]$.*

Proof This just follows from the proof of Proposition 2.9.   □

More generally, the same argument shows that if K is any field, then an element α is algebraic over K (i.e., satisfies a polynomial equation with coefficients in K) if and only if $[K(\alpha) : K]$ is finite, and that then this degree is also the degree of the minimal polynomial of α over K.

We now use Proposition 2.9 to prove that the algebraic numbers form a *field*; that is, the sum, difference and product of any two algebraic numbers is again algebraic, as is the quotient of an algebraic number by a non-zero algebraic number.

We can adjoin more than one number to a field; for example, if both α and β are algebraic, define $\mathbb{Q}(\alpha, \beta)$ to be $\mathbb{Q}(\alpha)(\beta)$, that is, all polynomial expressions in β with coefficients in $\mathbb{Q}(\alpha)$. It is easy to see that this just gives all the polynomials in the two variables α and β.

Corollary 2.11 *Suppose that α and β are algebraic. Then $\alpha + \beta$, $\alpha - \beta$ and $\alpha\beta$ are algebraic; if also $\beta \neq 0$, then α/β is algebraic.*

Proof As α and β are algebraic, Proposition 2.9 states that $[\mathbb{Q}(\alpha) : \mathbb{Q}]$ and $[\mathbb{Q}(\beta) : \mathbb{Q}]$ are both finite. Write

$$m = [\mathbb{Q}(\alpha) : \mathbb{Q}],$$
$$n = [\mathbb{Q}(\beta) : \mathbb{Q}].$$

Let's explain that $[\mathbb{Q}(\alpha, \beta) : \mathbb{Q}]$ is finite. A typical element of $\mathbb{Q}(\alpha, \beta)$ is a polynomial expression $\sum_{i=0}^{k} \sum_{j=0}^{l} a_{ij}\alpha^i \beta^j$. However, every α^i with $i \geq m$ can be written as a polynomial in α of degree at most $m - 1$ (by Corollary 2.10), and similarly every β^j with $j \geq n$ can be written as a polynomial in β of degree at most $n - 1$. Substituting these in, we see that any element of $\mathbb{Q}(\alpha, \beta)$ can be written

$$\sum_{i=0}^{m-1} \sum_{j=0}^{n-1} a'_{ij}\alpha^i \beta^j$$

for some a'_{ij}. It therefore follows that $\mathbb{Q}(\alpha, \beta)$ is spanned by the set $\{\alpha^i \beta^j \mid 0 \leq i \leq m - 1, 0 \leq j \leq n - 1\}$. Thus $\mathbb{Q}(\alpha, \beta)$ has a finite spanning set as a $\mathbb{Q}$-vector space, and therefore $[\mathbb{Q}(\alpha, \beta) : \mathbb{Q}]$ is finite.

To prove that $\alpha + \beta$ is algebraic, we simply note that $\alpha + \beta \in \mathbb{Q}(\alpha, \beta)$, so that $\mathbb{Q}(\alpha + \beta) \subseteq \mathbb{Q}(\alpha, \beta)$, and so $[\mathbb{Q}(\alpha + \beta) : \mathbb{Q}] \leq [\mathbb{Q}(\alpha, \beta) : \mathbb{Q}]$. It follows that $[\mathbb{Q}(\alpha + \beta) : \mathbb{Q}]$ is finite, and, again applying Proposition 2.9, $\alpha + \beta$ must be algebraic.

The arguments for $\alpha - \beta$, $\alpha\beta$ and α/β are all similar, as each lies in $\mathbb{Q}(\alpha, \beta)$. □

While this proof is easy, it doesn't give any recipe for writing down a polynomial with $\alpha + \beta$ as a root, given polynomials with α and β as roots. We will discuss this at the end of Sect. 2.6.

This is exactly what we need to deduce our desired result:

Corollary 2.12 *The algebraic numbers $\mathcal{A}$ form a field.*

Exercise 2.5 Find an algebraic number α where $\mathbb{Q}(\alpha)$ is strictly larger than the set $\{a + b\alpha \mid a, b \in \mathbb{Q}\}$.

Exercise 2.6 Let $\alpha = \sqrt[3]{2}$. Write $\frac{\alpha^2 - 1}{\alpha + 2}$ as a polynomial in α with rational coefficients.

Exercise 2.7 Let α be a root of $X^4 + 2X + 1 = 0$. Write $\frac{\alpha + 1}{\alpha^2 - 2\alpha + 2}$ as a polynomial in α with rational coefficients.

2.4   Number Fields

Although $\mathcal{A}$ is countable, it is still very much larger than the rational numbers $\mathbb{Q}$ (it has infinite degree over $\mathbb{Q}$, for example), and is too large to be really useful.

The fields in which we are going to generalise ideas of primes, factorisations, and so on, are the finite extensions of $\mathbb{Q}$:

Definition 2.13 A field K is a *number field* if it is a finite extension of $\mathbb{Q}$. The *degree* of K is the degree of the field extension $[K : \mathbb{Q}]$, i.e., the dimension of K as a vector space over $\mathbb{Q}$.

In particular, every element in K lies inside a finite extension of $\mathbb{Q}$, and, by Proposition 2.9, is necessarily algebraic.

Example 2.14

1. $\mathbb{Q}$ itself is a number field. Indeed, it will serve as the inspiration for our general theory.
2. $\mathbb{Q}(\sqrt{2}) = \{a + b\sqrt{2} \mid a, b \in \mathbb{Q}\}$ is a number field, since every element is a $\mathbb{Q}$-linear combination of 1 and $\sqrt{2}$, so $[\mathbb{Q}(\sqrt{2}) : \mathbb{Q}] = 2$, which is finite.
3. Similarly, $\mathbb{Q}(i)$ is a number field, as is $\mathbb{Q}(\sqrt{d})$ for any integer d. Note that we may assume that d is not divisible by a square ("squarefree"), because if $d = m^2 d'$, $\mathbb{Q}(\sqrt{d}) = \mathbb{Q}(\sqrt{d'})$.
   Indeed, it is easy to see (from the quadratic formula) that every quadratic field $\mathbb{Q}(\alpha)$ is of this form. So every quadratic number field is $\mathbb{Q}(\sqrt{d})$ for some square-free d.
4. $\mathbb{Q}(\sqrt[3]{2})$ is a number field, as $[\mathbb{Q}(\sqrt[3]{2}) : \mathbb{Q}] = 3$, which is finite. Every element can be written in the form

$$a + b\sqrt[3]{2} + c(\sqrt[3]{2})^2,$$

   for $a, b, c \in \mathbb{Q}$, so 1, $\sqrt[3]{2}$ and $(\sqrt[3]{2})^2$ form a basis for $\mathbb{Q}(\sqrt[3]{2})$ as a vector space over $\mathbb{Q}$.

5. $\mathbb{Q}(\sqrt{2}, \sqrt{3})$ is also a number field; every element can be written in the form
   $a+b\sqrt{2}+c\sqrt{3}+d\sqrt{6}$ for rational numbers a, b, c and d, so that $\{1, \sqrt{2}, \sqrt{3}, \sqrt{6}\}$
   forms a basis for $\mathbb{Q}(\sqrt{2}, \sqrt{3})$ over $\mathbb{Q}$—it follows that $[\mathbb{Q}(\sqrt{2}, \sqrt{3}) : \mathbb{Q}] = 4$ (we
   showed the linear independence in Exercise 2.4).
6. $\mathbb{Q}(\pi)$ is *not* a number field; π does not satisfy any polynomial equation over $\mathbb{Q}$
   (as it is transcendental); therefore $[\mathbb{Q}(\pi) : \mathbb{Q}]$ is infinite.

Notice that every number field contains the rationals, so is infinite and is of characteristic 0.

Recall that the *characteristic* of a field is 0 if $1 + 1 + \cdots + 1$ is never equal to 0, and is p if p is the smallest number such that $1 + 1 + \cdots + 1 = 0$, where p is the number of 1s in the left-hand sum. Fields of characteristic 0 always contain $\mathbb{Q}$. Fields of characteristic p exist for any prime number p. They always contain the integers modulo p, $\{0, 1, \ldots, p-1\}$, which is the smallest field of characteristic p, and which we denote by $\mathbb{F}_p$.

On the other hand, every element of a number field is algebraic, so is a root of a polynomial with rational coefficients. As all roots of such polynomials are complex numbers, this means that we can view every number field as a subfield of the complex numbers $\mathbb{C}$. However, it is sometimes important to realise that there is not usually a natural way to do this; if a number field contains a square root $\sqrt{-1}$ of -1, we have a choice whether to view this as i or as $-i$ inside the complex numbers. In Chap. 3 we will think more about this issue.

Occasionally it will be useful to know that every extension is *simple*, that is, it is generated by a single element.

We need a preliminary result.

Lemma 2.15 *Suppose that $f(X) \in \mathbb{Q}[X]$ is an irreducible polynomial. Then it has distinct roots in $\mathbb{C}$.*

Proof Over $\mathbb{C}$, factorise $f(X)$ as $c \prod_{i=1}^{r} (X - \gamma_i)^{d_i}$.

If the lemma were false, $d_i > 1$ for some i, and so $f(X)$ would have a factor $(X - \gamma_i)^2$.

Writing $f(X) = (X - \gamma_i)^2 g(X)$, we see that $(X - \gamma_i)$ is also a factor of the derivative $f'(X)$. So $(X - \gamma_i)$ is a common factor of f and f', thus showing that the highest common factor h of f and f' must be of degree at least 1. But the highest common factor of f and f' is obtained by Euclid's algorithm in $\mathbb{Q}[X]$, and is a polynomial with rational coefficients that divides into both f and f'. However, f is irreducible, so its only factors are 1 and f. Since h has degree at least 1, we conclude that $h = f$. But then $f | f'$, which is absurd, since the degree of f is bigger than the degree of f', which is a nonzero polynomial (as f has degree $d \geq 1$). $\qquad \Box$

Remark 2.16 More generally, the proof shows that any irreducible polynomial over a field of characteristic 0 has distinct roots. In characteristic p, it can happen for an irreducible polynomial that its derivative is 0, since it may only involve terms in X^p, whose derivatives are divisible by p, and which therefore vanish; consider the polynomial $f(X) = X^p$ in a field of characteristic p; although this is not irreducible, it is an example where f divides f', as $f' = 0$.

Theorem 2.17 (Primitive Element) *Suppose $K \subseteq L$ is a finite extension of fields of characteristic 0 (e.g., number fields). Then $L = K(\gamma)$ for some element $\gamma \in L$.*

Proof Suppose L is generated over K by m elements. We first treat the case $m = 2$. So suppose $L = K(\alpha, \beta)$, and let f and g denote the minimal polynomials of α and β over K. Let $\alpha_1 = \alpha, \alpha_2, \ldots, \alpha_s$ be the roots of f in $\mathbb{C}$, and let $\beta_1 = \beta, \beta_2, \ldots, \beta_t$ be the roots of g. Irreducible polynomials always have distinct roots (Lemma 2.15). Thus $X = \frac{\alpha_i - \alpha_1}{\beta_1 - \beta_j}$ is the only solution (if $j \neq 1$) to

$$\alpha_i + X\beta_j = \alpha_1 + X\beta_1.$$

Choosing a $c \in K$ different from each of these X's, then each $\alpha_i + c\beta_j$ is different from $\alpha + c\beta$. We claim that $\gamma = \alpha + c\beta$ generates L over K. Certainly $\gamma \in K(\alpha, \beta) = L$. Now it suffices to verify that $\alpha, \beta \in K(\gamma)$.

The polynomials $g(X)$ and $f(\gamma - cX)$ both have coefficients in $K(\gamma)$, and have β as a root. The other roots of $g(X)$ are $\beta_2, \ldots, \beta_t$, and, as $\gamma - c\beta_j$ is not any α_i, unless $i = j = 1$, β is the only common root of $g(X)$ and $f(\gamma - cX)$. Thus, $(X - \beta)$ is the highest common factor of $g(X)$ and $f(\gamma - cX)$. But the highest common factor is a polynomial defined over any field containing the coefficients of the original two polynomials (think about how the Euclidean algorithm works for polynomials). In particular, it follows that $X - \beta$ has coefficients in $K(\gamma)$, so that $\beta \in K(\gamma)$. Then $\alpha = \gamma - c\beta \in K(\gamma)$. The result follows for $m = 2$.

Now we turn to the case where $m > 2$. We can prove this using the result we have just proven. After all, if $L = K(\alpha_1, \ldots, \alpha_m)$, we can view this as $K(\alpha_1, \ldots, \alpha_{m-2})(\alpha_{m-1}, \alpha_m)$, and the case $m = 2$ allows us to write this as $K(\alpha_1, \ldots, \alpha_{m-2})(\gamma_{m-1})$. Rewriting this as $K(\alpha_1, \ldots, \alpha_{m-3})(\alpha_{m-2}, \gamma_{m-1})$, and using the case $m = 2$ again reduces the number further still. Continuing in this way, we eventually get down to just one element. $\square$

This proof uses properties of fields of characteristic 0 in two places. Firstly, we used the fact that irreducible polynomials always have distinct roots, which is true for any field of characteristic 0. And then we chose a value of c different from all values in some finite set, which we can do because fields of characteristic 0 contain $\mathbb{Q}$, and so are infinite.

Corollary 2.18 *Let K be a number field. Then $K = \mathbb{Q}(\gamma)$ for some element γ.*

Proof Simply apply Theorem 2.17. $\square$

Let's illustrate this argument with one example.

Example 2.19 By the previous corollary, it should be possible to express the number field $\mathbb{Q}(\sqrt{2}, \sqrt{3})$ as $\mathbb{Q}(\gamma)$ for some element γ. By looking at the proof of Theorem 2.17, it seems that we should be able to take $\gamma = \sqrt{2} + c\sqrt{3}$ for almost any choice of c (only finitely many values might be excluded). Let's try $c = 1$, so that $\gamma = \sqrt{2} + \sqrt{3} \in \mathbb{Q}(\sqrt{2}, \sqrt{3})$. Then

$$
\begin{aligned}
1 &= 1 \\
\gamma &= \quad \sqrt{2} \ +\sqrt{3} \\
\gamma^2 &= 5 \qquad\qquad\qquad +2\sqrt{6} \\
\gamma^3 &= \quad 11\sqrt{2} +9\sqrt{3}
\end{aligned}
$$

and we see that $\sqrt{2} = (\gamma^3 - 9\gamma)/2$ and $\sqrt{3} = (11\gamma - \gamma^3)/2$. It follows that both $\sqrt{2}$ and $\sqrt{3}$ can be written as polynomials in γ, so that $\sqrt{2}, \sqrt{3} \in \mathbb{Q}(\gamma)$. Therefore $\mathbb{Q}(\sqrt{2}, \sqrt{3}) \subseteq \mathbb{Q}(\gamma)$. On the other hand, $\gamma \in \mathbb{Q}(\sqrt{2}, \sqrt{3})$, which gives the other inclusion $\mathbb{Q}(\gamma) \subseteq \mathbb{Q}(\sqrt{2}, \sqrt{3})$, and shows that $\mathbb{Q}(\sqrt{2}, \sqrt{3}) = \mathbb{Q}(\gamma)$, as required.

Exercise 2.8 What is the degree of $\mathbb{Q}(\sqrt[3]{2}, \sqrt{2})$ over $\mathbb{Q}$?

Exercise 2.9 Show that $\mathbb{Q}(\sqrt{3}, \sqrt{5}) = \mathbb{Q}(\sqrt{3} + \sqrt{5})$.

2.5 Integrality

We are going to do number theory in number fields, enlarged versions of the rational numbers. That is, we are going to study prime numbers, divisibility, and so on, in these larger fields.

Recall that prime numbers are defined as those positive integers which have no divisors other than themselves and 1. Even to talk about divisibility needs some notion of integrality; in $\mathbb{Q}$, any rational number is divisible by any non-zero rational number. It is only in the integers that divisibility and prime numbers are properly defined.

When we "do number theory", we almost always refer to properties of the integers $\mathbb{Z}$, rather than $\mathbb{Q}$. So to work in a number field K, we need to define a subset $\mathbb{Z}_K$ of "integers in K".

It would be nice if this subset satisfied the same algebraic properties as $\mathbb{Z}$—namely, $\mathbb{Z}_K$ should be a ring, so that we can add, subtract and multiply within $\mathbb{Z}_K$. Clearly we would like the integers in $\mathbb{Q}$ to turn out to be $\mathbb{Z}$!

It would also be desirable to arrange that, given two number fields $K \subseteq L$ and an element $\alpha \in K$, that α is an integer in K if and only if it is an integer in L. That is, if $K \subseteq L$ is an extension of number fields, we require $\mathbb{Z}_L \cap K = \mathbb{Z}_K$.

At the end of Chap. 1, we saw our first example of working in a more general number field. There, we looked at the Gaussian integers,

$$
\mathbb{Z}[i] = \{a + bi \mid a, b \in \mathbb{Z}\}
$$

which seemed to have the appropriate properties in $\mathbb{Q}(i)$. It seems reasonable to hope that our definition of integers should give $\mathbb{Z}[i]$ as the integers for $\mathbb{Q}(i)$.

We have also seen that every number field can be written in the form $\mathbb{Q}(\gamma)$, and at first glance, it might seem reasonable to suggest that we define its integers to be $\mathbb{Z}[\gamma]$. This is, after all, a ring; the elements are polynomials in γ with integer coefficients, and two of these can be added, subtracted or multiplied. In addition, it gives the right answer for $\mathbb{Q}(i)$.

Unfortunately, a moment's reflection will reveal that this is not a good definition. Indeed, it isn't even well-defined! That is, we may be able to write our number field in more than one way as $\mathbb{Q}(\gamma)$, but these may give different answers for the integers. For example, as $\sqrt{8} = 2\sqrt{2}$, we see that $\mathbb{Q}(\sqrt{8}) = \mathbb{Q}(\sqrt{2})$; on the other hand, $\mathbb{Z}[\sqrt{8}] \neq \mathbb{Z}[\sqrt{2}]$, since $\sqrt{2} \notin \mathbb{Z}[\sqrt{8}]$.

We need some more intrinsic way to determine which element of a given number field is an integer.

Associated to α is its minimal polynomial over $\mathbb{Q}$, the monic polynomial with rational coefficients of smallest degree which has α as a root. We're going to use this to give our definition of an integer:

Definition 2.20 Let α be an algebraic number. We say that α is an *algebraic integer* if the minimal polynomial of α over $\mathbb{Q}$ has coefficients in $\mathbb{Z}$.

Before we explain that the algebraic integers form a *ring*, that is, they are closed under addition, subtraction and multiplication, let's look at some examples:

Example 2.21

1. Every integer n is an algebraic integer. Its minimal polynomial over $\mathbb{Q}$ is $X - n$, and the coefficients of this polynomial are indeed integral.
2. i is an algebraic integer, as its minimal polynomial is $X^2 + 1$, which is in $\mathbb{Z}[X]$.
3. $\sqrt{2}$ is an algebraic integer, as its minimal polynomial is $X^2 - 2$, again in $\mathbb{Z}[X]$.
4. $\omega = (-1 + \sqrt{-3})/2$ is an algebraic integer, perhaps surprisingly; it is a root of the polynomial $X^2 + X + 1$—since this polynomial is irreducible, this must be the minimal polynomial of ω.
5. $(-1 + \sqrt{3})/2$ is *not* an algebraic integer, as its minimal polynomial is $X^2 + X - \frac{1}{2}$, which involves fractional coefficients.
6. π is not an algebraic integer, since it is not even an algebraic number.

You might be surprised that $(-1 + \sqrt{-3})/2$ should be an integer, but that $(-1 + \sqrt{3})/2$ isn't, but apart from that, I hope that you agree that the definition looks reasonable.

When it comes to checking whether or not a given algebraic number α is an algebraic integer, it is sometimes convenient to be able to check a weaker condition.

Lemma 2.22 *Suppose that α satisfies any monic polynomial with coefficients in $\mathbb{Z}$. Then α is an algebraic integer.*

Proof Suppose α is a root of the monic polynomial $f(X) \in \mathbb{Z}[X]$. Let $m(X) \in \mathbb{Q}[X]$ denote the minimal polynomial of α over $\mathbb{Q}$. We will show that $m(X) \in \mathbb{Z}[X]$. We have already seen that $m(X) | f(X)$, so that $f(X) = q(X)m(X)$ for some polynomial $q(X) \in \mathbb{Q}[X]$. Since $f(X)$ and $m(X)$ are both monic, clearly $q(X)$ is also.

So $f(X) = q(X)m(X)$ expresses $f(X) \in \mathbb{Z}[X]$ as a product of two monic polynomials $q(X)$ and $m(X)$ with rational coefficients. We explain that this implies $q(X)$ and $m(X)$ are both in $\mathbb{Z}[X]$.

Choose positive integers a and b so that $aq(X)$ and $bm(X)$ are polynomials with integer coefficients, and where the highest common factors of the coefficients of $aq(X)$ and $bm(X)$ are both 1. (Indeed, a and b are just the least common multiples of the denominators of the coefficients of q and m respectively.) Then

$$(ab)f(X) = aq(X).bm(X).$$

If $ab \neq 1$, choose a prime number $p|ab$. There are coefficients of $aq(X)$ and $bm(X)$ not divisible by p. So there are also terms in the product whose coefficients are not divisible by p (consider the term in the product coming from the first term of $aq(X)$ with coefficient not divisible by p with the first term of $bm(X)$ with coefficient not divisible by p). On the other hand, the product is $(ab)f(X)$, so all the coefficients must be divisible by the integer ab, and therefore by p. This contradiction shows that $ab = 1$, and therefore $a = b = 1$, as a and b are positive integers.

Therefore both $q(X)$ and $m(X)$ are already in $\mathbb{Z}[X]$. In particular, $m(X) \in \mathbb{Z}[X]$, and so the minimal polynomial of α has integral coefficients. $\square$

Remark 2.23 Here, we have essentially proven Gauss's Lemma: If a polynomial $f(X) \in \mathbb{Z}[X]$ is reducible in $\mathbb{Q}[X]$ then it is reducible in $\mathbb{Z}[X]$ (that is, if $f(X)$ factorises into polynomials with rational coefficients then it factorises into polynomials with integer coefficients).

Remark 2.24 Suppose that α is an algebraic number. Then α is the root of some monic polynomial with coefficients in $\mathbb{Q}$:

$$X^n + a_{n-1}X^{n-1} + a_{n-2}X^{n-2} + \cdots + a_0 = 0.$$

Let d be an integer which is a common multiple of all the denominators of $a_{n-1}, \ldots, a_0$. Then $d\alpha$ is a root of

$$X^n + a_{n-1}dX^{n-1} + a_{n-2}d^2X^{n-2} + \cdots + a_0d^n = 0,$$

which is a monic polynomial with integer coefficients. Therefore $d\alpha$ is an algebraic integer. This shows that every algebraic number has an integer multiple which is an algebraic integer. Equivalently, every algebraic number can be expressed as the quotient of an algebraic integer by an element of $\mathbb{Z}$.

Exercise 2.10 Show that $\frac{1+\sqrt{5}}{2}$ is an algebraic integer.

Exercise 2.11 Show that $\frac{1+\sqrt{3}}{\sqrt{2}}$ is an algebraic integer.

Exercise 2.12 Let a be an integer. Show that $\alpha = (1 + a^{1/3} + a^{2/3})/3$ is a root of

$$X^3 - X^2 + \frac{1-a}{3}X - \frac{(1-a)^2}{27} = 0.$$

[*Hint: Expand* $(\alpha - 1/3)^3$.] Deduce that if $a \equiv 1 \pmod 9$, then α is an algebraic integer.

2.6 The Ring of All Algebraic Integers

We will want to study factorisation and so on in number fields. This will require a definition of integers and primes in these fields. Amongst the properties that we would like to hold is that the integers in a number field have the same algebraic structure as the integers $\mathbb{Z}$; in particular, that they form a ring, so that we can add, subtract and multiply two integers.

Given two integers α and β, we will need to prove, for example, that $\alpha + \beta$ is an integer. From the definition, it looks as if this will mean finding a monic polynomial with integer coefficients that has $\alpha + \beta$ as a root.

Our approach will resemble the method we used earlier to show that the algebraic numbers form a field: we will reformulate the condition on integrality into one involving abstract algebra and which resembles Proposition 2.9. By a process rather similar to Corollary 2.11, we will show that if α and β are algebraic integers, so are $\alpha + \beta$, $\alpha - \beta$ and $\alpha\beta$.

Looking back at Proposition 2.9, we reformulated the property of being an algebraic number in terms of field extensions of $\mathbb{Q}$ of finite degree. We will do something similar for $\mathbb{Z}$, and our reformulation will involve the ring $\mathbb{Z}[\alpha]$, consisting of all polynomial expressions in α with integer coefficients. For algebraic numbers, we then used results and terminology from vector spaces over fields; the analogous concept for rings is called a *module*.

Recall that a *module M* over a ring R is like a vector space over a field; we should be able to add two elements of M together to get another element of M, and to multiply an element of M by an element of R, in such a way that the same rules are satisfied as for vector spaces.

The theory of modules over rings is a little more complicated than vector spaces over fields, but for now at least, we just need the concept which is analogous to "finite dimensional" for vector spaces. The appropriate condition is that the module $\mathbb{Z}[\alpha]$ is *finitely generated* over $\mathbb{Z}$. This means that there are finitely many elements $\omega_1, \ldots, \omega_n \in \mathbb{Z}[\alpha]$ such that every element of $\mathbb{Z}[\alpha]$ can be written as a sum $a_1\omega_1 + \cdots + a_n\omega_n$ for suitable integers $a_1, \ldots, a_n \in \mathbb{Z}$.

Proposition 2.25 *Let $\alpha \in \mathbb{C}$. The following are equivalent:*

1. *α is an algebraic integer;*
2. *$\mathbb{Z}[\alpha]$ is a finitely generated module over $\mathbb{Z}$.*

Proof (1) $\Rightarrow$ (2). Suppose that α is an algebraic integer. Then it is a root of a monic polynomial $f(X) \in \mathbb{Z}[X]$ of some degree n. Given any polynomial $g(X) \in \mathbb{Z}[X]$, write

$$g(X) = q(X)f(X) + r(X)$$

for $q(X), r(X) \in \mathbb{Z}[X]$, and where $r(X) = 0$ or the degree of $r(X)$ is less than n. (If $f(X)$ were not monic, we could only deduce that $q(X)$ and $r(X)$ would have *rational* coefficients.)

Substitute in $X = \alpha$; then $g(\alpha) = r(\alpha)$ as α is a root of f. This shows that $g(\alpha)$ can also be expressed as a polynomial expression of degree less than n, so $g(\alpha)$ can be written as a linear combination of $1, \alpha, \ldots, \alpha^{n-1}$ with integer coefficients.

We conclude that any polynomial expression in α with integer coefficients can be expressed as an integer linear combination of $1, \alpha, \ldots, \alpha^{n-1}$. Therefore $\mathbb{Z}[\alpha]$ is finitely generated as a $\mathbb{Z}$-module.

(2) $\Rightarrow$ (1). Suppose that $\mathbb{Z}[\alpha] = \mathbb{Z}\omega_1 + \cdots + \mathbb{Z}\omega_n$. For each i, the product $\alpha\omega_i$ is again in $\mathbb{Z}[\alpha]$, so can be written as a linear combination of the spanning set:

$$\alpha\omega_i = \sum_{j=1}^{n} a_{ij}\omega_j \tag{2.3}$$

with each $a_{ij} \in \mathbb{Z}$. Consider the column vector $\mathbf{v} = (\omega_1 \cdots \omega_n)^t$. Then (2.3) implies that $\alpha\mathbf{v} = A\mathbf{v}$ where $A = (a_{ij})$. That is, $\mathbf{v}$ is an eigenvector of A with eigenvalue α. As α is an eigenvalue, it is a root of the characteristic polynomial of A. Characteristic polynomials are always monic; also, as the entries of A are integral, its characteristic polynomial has coefficients in $\mathbb{Z}$. Thus α is a root of a monic polynomial with integer coefficients, and so α is integral. $\square$

The next result is really a corollary to the proof of the previous proposition, and is a mild generalisation:

Corollary 2.26 *Let R be a ring containing $\mathbb{Z}$. If R is finitely generated as a $\mathbb{Z}$-module, then every element $\alpha \in R$ is the root of a monic polynomial with coefficients in $\mathbb{Z}$.*

Proof We argue exactly as above; since R is finitely generated, $R = \mathbb{Z}\omega_1 + \cdots + \mathbb{Z}\omega_n$. For each i, we have $\alpha\omega_i = \sum_{j=1}^{n} a_{ij}\omega_j$ for some integers $a_{ij} \in \mathbb{Z}$, and then α is a root of the characteristic polynomial for the matrix (a_{ij}), as required. $\square$

Next, consider what happens for two algebraic integers α and β:

Proposition 2.27 *Suppose that α and β are algebraic integers. Then $\mathbb{Z}[\alpha, \beta]$ is finitely generated as a $\mathbb{Z}$-module.*

Proof By Proposition 2.25, $\mathbb{Z}[\alpha]$ and $\mathbb{Z}[\beta]$ are both finitely generated as $\mathbb{Z}$-modules. That is, there are elements $\omega_1, \ldots, \omega_m \in \mathbb{Z}[\alpha]$ such that every element of $\mathbb{Z}[\alpha]$ can be written as a $\mathbb{Z}$-linear combination of these elements. Similarly, there are elements $\theta_1, \ldots, \theta_n \in \mathbb{Z}[\beta]$ such that every element of $\mathbb{Z}[\beta]$ is a $\mathbb{Z}$-linear combination of these elements. Let's show that every element of $\mathbb{Z}[\alpha, \beta]$ is a $\mathbb{Z}$-linear combination of the finite set $\{\omega_i\theta_j \mid 1 \leq i \leq m, 1 \leq j \leq n\}$.

Every element of $\mathbb{Z}[\alpha, \beta]$ can be written as a polynomial $\sum_{k,l} a_{kl}\alpha^k\beta^l$, with $a_{kl} \in \mathbb{Z}$. Since each $\alpha^k \in \mathbb{Z}[\alpha]$, it can be written as some $\mathbb{Z}$-linear combination of $\{\omega_i \mid 1 \leq i \leq m\}$. Similarly, β^j can be written as a $\mathbb{Z}$-linear combination of $\{\theta_j \mid 1 \leq j \leq n\}$. Substituting these in, we see that every element can be written as a $\mathbb{Z}$-linear combination of the set $\{\omega_i \theta_j \mid 1 \leq i \leq m, 1 \leq j \leq n\}$, as required. $\square$

After this reformulation, it is easy to prove that algebraic integers form a ring:

Corollary 2.28 *The set of all algebraic integers forms a ring.*

Proof Let α and β be algebraic integers. We need to check that $\alpha + \beta$, $\alpha - \beta$ and $\alpha\beta$ are algebraic integers. Since $\alpha + \beta \in \mathbb{Z}[\alpha, \beta]$, and Proposition 2.27 shows that $\mathbb{Z}[\alpha, \beta]$ is finitely generated as a $\mathbb{Z}$-module, Proposition 2.25 implies that $\alpha + \beta$ is an algebraic integer.

As $\alpha - \beta$ and $\alpha\beta$ are also in $\mathbb{Z}[\alpha, \beta]$, the same argument applies to show that they are also integral. $\square$

Not only does this prove the result we want, but the argument of Proposition 2.25 also suggests a way to construct polynomials satisfied by the sum (or difference, or product) of two algebraic numbers.

Example 2.29 To explain the procedure, let's show that the sum

$$\theta = \left(\frac{1 + \sqrt{5}}{2}\right) + \left(\frac{-1 + \sqrt{-3}}{2}\right) = (\sqrt{5} + \sqrt{-3})/2$$

is an algebraic integer, by computing its minimal polynomial.

Write $\alpha = (1 + \sqrt{5})/2$ and $\beta = (-1 + \sqrt{-3})/2$. Then α has minimal polynomial $X^2 - X - 1$ and β has minimal polynomial $X^2 + X + 1$. One way to proceed is as follows.

Form the vector $\mathbf{v} = (1 \ \alpha \ \beta \ \alpha\beta)^t$. We are going to find matrices A and B with entries in $\mathbb{Z}$ such that $A\mathbf{v} = \alpha\mathbf{v}$ and $B\mathbf{v} = \beta\mathbf{v}$. That is, α is an eigenvalue of A, and β is an eigenvalue of B. Then $(A + B)\mathbf{v} = (\alpha + \beta)\mathbf{v}$, and so $\alpha + \beta$ is an eigenvalue of $A + B$. It is therefore a root of the characteristic polynomial of $A + B$, which is defined over $\mathbb{Z}$, since the entries of $A + B$ are integers. This gives a polynomial with $\alpha + \beta$ as a root.

Let's first try to construct the matrix A. It should be a 4×4 matrix such that

$$A \begin{pmatrix} 1 \\ \alpha \\ \beta \\ \alpha\beta \end{pmatrix} = \alpha \begin{pmatrix} 1 \\ \alpha \\ \beta \\ \alpha\beta \end{pmatrix} = \begin{pmatrix} \alpha \\ \alpha^2 \\ \alpha\beta \\ \alpha^2\beta \end{pmatrix}.$$

But α is a root of $X^2 = X + 1$, so $\alpha^2 = \alpha + 1$, and so we need to solve

$$A \begin{pmatrix} 1 \\ \alpha \\ \beta \\ \alpha\beta \end{pmatrix} = \begin{pmatrix} \alpha \\ \alpha + 1 \\ \alpha\beta \\ (\alpha + 1)\beta \end{pmatrix},$$

and it is easy to see that

$$A = \begin{pmatrix} 0 & 1 & 0 & 0 \\ 1 & 1 & 0 & 0 \\ 0 & 0 & 0 & 1 \\ 0 & 0 & 1 & 1 \end{pmatrix}.$$

Similarly, we can find a matrix B with the property that

$$B \begin{pmatrix} 1 \\ \alpha \\ \beta \\ \alpha\beta \end{pmatrix} = \beta \begin{pmatrix} 1 \\ \alpha \\ \beta \\ \alpha\beta \end{pmatrix} = \begin{pmatrix} \beta \\ \alpha\beta \\ \beta^2 \\ \alpha\beta^2 \end{pmatrix} = \begin{pmatrix} \beta \\ \alpha\beta \\ -(\beta+1) \\ -\alpha(\beta+1) \end{pmatrix};$$

take

$$B = \begin{pmatrix} 0 & 0 & 1 & 0 \\ 0 & 0 & 0 & 1 \\ -1 & 0 & -1 & 0 \\ 0 & -1 & 0 & -1 \end{pmatrix}.$$

Then

$$A + B = \begin{pmatrix} 0 & 1 & 1 & 0 \\ 1 & 1 & 0 & 1 \\ -1 & 0 & -1 & 1 \\ 0 & -1 & 1 & 0 \end{pmatrix},$$

and the argument above shows that θ should be a root of the characteristic polynomial of $A + B$.

Exercise 2.13 Show that this characteristic polynomial is $X^4 - X^2 + 4$, and verify explicitly that θ is a root of this polynomial.

In the same way, as $AB\mathbf{v} = A(B\mathbf{v}) = A(\beta\mathbf{v}) = \beta(A\mathbf{v}) = \alpha\beta\mathbf{v}$, $\alpha\beta$ is an eigenvalue of AB, and is therefore a root of the characteristic polynomial of AB.

More generally, if α is a root of an equation of degree m, and β is a root of an equation of degree n, form the vector of length mn:

$$\mathbf{v} = (1, \ldots, \alpha^{m-1}, \beta, \ldots, \alpha^{m-1}\beta, \ldots \ldots; \beta^{n-1}, \ldots, \alpha^{m-1}\beta^{n-1})^t.$$

As above, we can find $mn \times mn$-matrices A and B such that $A\mathbf{v} = \alpha\mathbf{v}$ and $B\mathbf{v} = \beta\mathbf{v}$. Then A and B will be $mn \times mn$-matrices, $\alpha + \beta$, $\alpha - \beta$ and $\alpha\beta$ are easily seen to be eigenvalues of $A + B$, $A - B$ and AB respectively (with $\mathbf{v}$ as eigenvector), and the characteristic polynomials of $A + B$, $A - B$ and AB have degree mn.

Further, notice that if α and β are both algebraic integers, then the matrices A and B have entries in $\mathbb{Z}$, and so the entries of $A + B$, $A - B$ and AB are all also in $\mathbb{Z}$. Therefore the characteristic polynomials of these three matrices are all integral, and are monic by definition, so this gives another proof that the eigenvalues $\alpha + \beta$, $\alpha - \beta$ and $\alpha\beta$ are all algebraic integers.

Exercise 2.14 Use this method to find a degree 6 polynomial satisfied by $\sqrt{2} + \sqrt[3]{2}$.

Of course, the same method also shows that the sum, difference and product of any two algebraic numbers is again algebraic; the two matrices A and B will in general no longer be integral, but have rational entries. We can extend the method to the case of quotients; if $\beta \neq 0$, then B will be invertible, and then $\mathbf{v}$ is an eigenvector of AB^{-1} with eigenvalue α/β. This quotient is a root of the characteristic polynomial of the rational matrix AB^{-1}, as required.

2.7 Rings of Integers of Number Fields

Now we have an obvious definition for the integers in a number field.

Definition 2.30 Let K be a number field. Then the integers in K are

$$\mathbb{Z}_K = \{\alpha \in K \mid \alpha \text{ is an algebraic integer}\}.$$

Probably the first check to make is that this gives the right answer for the rational number field $\mathbb{Q}$. Luckily, this is straightforward; a rational $a \in \mathbb{Q}$ has minimal polynomial $X - a$, and the coefficients are in $\mathbb{Z}$ if and only if $a \in \mathbb{Z}$. So the integers in $\mathbb{Q}$ using Definition 2.30 are indeed $\mathbb{Z}$, as one hopes.

Also, if $K \subseteq L$ is an extension of number fields and $\alpha \in K$, then α is an integer in K if and only if it is an integer in L. This follows simply because the condition determining whether or not α is an algebraic integer makes no reference to any field K.

Corollary 2.31 *Let K be a number field. Then $\mathbb{Z}_K$ is a ring.*

Proof Given $\alpha, \beta \in \mathbb{Z}_K$, we need to check that $\alpha + \beta$, $\alpha - \beta$ and $\alpha\beta$ all lie in $\mathbb{Z}_K$. But they certainly all lie in K, and Corollary 2.28 implies that they are all algebraic integers, so they lie in $\mathbb{Z}_K$, as required. □

Remark 2.32 $\mathbb{Z}_K$ is even an integral domain, since $\mathbb{Z}_K \subset K$, and as K is a field, it has no zero-divisors.

We say that $\mathbb{Z}_K$ is the *ring of integers* of K. In the literature, you will often see the ring of integers written as $\mathcal{O}_K$, for historical reasons (an older terminology for ring of integers is *order*—this word is still used to refer to certain subrings of $\mathbb{Z}_K$).

The following generalisation of Proposition 2.25 allows us to characterise the ring of integers $\mathbb{Z}_K$ as the largest subring of K which is a finitely generated $\mathbb{Z}$-module:

Proposition 2.33 *Suppose R is a subring of a number field K, and that R is finitely generated as a $\mathbb{Z}$-module. Then $R \subseteq \mathbb{Z}_K$.*

Proof This is immediate from Corollary 2.26. □

Earlier, we suggested that the ring of integers in $\mathbb{Q}(i)$ should be $\mathbb{Z}[i]$. We will now compute the rings of integers in all quadratic fields $\mathbb{Q}(\sqrt{d})$.

Proposition 2.34 *Suppose that d is a squarefree integer (i.e., not divisible by the square of any prime). Then*

1. *If $d \equiv 2$ or 3 (mod 4), then the ring of integers in $\mathbb{Q}(\sqrt{d})$ is*

$$\mathbb{Z}[\sqrt{d}] = \{a + b\sqrt{d} \mid a, b \in \mathbb{Z}\}.$$

2. *If $d \equiv 1$ (mod 4), then the ring of integers in $\mathbb{Q}(\sqrt{d})$ is*

$$\mathbb{Z}[\rho_d] = \{a + b\rho_d \mid a, b \in \mathbb{Z}\}$$

where $\rho_d = \frac{1+\sqrt{d}}{2}$.

Proof Let $\alpha = a + b\sqrt{d}$ with $a, b \in \mathbb{Q}$. Then α satisfies the equation $(X-a)^2 = b^2 d$, or

$$X^2 - 2aX + (a^2 - b^2 d) = 0.$$

We seek conditions on a and b to make this have integer coefficients. This implies that

$$2a \in \mathbb{Z}$$
$$a^2 - b^2 d \in \mathbb{Z}$$

Clearly the first condition implies $a \in \mathbb{Z}$ or $a = \frac{A}{2}$ where A is an odd integer. In the first case, the second condition becomes $b^2 d \in \mathbb{Z}$, and, as d is squarefree, this requires $b \in \mathbb{Z}$. So the set $\{a + b\sqrt{d} \mid a, b \in \mathbb{Z}\}$ is always contained in the ring of integers.

Let's examine when the second case can arise. Here $a = \frac{A}{2}$, and we need

$$\frac{A^2}{4} - b^2 d \in \mathbb{Z},$$

or

$$A^2 - 4b^2 d \equiv 0 \,(\mathrm{mod}\, 4).$$

This certainly requires $4b^2 d \in \mathbb{Z}$; again, as d is squarefree, $2b$ must be an integer, B say. Further, b itself cannot be in $\mathbb{Z}$; otherwise

$$\frac{A^2}{4} - b^2 d \notin \mathbb{Z}.$$

Thus B is an odd integer. Then

$$A^2 - B^2 d \equiv 0 \,(\mathrm{mod}\ 4)$$

with A and B odd integers. But the squares of odd numbers are all 1 (mod 4). Thus

$$1 - d \equiv 0 \,(\mathrm{mod}\ 4).$$

If $d \equiv 1$ (mod 4), the second case can arise, and the integers are

$$\{a + b\sqrt{d} \mid \text{either } a, b \in \mathbb{Z}, \text{ or both } a \text{ and } b \text{ are halves of odd integers}\},$$

a set which is easily seen to be the same as that of the statement. On the other hand, if $d \not\equiv 1$ (mod 4), then the only integers are $\{a + b\sqrt{d} \mid a, b \in \mathbb{Z}\}$ as claimed.   □

In particular, if $d = -1$, so that $d \equiv 3 (\mathrm{mod}\ 4)$, this result shows that the ring of integers of $\mathbb{Q}(i)$ is $\mathbb{Z}[i]$.

However, as already remarked, the ring of integers of $\mathbb{Q}(\sqrt{d})$ is not always just $\mathbb{Z}[\sqrt{d}]$. Although every element in $\mathbb{Z}[\sqrt{d}]$ is an algebraic integer, there are sometimes additional integers; if $d = -3$, for example, then $(-1 + \sqrt{-3})/2$ is an integer, as it is a root of $X^2 + X + 1$. Similarly, if $d = 5$, then $(1 + \sqrt{5})/2$ is an integer, as it is a root of $X^2 - X - 1$.

Exercise 2.15 Show that the square of the modulus of the complex number $a + b(\frac{1+\sqrt{-3}}{2}) \in \mathbb{Q}(\sqrt{-3})$ is $a^2 + ab + b^2$.

[*Hint: As usual, write down the real and imaginary parts, and consider the sum of their squares.*]

Find the elements in the ring of integers of $\mathbb{Q}(\sqrt{-3})$ with squared modulus 19. And which elements in the ring of integers of $\mathbb{Q}(\sqrt{-2})$ have squared modulus 19?

Exercise 2.16 Use Exercise 2.11 to see that the ring of integers of $\mathbb{Q}(\sqrt{2}, \sqrt{3})$ is bigger than $\mathbb{Z}[\sqrt{2}, \sqrt{3}]$.

Chapter 3
Fields, Discriminants and Integral Bases

By definition, every number field K is a finite extension of $\mathbb{Q}$. In particular, if K has degree n, then there must be elements $\alpha_1, \ldots, \alpha_n \in K$ such that every element of K can be written as a linear combination

$$x_1\alpha_1 + x_2\alpha_2 + \cdots + x_n\alpha_n$$

where $x_1, \ldots, x_n \in \mathbb{Q}$.

We've even seen (Corollary 2.18) that there is a particular element γ such that $K = \mathbb{Q}(\gamma)$, so that every element of K can be written as

$$x_1\gamma^{n-1} + x_2\gamma^{n-2} + \cdots + x_n.1,$$

a special case of the above, where our basis has a particular form.

We can ask exactly analogous questions about the ring of integers $\mathbb{Z}_K$.

1. Do there exist elements $\alpha_1, \ldots, \alpha_n \in \mathbb{Z}_K$ such that every element of $\mathbb{Z}_K$ is of the form
$$x_1\alpha_1 + x_2\alpha_2 + \cdots + x_n\alpha_n$$
   for some $x_i \in \mathbb{Z}$?
2. Does there exist a single element $\gamma \in \mathbb{Z}_K$ such that every element of $\mathbb{Z}_K$ is of the form
$$x_1\gamma^{n-1} + x_2\gamma^{n-2} + \cdots + x_n.1,$$
   for some $x_i \in \mathbb{Z}$?

It will turn out that the first question has a positive answer, but the second does not, in general.

In general, the versions of the questions for $\mathbb{Z}_K$ are somewhat harder than for K, and some of the material of this chapter could be omitted at a first reading. This is particularly true for Sect. 3.5.

F. Jarvis, *Algebraic Number Theory*, Springer Undergraduate
Mathematics Series, DOI: 10.1007/978-3-319-07545-7_3,
© Springer International Publishing Switzerland 2014

3.1 Embeddings

Suppose then that K is a number field and that $[K : \mathbb{Q}] = n$. By Corollary 2.18, there exists an element $\gamma \in K$ such that $K = \mathbb{Q}(\gamma)$. Let f denote the minimal polynomial of γ over $\mathbb{Q}$; it follows from Corollary 2.10 that f has degree n.

As $\mathbb{C}$ is algebraically closed, we can factor $f(X)$ completely over $\mathbb{C}$, and write it as

$$f(X) = \prod_{i=1}^{n}(X - \gamma_i),$$

where $\gamma_1, \ldots, \gamma_n \in \mathbb{C}$ are the (complex) roots of f. Of course, one of these is γ itself, so we will assume $\gamma_1 = \gamma$.

Definition 3.1 If $\gamma \in K$ has $f(X) \in \mathbb{Q}[X]$ as its minimal polynomial as above, then the roots $\gamma_1, \ldots, \gamma_n$ are the *conjugates* of γ.

Notice that conjugate elements have the same minimal polynomial; indeed, $\gamma_1, \ldots, \gamma_n$ are all roots of the monic irreducible polynomial f, and so f is the minimal polynomial for each of them.

By Lemma 2.15, the conjugates of an algebraic number are all distinct.

Example 3.2 Suppose that $\alpha = i$. Then its minimal polynomial is $X^2 + 1$, and the two complex roots of this are $\pm i$. Thus the two conjugates of i are i and $-i$.

The next exercise gives some justification for the terminology:

Exercise 3.1 Suppose that $\alpha = a + bi \in \mathbb{Q}(i)$. Show that its conjugates (in the sense above) are just α and $\overline{\alpha}$.

Thus the conjugates of a complex number (in this sense) are the same as the conjugates (in the familiar sense). But the concept is more general, and applies in other situations.

Exercise 3.2 Find the conjugates of $\sqrt{2}$.

Exercise 3.3 Find the conjugates of $\sqrt{2} + \sqrt{3}$.

Exercise 3.4 Find the conjugates of $\sqrt[3]{2}$.

Remark 3.3 Clearly this concept of conjugacy generalises somewhat; given an extension $L \subseteq K$ of fields, if $\alpha \in K$ has minimal polynomial $f(X) \in L[X]$ over L, then the *conjugates of α over L* are the roots of f.

Given any element of K, we can write it as a polynomial expression in γ with coefficients in $\mathbb{Q}$, simply because $K = \mathbb{Q}(\gamma)$. For each $k = 1, \ldots, n$, the map $\sigma_k : \gamma \mapsto \gamma_k$ induces a field homomorphism

$$\sigma_k : \mathbb{Q}(\gamma) \longrightarrow \mathbb{Q}(\gamma_k) \subset \mathbb{C}.$$

$$\sum_{i=0}^{n-1} x_i \gamma^i \mapsto \sum_{i=0}^{n-1} x_i \gamma_k^i$$

Remark 3.4 This map is *well-defined*—that is, if the same element of $\mathbb{Q}(\gamma)$ can be written in two different ways as a polynomial expression of γ, then applying σ_k to either expression gives the same answer.

Indeed, if $g_1(\gamma) = g_2(\gamma)$, then γ is a root of $g_1 - g_2$, and so the minimal polynomial of γ divides $g_1 - g_2$. But this minimal polynomial is just f. As γ_k is also a root of f, we see that $f(\gamma_k) = 0$, and so $g_1(\gamma_k) = g_2(\gamma_k)$.

Remark 3.5 A similar argument shows that all these maps are *injective*. If $g_1(\gamma)$ and $g_2(\gamma)$ are two elements of $K = \mathbb{Q}(\gamma)$ that map to the same element under σ_k, then $g_1(\gamma_k) = g_2(\gamma_k)$, and so γ_k must be a root of $g_1 - g_2$. Therefore the minimal polynomial of γ_k divides $g_1 - g_2$. But this minimal polynomial is exactly f, and so $f | g_1 - g_2$, from which we conclude that $g_1(\gamma) = g_2(\gamma)$.

Definition 3.6 We will use the word *embedding* to mean an injective field homomorphism; thus $\sigma_1, \ldots, \sigma_n$ are all embeddings.

Proposition 3.7 *If K is a number field of degree n, then the maps $\sigma_1, \ldots, \sigma_n$ are all of the n distinct field embeddings $K \longrightarrow \mathbb{C}$.*

Proof The arguments just given show that they are all well-defined injective field homomorphisms.

Conversely, if $\sigma : K \longrightarrow \mathbb{C}$ is a field homomorphism, and $K = \mathbb{Q}(\gamma)$, then σ must be determined by its effect on γ, as

$$\sigma \left(\sum_{i=0}^{n-1} x_i \gamma^i \right) = \sum_{i=0}^{n-1} x_i \sigma(\gamma)^i.$$

Further, applying σ to the equality $f(\gamma) = 0$ gives

$$f(\sigma(\gamma)) = \sigma(f(\gamma)) = \sigma(0) = 0,$$

and so $\sigma(\gamma)$ is a root of f, and is therefore γ_k for some k. It is then clear that $\sigma = \sigma_k$. □

As an example, let's consider the field $K = \mathbb{Q}(i)$. We have already seen in Example 3.2 that the conjugates of i are i and $-i$, so we get two embeddings from K into $\mathbb{C}$, given by $\sigma_1(a + bi) = a + bi$ and $\sigma_2(a + bi) = a - bi$. This gives us two ways to think of $\mathbb{Q}(i)$ as a subfield of $\mathbb{C}$.

Remark 3.8 It is sometimes important to be careful when writing $\mathbb{Q}(\sqrt{2})$, say, to keep in mind that the element "$\sqrt{2}$" should be regarded as just an abstract square root of 2, and is not necessarily to be identified with the positive real number 1.4142.... We are writing $\mathbb{Q}(\sqrt{2})$ as a shorthand for "$\mathbb{Q}(\alpha)$ where α is some number with $\alpha^2 = 2$".

Then choosing an embedding from $\mathbb{Q}(\sqrt{2})$ into $\mathbb{C}$ is tantamount to identifying the abstract element $\sqrt{2}$ with the particular number $1.4142\ldots$ or $-1.4142\ldots$.

Exercise 3.5 Write down the embeddings from $K = \mathbb{Q}(\sqrt{2}, \sqrt{3})$ into $\mathbb{C}$.
  [*Hint: Since $[K : \mathbb{Q}] = 4$, you should find 4 embeddings, $\sigma_1, \sigma_2, \sigma_3$ and σ_4, say.*]

1. Find an element $\alpha \in K$ such that $\sigma_1(\alpha) = \sigma_2(\alpha) = \sigma_3(\alpha) = \sigma_4(\alpha)$.
2. Find an element $\beta \in K$ such that $\sigma_1(\beta), \sigma_2(\beta), \sigma_3(\beta)$ and $\sigma_4(\beta)$ are all different.
3. Find an element $\gamma \in K$ such that $\sigma_1(\gamma) = \sigma_2(\gamma)$ and $\sigma_3(\gamma) = \sigma_4(\gamma)$, but $\sigma_1(\gamma) \neq \sigma_3(\gamma)$. (Your answer will depend on the order you wrote down your embeddings.)

Slightly more generally, the same argument as Proposition 3.7 shows that

Proposition 3.9 *Suppose that $K \subseteq L$ is a finite extension of fields, and that we have a fixed embedding $\iota : K \longrightarrow \mathbb{C}$. Then there are $[L : K]$ ways to extend the embedding ι to an embedding $L \longrightarrow \mathbb{C}$ (that is, to define embeddings $L \longrightarrow \mathbb{C}$ which agree with ι on the elements of L that belong to K).*

Proof By the Theorem of the Primitive Element (Theorem 2.17), we can write $L = K(\gamma)$, where γ has minimal polynomial over K of degree $n = [L : K]$. Then we let $\gamma_1, \ldots, \gamma_n$ denote the roots of the minimal polynomial, and define extensions $\sigma_k : L \longrightarrow \mathbb{C}$ by insisting that

$$\sigma_k \left(\sum_{i=0}^{n-1} x_i \gamma^i \right) = \sum_{i=0}^{n-1} \iota(x_i) \gamma_k^i.$$

The verification that these are all the embeddings is then identical to the previous arguments, and is left to the reader as an exercise.   □

Suppose that K is a number field, and that $\alpha \in K$. We will next try to understand the images of α under each of the embeddings.

Let's do an example. Suppose that $K = \mathbb{Q}(\sqrt{2}, \sqrt{3})$, and that $\alpha = \sqrt{6}$. In Exercise 3.5, you should find four embeddings, given by:

$$\sigma_1 \left(a + b\sqrt{2} + c\sqrt{3} + d\sqrt{6} \right) = a + b\sqrt{2} + c\sqrt{3} + d\sqrt{6}$$

$$\sigma_2 \left(a + b\sqrt{2} + c\sqrt{3} + d\sqrt{6} \right) = a + b\sqrt{2} - c\sqrt{3} - d\sqrt{6}$$

$$\sigma_3 \left(a + b\sqrt{2} + c\sqrt{3} + d\sqrt{6} \right) = a - b\sqrt{2} + c\sqrt{3} - d\sqrt{6}$$

$$\sigma_4 \left(a + b\sqrt{2} + c\sqrt{3} + d\sqrt{6} \right) = a - b\sqrt{2} - c\sqrt{3} + d\sqrt{6}.$$

Then $\sigma_1(\sqrt{6}) = \sigma_4(\sqrt{6}) = \sqrt{6}$, and $\sigma_2(\sqrt{6}) = \sigma_3(\sqrt{6}) = -\sqrt{6}$. Unsurprisingly, these images are just the conjugates of $\sqrt{6}$, but each occurs twice.

To study the general case, where K is any number field, we need one result from field theory:

Theorem 3.10 *Suppose that $K \subseteq L \subseteq M$ is a "tower" of fields. Then, assuming M is a finite extension of L, and L is a finite extension of K, we have $[M : K] = [M : L][L : K]$.*

Proof Suppose that $[M : L] = m$ and $[L : K] = n$. Then there are elements $\omega_1, \ldots, \omega_n$ such that every element of L is a linear combination of $\omega_1, \ldots, \omega_n$ with coefficients in K, and elements $\theta_1, \ldots, \theta_m$ such that every element of M is a linear combination of $\theta_1, \ldots, \theta_m$ with coefficients in L. We claim that $\{\theta_i \omega_j\}$ is a basis for M as a K-vector space.

Given $\mu \in M$, express it first as a linear combination of $\theta_1, \ldots, \theta_m$ with coefficients in L, and then express each of these coefficients as linear combinations of $\omega_1, \ldots, \omega_n$ with coefficients in K. This shows that μ can be written as a linear combination of $\{\theta_i \omega_j\}$ with coefficients in K.

Furthermore, these elements form a linearly independent set. To see this, we take a linear combination which is 0:

$$\alpha_{11}\theta_1\omega_1 + \alpha_{12}\theta_1\omega_2 + \cdots + \alpha_{1n}\theta_1\omega_n + \alpha_{21}\theta_2\omega_1 + \cdots + \alpha_{mn}\theta_m\omega_n = 0.$$

Rearrange this as

$$(\alpha_{11}\omega_1 + \cdots + \alpha_{1n}\omega_n)\theta_1 + \cdots + (\alpha_{m1}\omega_1 + \cdots + \alpha_{mn}\omega_n)\theta_m = 0.$$

Now this is a linear combination of $\theta_1, \ldots, \theta_m$ with coefficients in L, and since they form a basis, each of the coefficients must vanish. Thus, for each i,

$$\alpha_{i1}\omega_1 + \cdots + \alpha_{in}\omega_n = 0,$$

and as $\omega_1, \ldots, \omega_n$ forms a basis for L as a vector space over K, we again conclude that each $\alpha_{ij} = 0$, as required.

Thus $\{\theta_i \omega_j\}$ form a basis for M over K, and so $[M : K] = mn$, as required. $\square$

Let's return to the general case, where K is a number field of degree n over $\mathbb{Q}$.

Suppose $\alpha \in K$ with minimal polynomial $g(X) \in \mathbb{Q}[X]$. Then α generates a field $\mathbb{Q}(\alpha)$ contained in K. If g has degree d_α, then $[\mathbb{Q}(\alpha) : \mathbb{Q}] = d_\alpha$. Suppose that the conjugates of α are written $\alpha_1 = \alpha, \alpha_2, \ldots, \alpha_{d_\alpha}$.

From the tower of fields $\mathbb{Q} \subseteq \mathbb{Q}(\alpha) \subseteq K$, we know that

$$[K : \mathbb{Q}] = [K : \mathbb{Q}(\alpha)][\mathbb{Q}(\alpha) : \mathbb{Q}],$$

and so we see that $d_\alpha | n$. Write $r = r_\alpha$ for n/d_α.

Proposition 3.11 *The images* $\sigma_i(\alpha)$ *are the conjugates* $\{\alpha_1, \ldots, \alpha_{d_\alpha}\}$, *each occurring with multiplicity* r_α.

Proof We have extension fields $\mathbb{Q} \subseteq \mathbb{Q}(\alpha) \subseteq K$. By Proposition 3.7, we know that there are d_α embeddings $\iota_k : \mathbb{Q}(\alpha) \longrightarrow \mathbb{C}$. The embedding ι_k is determined by the property that $\iota_k(\alpha) = \alpha_k$.

Make a choice of any of these embeddings $\iota_k : \mathbb{Q}(\alpha) \longrightarrow \mathbb{C}$. As the extension $\mathbb{Q}(\alpha) \subseteq K$ has degree r_α, we know by Proposition 3.9 that the embedding ι_k extends to an embedding $K \longrightarrow \mathbb{C}$ in r_α ways; by definition of an extension of embeddings, each extension maps α to α_k.

We can do this for each of the d_α embeddings ι_k, extending each in r_α ways. We thus obtain $d_\alpha r_\alpha = n$ embeddings from K to $\mathbb{C}$. But we know that there should be exactly n embeddings from K into $\mathbb{C}$, again by Proposition 3.7.

We therefore conclude that all of the embeddings $\sigma_i : K \longrightarrow \mathbb{C}$ have been obtained, and we have seen that α is taken to each of its conjugates $\{\alpha_1, \ldots, \alpha_{d_\alpha}\}$ with multiplicity r_α.   □

Corollary 3.12 *Suppose* α *in* K *has minimal polynomial* g *of degree* d_α, *and that* $r_\alpha = n/d_\alpha$. *Then*

$$\prod_{i=1}^{n} (X - \sigma_k(\alpha)) = g(X)^{r_\alpha}.$$

Proof Both sides are monic polynomials with the same roots.   □

Exercise 3.6 Verify Corollary 3.12 explicitly for each of the three elements α, β and γ you found in Exercise 3.5.

3.2 Norms and Traces

Again let K be a number field, with $[K : \mathbb{Q}] = n$. Suppose that $\alpha \in K$. Then multiplication by α gives a map

$$m_\alpha : K \longrightarrow K$$

$$x \longmapsto \alpha x$$

This map is $\mathbb{Q}$-linear: it is easy to see that

$$m_\alpha(x + x') = \alpha(x + x') = \alpha x + \alpha x' = m_\alpha(x) + m_\alpha(x')$$

and that

$$m_\alpha(tx) = \alpha(tx) = t(\alpha x) = tm_\alpha(x)$$

for $t \in \mathbb{Q}$. (Of course, the map is even K-linear, since $m_\alpha(tx) = tm_\alpha(x)$ even for $t \in K$, but we won't need that.)

After choosing a basis for K over $\mathbb{Q}$, the map is represented by a $n \times n$-matrix. We define the *trace* of α, written $T_{K/\mathbb{Q}}(\alpha)$, to be the trace of this matrix, and the *norm*, written $N_{K/\mathbb{Q}}(\alpha)$, to be its determinant. Choosing a different basis would give a conjugate $n \times n$-matrix representing the map; it is a well-known result from linear algebra that the trace and determinant of an endomorphism do not depend on the choice of basis. When the field K is clearly understood, we may simply write $N(\alpha)$ and $T(\alpha)$ for the norm and trace.

In the same way, if L/K is an extension of number fields, there is a notion of $T_{L/K}$ and $N_{L/K}$.

Example 3.13 Let's look at one example. Suppose that $K = \mathbb{Q}(\sqrt{2}, \sqrt{3})$, and take $\alpha = \sqrt{2} + \sqrt{3}$. We can choose a basis $\{1, \sqrt{2}, \sqrt{3}, \sqrt{6}\}$ for K, and see how multiplying by α affects an element:

$$\alpha(a + b\sqrt{2} + c\sqrt{3} + d\sqrt{6}) = (2b + 3c) + (a + 3d)\sqrt{2} + (a + 2d)\sqrt{3} + (b + c)\sqrt{6},$$

which we interpret as a map on coefficients

$$\begin{pmatrix} a \\ b \\ c \\ d \end{pmatrix} \mapsto \begin{pmatrix} 2b + 3c \\ a + 3d \\ a + 2d \\ b + c \end{pmatrix},$$

which is the map given by multiplication by

$$\begin{pmatrix} 0 & 2 & 3 & 0 \\ 1 & 0 & 0 & 3 \\ 1 & 0 & 0 & 2 \\ 0 & 1 & 1 & 0 \end{pmatrix};$$

the trace is the sum of the diagonal entries, which is 0, and the norm of α is the determinant of the matrix, which is 1.

In fact, even more is true. We have defined $N_{K/\mathbb{Q}}(\alpha)$ and $T_{K/\mathbb{Q}}(\alpha)$ to be the determinant and trace of the matrix given by multiplication, but in the fact the minimal polynomial of α is exactly the characteristic polynomial of this matrix; recall that the determinant and trace are just two coefficients of this polynomial:

Proposition 3.14 *Suppose that α is an algebraic number with minimal polynomial $g(X) \in \mathbb{Q}[X]$, and form the map m_α as above. Then the characteristic polynomial of the matrix of m_α is $g(X)$.*

Proof Suppose that the minimal polynomial for α is given by $x^n + c_1 x^{n-1} + \cdots + c_n = 0$.

As just mentioned, we can compute the characteristic polynomial after choosing any basis. One basis for $\mathbb{Q}(\alpha)$ over $\mathbb{Q}$ is given by $\{1, \alpha, \alpha^2, \ldots, \alpha^{n-1}\}$, where α has degree n.

Since $\alpha.\alpha^k = \alpha^{k+1}$ for $k = 0, \ldots, n-2$, and $\alpha.\alpha^{n-1} = \alpha^n = -c_1\alpha^{n-1} - \cdots - c_n$, the map m_α is given by

$$
\begin{aligned}
m_\alpha\left(a_0 + a_1\alpha + \cdots + a_{n-1}\alpha^{n-1}\right) &= \alpha\left(a_0 + a_1\alpha + \cdots + a_{n-1}\alpha^{n-1}\right) \\
&= a_0\alpha + \cdots + a_{n-2}\alpha^{n-1} + a_{n-1}\alpha^n \\
&= a_0\alpha + \cdots + a_{n-2}\alpha^{n-1} \\
&\quad + a_{n-1}\left(-c_1\alpha^{n-1} - \cdots - c_n\right) \\
&= -a_{n-1}c_n + (a_0 - a_{n-1}c_{n-1})\alpha \\
&\quad + \cdots + (a_{n-2} - a_{n-1}c_1)\alpha^{n-1}
\end{aligned}
$$

and so the map of m_α using this basis is given by

$$
\begin{pmatrix} a_0 \\ a_1 \\ \vdots \\ a_{n-1} \end{pmatrix} \mapsto \begin{pmatrix} -a_{n-1}c_n \\ a_0 - a_{n-1}c_{n-2} \\ \vdots \\ a_{n-2} - a_{n-1}c_1 \end{pmatrix},
$$

which is the same as multiplication by the matrix

$$
\begin{pmatrix} & & & -c_n \\ 1 & & & -c_{n-1} \\ & 1 & & -c_{n-2} \\ & & \ddots & \vdots \\ & & 1 & -c_1 \end{pmatrix},
$$

and it is easy to check that this matrix has characteristic polynomial given by $x^n + c_1x^{n-1} + \cdots + c_n = 0$. $\qquad\square$

Now let's explore some properties of norms and traces.

Lemma 3.15 *Suppose $\alpha \in K$. Then $N_{K/\mathbb{Q}}(\alpha)$ and $T_{K/\mathbb{Q}}(\alpha)$ are both in $\mathbb{Q}$.*

Proof This simply follows because they are the trace and determinant of a matrix with entries in $\mathbb{Q}$. $\qquad\square$

This is a rather abstract definition of the trace and norm of an element, but we can make it a little more concrete.

Proposition 3.16 *Write $\sigma_1, \ldots, \sigma_n$ for the embeddings of K into $\mathbb{C}$. If $\alpha \in K$, then*

$$N_{K/\mathbb{Q}}(\alpha) = \prod_{k=1}^{n} \sigma_k(\alpha) \quad and \quad T_{K/\mathbb{Q}}(\alpha) = \sum_{k=1}^{n} \sigma_k(\alpha).$$

Proof Let g denote the minimal polynomial of α over $\mathbb{Q}$. Note that $\mathbb{Q}(\alpha)$ may be smaller than K (for example, we might even have $\alpha \in \mathbb{Q}$), so the degree of g may be strictly smaller than n. As g is irreducible, $[\mathbb{Q}(\alpha) : \mathbb{Q}] = \deg g$, and we will write d_α for this degree.

We have field extensions $\mathbb{Q} \subseteq \mathbb{Q}(\alpha) \subseteq K$; let $\{\beta_1, \ldots, \beta_{r_\alpha}\}$ denote a basis for K over $\mathbb{Q}(\alpha)$, where $[K : \mathbb{Q}(\alpha)] = r_\alpha = n/d_\alpha$. Clearly $\{1, \alpha, \ldots, \alpha^{d_\alpha-1}\}$ is a basis for $\mathbb{Q}(\alpha)$ over $\mathbb{Q}$. Standard results in field theory (see the proof of Theorem 3.10) now show that the set of products $\{\beta_i \alpha^j \mid 1 \le i \le r_\alpha, 0 \le j < d_\alpha\}$ forms a basis for K over $\mathbb{Q}$.

Choose this basis, and fix one of the β_i. Consider the multiplication-by-α map m_α on the block spanned by $\{\beta_i, \beta_i\alpha, \ldots, \beta_i\alpha^{d_\alpha-1}\}$. It is easy to see the matrix of this map on this block is the same for all choices of β_i, and that it is the same as the matrix of the map m_α on $\mathbb{Q}(\alpha)$, where we use the basis $\{1, \alpha, \ldots, \alpha^{d_\alpha-1}\}$. We have already noted that this matrix has characteristic polynomial g.

It follows that the characteristic polynomial of m_α on K is given by $g(X)^{r_\alpha}$. But the roots of g, by definition, are exactly the conjugates of α. The roots of $g(X)^{r_\alpha}$ are therefore the conjugates of α, taken with multiplicity r_α.

By Proposition 3.11, these are exactly the images of α under all the embeddings $\sigma_i : K \longrightarrow \mathbb{C}$, and the result then follows. $\qquad\square$

Corollary 3.17 *If $\alpha \in \mathbb{Z}_K$, then $N_{K/\mathbb{Q}}(\alpha)$ and $T_{K/\mathbb{Q}}(\alpha)$ are both in $\mathbb{Z}$.*

Proof As $\alpha \in \mathbb{Z}_K$, its minimal polynomial $g(X) \in \mathbb{Z}[X]$. With the notation of Corollary 3.12, we see that $g(X)^{r_\alpha} \in \mathbb{Z}[X]$. But this implies that the product $\prod_{i=1}^{n}(X - \sigma_i(\alpha)) \in \mathbb{Z}[X]$; the constant coefficient of this polynomial is $(-1)^n N_{K/\mathbb{Q}}(\alpha)$, and the coefficient of X^{n-1} is $-T_{K/\mathbb{Q}}(\alpha)$. $\qquad\square$

Exercise 3.7 Compute the norm and trace of $a + bi \in \mathbb{Q}(i)$, using both the definition as the determinant and trace of the multiplication map, and also Proposition 3.16.

Exercise 3.8 Compute the norm and trace of $\sqrt{2} + \sqrt{3} \in \mathbb{Q}(\sqrt{2}, \sqrt{3})$, again both from the definition and from Proposition 3.16.

3.3 The Discriminant

As before, suppose that K is a number field of degree n over $\mathbb{Q}$. We have seen that this means that:

1. K is generated over $\mathbb{Q}$ by n elements (the definition of the degree);
2. there are n embeddings $\sigma_1, \ldots, \sigma_n$ from K into $\mathbb{C}$ (see Proposition 3.7).

Suppose that $\{\omega_1, \ldots, \omega_n\}$ lie in K. For the moment, we won't assume that these are a basis as in (1).

Consider the matrix:

$$M = \begin{pmatrix} \sigma_1(\omega_1) & \sigma_1(\omega_2) & \cdots & \sigma_1(\omega_n) \\ \sigma_2(\omega_1) & \sigma_2(\omega_2) & \cdots & \sigma_2(\omega_n) \\ \vdots & \vdots & \ddots & \vdots \\ \sigma_n(\omega_1) & \sigma_n(\omega_2) & \cdots & \sigma_n(\omega_n) \end{pmatrix}.$$

We will use the determinant of M as a measure of how "widely spaced" the set $\{\omega_1, \ldots, \omega_n\}$ is. (More explanation will be given in Remark 3.25, and again in Chap. 10.) One reason why this is not quite satisfactory is that this determinant is defined only up to sign; taking the same set, but in a different order may multiply the determinant by -1. To avoid this issue, we will use the square of this determinant.

Definition 3.18 Define the *discriminant* of $\{\omega_1, \ldots, \omega_n\}$ to be $\Delta\{\omega_1, \ldots, \omega_n\} = (\det M)^2$.

Here is a reformulation, sometimes more useful for computation:

Lemma 3.19 *With the notation as above, form the matrix T, where $T_{ij} = T_{K/\mathbb{Q}}(\omega_i\omega_j)$. Then $\Delta\{\omega_1, \ldots, \omega_n\} = \det T$.*

Proof Simply notice that $\det M = \det M^t$, and so

$$\Delta\{\omega_1, \ldots, \omega_n\} = (\det M)^2 = \det(M^t M).$$

But

$$(M^t M)_{ij} = \sum_{k=1}^{n} M^t_{ik} M_{kj} = \sum_{k=1}^{n} M_{ki} M_{kj} = \sum_{k=1}^{n} \sigma_k(\omega_i)\sigma_k(\omega_j) = \sum_{k=1}^{n} \sigma_k(\omega_i\omega_j),$$

which is equal to $T_{K/\mathbb{Q}}(\omega_i\omega_j)$ by Proposition 3.16. The result follows. $\square$

Exercise 3.9 Let $K = \mathbb{Q}(\sqrt{2}, \sqrt{3})$. Compute $\Delta\{1, \sqrt{2}, \sqrt{3}, \sqrt{6}\}$ using the formula from Definition 3.18, and verify that it agrees with the formula given in Lemma 3.19.

Corollary 3.20 *Suppose that $\{\omega_1, \ldots, \omega_n\}$ consists of elements of $\mathbb{Z}_K$. Then $\Delta\{\omega_1, \ldots, \omega_n\} \in \mathbb{Z}$.*

Proof If each $\omega_i \in \mathbb{Z}_K$, then $\omega_i\omega_j \in \mathbb{Z}_K$, as $\mathbb{Z}_K$ is closed under multiplication. By Corollary 3.17, this means that $T_{K/\mathbb{Q}}(\omega_i\omega_j) \in \mathbb{Z}$. Finally, $\Delta\{\omega_1, \ldots, \omega_n\}$ is the square of the determinant of a matrix with entries in $\mathbb{Z}$, so is itself in $\mathbb{Z}$. $\square$

Example 3.21 Let's consider one special case. As $K = \mathbb{Q}(\gamma)$ for some γ, one natural basis for K over $\mathbb{Q}$ is given by $\{1, \gamma, \gamma^2, \ldots, \gamma^{n-1}\}$. As usual, write $\gamma_1, \ldots, \gamma_n$ for the conjugates of γ; then the discriminant $\Delta\{1, \gamma, \gamma^2, \ldots, \gamma^{n-1}\}$ is given by

$$\begin{vmatrix} 1 & \gamma_1 & \cdots & \gamma_1^{n-1} \\ 1 & \gamma_2 & \cdots & \gamma_2^{n-1} \\ \vdots & \vdots & \ddots & \vdots \\ 1 & \gamma_n & \cdots & \gamma_n^{n-1} \end{vmatrix}^2.$$

This is a Vandermonde determinant, and is equal to $\prod_{i<j}(\gamma_i - \gamma_j)^2$. We have already remarked (Lemma 2.15) that the conjugates of γ are distinct, and so we conclude that the discriminant $\Delta\{1, \gamma, \gamma^2, \ldots, \gamma^{n-1}\}$ is nonzero.

Let's remark if $f(X)$ is the minimal polynomial of γ, its roots are $\gamma_1, \ldots, \gamma_n$, and its discriminant is defined to be exactly $\prod_{i<j}(\gamma_i - \gamma_j)^2$, so that the discriminant of $f(X)$ coincides with the discriminant $\Delta\{1, \gamma, \ldots, \gamma^{n-1}\}$, which justifies the terminology.

In fact, we shall see that the discriminant of any basis is nonzero. For this, we need one preliminary result.

Proposition 3.22 *Suppose that the elements of two sets $\{\omega_1, \ldots, \omega_n\}$ and $\{\omega_1', \ldots, \omega_n'\}$ are related by*

$$\omega_i' = c_{1i}\omega_1 + \cdots + c_{ni}\omega_n$$

for rational numbers $c_{ij} \in \mathbb{Q}$. Write C for the matrix (c_{ij}). Then

$$\Delta\{\omega_1', \ldots, \omega_n'\} = (\det C)^2 \Delta\{\omega_1, \ldots, \omega_n\}.$$

Proof Set

$$M' = \begin{pmatrix} \sigma_1(\omega_1') & \sigma_1(\omega_2') & \cdots & \sigma_1(\omega_n') \\ \sigma_2(\omega_1') & \sigma_2(\omega_2') & \cdots & \sigma_2(\omega_n') \\ \vdots & \vdots & \ddots & \vdots \\ \sigma_n(\omega_1') & \sigma_n(\omega_2') & \cdots & \sigma_n(\omega_n') \end{pmatrix},$$

so that $\Delta\{\omega_1', \ldots, \omega_n'\} = (\det M')^2$. Note that

$$\sigma_k(\omega_i') = c_{1i}\sigma_k(\omega_1) + \cdots + c_{ni}\sigma_k(\omega_n)$$

since σ_k is a homomorphism which is the identity on rational numbers.

It is easy to see that this implies that $M' = CM$, where $C = (c_{ij})$. The result follows from the multiplicativity of the determinant. $\qquad\square$

Exercise 3.10 In Example 2.19, we showed that if $K = \mathbb{Q}(\sqrt{2}, \sqrt{3})$, then $K = \mathbb{Q}(\gamma)$ for $\gamma = \sqrt{2} + \sqrt{3}$. Compute $\Delta\{1, \gamma, \gamma^2, \gamma^3\}$ using the formula from Example 3.21.

You have also computed $\Delta\{1, \sqrt{2}, \sqrt{3}, \sqrt{6}\}$ (Exercise 3.9). Write the set $\{1, \gamma, \gamma^2, \gamma^3\}$ in terms of the basis $\{1, \sqrt{2}, \sqrt{3}, \sqrt{6}\}$ and verify the formula of Proposition 3.22 in this example.

Now we can prove the claim we made earlier.

Proposition 3.23 *Suppose that* $\{\omega_1, \ldots, \omega_n\}$ *is a basis for* K *over* $\mathbb{Q}$. *Then* $\Delta\{\omega_1, \ldots, \omega_n\} \neq 0$.

Proof As usual, write $K = \mathbb{Q}(\gamma)$ for some element $\gamma \in K$. Then $\{1, \gamma, \ldots, \gamma^{n-1}\}$ is a basis for K over $\mathbb{Q}$. We can write the basis $\{\omega_1, \ldots, \omega_n\}$ in terms of $\{1, \gamma, \ldots, \gamma^{n-1}\}$ as

$$\omega_i = c_{1i}1 + c_{2i}\gamma + \cdots + c_{ni}\gamma^{n-1}.$$

The condition that $\{\omega_1, \ldots, \omega_n\}$ is also a basis means that $\det(c_{ij}) \neq 0$. Indeed, if it is a basis, we can write

$$\gamma^{i-1} = c'_{1i}\omega_1 + c'_{2i}\omega_2 + \cdots + c'_{ni}\omega_n,$$

for some c'_{ij}, and it is easy to see that this implies that $C'C = I$, where $C = (c_{ij})$ and $C' = (c'_{ij})$, so that C and C' are invertible. The previous proposition shows that

$$\Delta\{\omega_1, \ldots, \omega_n\} = (\det(c_{ij}))^2 \Delta\{1, \gamma, \ldots, \gamma^{n-1}\},$$

and the result follows.   □

There is a converse to this result also:

Proposition 3.24 *The set* $\{\omega_1, \ldots, \omega_n\}$ *is a basis for* K *over* $\mathbb{Q}$ *if and only if* $\Delta\{\omega_1, \ldots, \omega_n\} \neq 0$.

Proof We have already done the hard work, to see that the discriminant of a basis is nonzero.

Conversely, if $\{\omega_1, \ldots, \omega_n\}$ are linearly dependent over $\mathbb{Q}$, then there is some dependency

$$x_i\omega_1 + \cdots + x_n\omega_n = 0$$

for some $x_1, \ldots, x_n \in \mathbb{Q}$, not all zero. Apply the embedding σ_k to this equality; as σ is a field homomorphism and fixes each element of $\mathbb{Q}$, we get

$$x_i\sigma_k(\omega_1) + \cdots + x_n\sigma_k(\omega_n) = 0.$$

This gives a linear dependency between the columns of the matrix M above, with $M_{ij} = \sigma_i(\omega_j)$, and so $\det M = 0$. Thus $\Delta\{\omega_1, \ldots, \omega_n\} = 0$, as required.   □

Remark 3.25 As an aside, let's give a preliminary justification of the comments above that the discriminant measures how widely spaced the set is. We will cover this in more detail later in the book (Chap. 10).

Some of the n embeddings may map K into the real numbers $\mathbb{R} \subset \mathbb{C}$; we call these *real embeddings*. The other embeddings occur in complex conjugate pairs; if

$\sigma : K \longrightarrow \mathbb{C}$ is an embedding, then so is $\overline{\sigma}$, where $\overline{\sigma}(\omega) = \overline{\sigma(\omega)}$. These *complex embeddings* therefore occur as complex conjugate pairs.

Write r_1 for the number of real embeddings of K into $\mathbb{C}$, and r_2 for the number of complex conjugate pairs of embeddings. Since there are n embeddings in total, we have $r_1 + 2r_2 = n$. Every pair $(\sigma, \overline{\sigma})$ of complex embeddings together map K into $\mathbb{C}^2$, but it is easy to see that the image is actually contained in a *real* 2-dimensional subspace; after all, if $\sigma(\omega) = a + bi$, then $\overline{\sigma}(\omega) = a - bi$, so that the real and imaginary parts of $\overline{\sigma}(\omega)$ are already determined by the real and imaginary parts of $\sigma(\omega)$. Since each real embedding maps K into $\mathbb{R} \subset \mathbb{C}$ and each pair of complex embeddings map K into a 2-dimensional real subspace of $\mathbb{C}^2$, we see that the collection of all embeddings $\iota = (\sigma_1, \ldots, \sigma_n)$ maps K into a real subspace V of $\mathbb{C}^n$ of real dimension n.

Given our set $\{\omega_1, \ldots, \omega_n\}$, the image of $\mathbb{Z}\iota(\omega_1) + \cdots + \mathbb{Z}\iota(\omega_n)$ is contained in this subspace V. When the set is not a basis, the image will lie in a subspace of V of strictly smaller dimension, and the discriminant will vanish—but if it is a basis, the discriminant will measure the volume of a fundamental region (see Definition 7.1) for the image, and thus how sparsely these points are spaced.

3.4 Integral Bases

We say that the set $\{\omega_1, \ldots, \omega_n\}$ is an *integral basis* for the ring of integers $\mathbb{Z}_K$ when every element of $\mathbb{Z}_K$ is uniquely expressible as a $\mathbb{Z}$-linear combination of elements of the set.

Example 3.26 We have already seen examples of integral bases for quadratic fields (see Proposition 2.34). If $K = \mathbb{Q}(\sqrt{d})$, with d a squarefree integer, and $d \equiv 1 \pmod 4$, then $\mathbb{Z}_K = \mathbb{Z} + \mathbb{Z}.\frac{1+\sqrt{d}}{2}$, so an integral basis is $\{1, \frac{1+\sqrt{d}}{2}\}$. Similarly, if $d \equiv 2 \pmod 4$ or $d \equiv 3 \pmod 4$, then $\mathbb{Z}_K = \mathbb{Z} + \mathbb{Z}\sqrt{d}$, so an integral basis is $\{1, \sqrt{d}\}$.

In general, it is not obvious that such bases exist, but the main result of this section is that they do for all number fields K.

Equivalently, we will prove that the ring of integers of K is a *free abelian group of rank* $n = [K : \mathbb{Q}]$. Recall that a free abelian group A of rank n is one which is the direct sum of n subgroups, each infinite cyclic (so isomorphic to $\mathbb{Z}$). Then $A \cong \mathbb{Z}\omega_1 + \cdots + \mathbb{Z}\omega_n$, so that every element of A can be expressed uniquely as $x_1\omega_1 + \cdots + x_n\omega_n$ for some $x_i \in \mathbb{Z}$. This is exactly the property required of an integral basis.

Suppose that $[K : \mathbb{Q}] = n$. By definition of the degree, we can choose a basis $\{\omega_1, \ldots, \omega_n\}$ for K over $\mathbb{Q}$; thus every element of K can be written $x_1\omega_1 + \cdots + x_n\omega_n$ for $x_i \in \mathbb{Q}$.

Theorem 3.27 *Let K be a number field. Then the ring of integers $\mathbb{Z}_K$ has an integral basis.*

Proof Given any basis, it follows from Remark 2.24 that we can replace each element in our basis with a nonzero multiple so that every basis element is in $\mathbb{Z}_K$.

We also know (Proposition 3.23) that the discriminant of every basis consisting of elements of $\mathbb{Z}_K$ is an integer.

Choose a basis $\{\omega_1, \ldots, \omega_n\}$, consisting of elements of $\mathbb{Z}_K$, such that $|\Delta\{\omega_1, \ldots, \omega_n\}|$ is as small as possible (since it is always a positive integer, this bound is attained).

We claim that this set is indeed an integral basis for K. If not, there would be some element $\omega \in \mathbb{Z}_K$ whose expression in terms of this basis

$$\omega = x_1\omega_1 + \cdots + x_n\omega_n$$

has coefficients which are in $\mathbb{Q}$, but not all in $\mathbb{Z}$. Reordering our basis elements if necessary, suppose that $x_1 \notin \mathbb{Z}$. Then we can choose $a_1 \in \mathbb{Z}$ with $|x_1 - a_1| \leq \frac{1}{2}$.

Define $\omega_1' = \omega - a_1\omega_1 = (x_1 - a_1)\omega_1 + x_2\omega_2 + \cdots + x_n\omega_n$. Then ω_1' is again in $\mathbb{Z}_K$ (as $\omega \in \mathbb{Z}_K$, $\omega_1 \in \mathbb{Z}_K$, and $a_1 \in \mathbb{Z}$). Define also $\omega_2' = \omega_2, \ldots, \omega_n' = \omega_n$.

Then $\{\omega_1', \ldots, \omega_n'\}$ is another basis; it is easy to see that each of the elements of both sets can be expressed as a linear combination of the other (recall that $x_1 - a_1 \neq 0$).

We now apply Proposition 3.22. For the change of basis from $\{\omega_1, \ldots, \omega_n\}$ to $\{\omega_1', \ldots, \omega_n'\}$, the matrix C is given by

$$\begin{pmatrix} x_1 - a_1 & x_2 & x_3 & \ldots & x_n \\ 0 & 1 & 0 & \ldots & 0 \\ 0 & 0 & 1 & \ldots & 0 \\ \vdots & \vdots & \vdots & \ddots & \vdots \\ 0 & 0 & 0 & \ldots & 1 \end{pmatrix},$$

and Proposition 3.22 gives

$$\Delta\{\omega_1', \ldots, \omega_n'\} = (x_1 - a_1)^2 \Delta\{\omega_1, \ldots, \omega_n\}.$$

But $|x_1 - a_1| \leq \frac{1}{2}$, so this means that

$$\Delta\{\omega_1', \ldots, \omega_n'\} < \Delta\{\omega_1, \ldots, \omega_n\}$$

and this contradicts the minimality of the discriminant of the basis $\{\omega_1, \ldots, \omega_n\}$. $\square$

So integral bases exist; in particular, the ring of integers of a number field of degree n is a free abelian group of rank n.

Proposition 3.28 *If $\{\omega_1, \ldots, \omega_n\}$ and $\{\omega_1', \ldots, \omega_n'\}$ are two integral bases for a number field K, then*

$$\Delta\{\omega_1', \ldots, \omega_n'\} = \Delta\{\omega_1, \ldots, \omega_n\}.$$

Proof If $\{\omega_1, \ldots, \omega_n\}$ and $\{\omega_1', \ldots, \omega_n'\}$ are two integral bases, then each element of the second can be written as an integral linear combination of those in the first; that is, in the notation of Proposition 3.22, each $c_{ij} \in \mathbb{Z}$. The equality

$$\Delta\{\omega_1', \ldots, \omega_n'\} = (\det C)^2 \Delta\{\omega_1, \ldots, \omega_n\}$$

then implies that the integer $\Delta\{\omega_1, \ldots, \omega_n\}$ divides the integer $\Delta\{\omega_1', \ldots, \omega_n'\}$. But the same argument applies in the other direction too; each element of the first basis can be written as an integral linear combination of the second basis, and a similar argument shows that the integer $\Delta\{\omega_1', \ldots, \omega_n'\}$ divides the integer $\Delta\{\omega_1, \ldots, \omega_n\}$. From this we see that

$$\Delta\{\omega_1', \ldots, \omega_n'\} = \pm\Delta\{\omega_1, \ldots, \omega_n\}.$$

Also each $c_{ij} \in \mathbb{Z}$, so that $\det C \in \mathbb{Z}$, and $(\det C)^2 > 0$. Thus

$$\Delta\{\omega_1', \ldots, \omega_n'\} = \Delta\{\omega_1, \ldots, \omega_n\},$$

as required. $\square$

Definition 3.29 Suppose that K is a number field. The *discriminant* D_K of K is defined to be the discriminant of any integral basis for K. It exists by Proposition 3.28.

Example 3.30 Consider the case $K = \mathbb{Q}(\sqrt{d})$ with d squarefree and $d \equiv 1 \pmod 4$. Then an integral basis is $\{1, \frac{1+\sqrt{d}}{2}\}$. There are two embeddings into $\mathbb{C}$, given by $\sigma_1(a + b\sqrt{d}) = a + b\sqrt{d}$ and $\sigma_2(a + b\sqrt{d}) = a - b\sqrt{d}$. The discriminant is

$$\begin{vmatrix} \sigma_1(1) & \sigma_1\left(\frac{1+\sqrt{d}}{2}\right) \\ \sigma_2(1) & \sigma_2\left(\frac{1+\sqrt{d}}{2}\right) \end{vmatrix}^2 = \begin{vmatrix} 1 & \frac{1+\sqrt{d}}{2} \\ 1 & \frac{1-\sqrt{d}}{2} \end{vmatrix}^2 = \left(-\sqrt{d}\right)^2 = d.$$

Thus if $K = \mathbb{Q}(\sqrt{d})$ as above, $D_K = d$.

Exercise 3.11 Recall that an integral basis for $K = \mathbb{Q}(\sqrt{d})$ with d squarefree and $d \equiv 2 \pmod 4$ or $d \equiv 3 \pmod 4$ is $\{1, \sqrt{d}\}$. Show that in this case $D_K = 4d$.

3.5 Further Theory of the Discriminant

This section contains some further results on the discriminant that are used later in the book. The reader may wish to skip this section, or at least the proofs, on first reading.

Proposition 3.31 *Suppose that $K = \mathbb{Q}(\gamma)$, and that the minimal polynomial of γ over $\mathbb{Q}$ is $f(X) \in \mathbb{Q}[X]$ of degree n. Then*

$$\Delta\left\{1, \gamma, \ldots, \gamma^{n-1}\right\} = (-1)^{n(n-1)/2} N_{K/\mathbb{Q}}(f'(\gamma)).$$

Proof By Example 3.21, the discriminant $\Delta\{1, \gamma, \ldots, \gamma^{n-1}\} = \prod_{i<j}(\gamma_i - \gamma_j)^2$, where the conjugates of γ are $\gamma_1, \ldots, \gamma_n$. Recall that the conjugates are the roots in $\mathbb{C}$ of the minimal polynomial $f(X)$, and that minimal polynomials are monic. So $f(X) = \prod_{i=1}^{n}(X - \gamma_i)$. Using the product rule,

$$f'(X) = \sum_{k=1}^{n} \prod_{i \neq k}(X - \gamma_i), \tag{3.1}$$

and so

$$f'(\gamma_j) = \prod_{i \neq j}(\gamma_j - \gamma_i),$$

since only the term with $k = j$ in (3.1) doesn't have a factor $(X - \gamma_j)$. Then

$$N_{K/\mathbb{Q}}(f'(\gamma)) = \prod_{j=1}^{n} f'(\gamma_j)$$

$$= \prod_{j=1}^{n} \prod_{i \neq j}(\gamma_j - \gamma_i),$$

and notice that if $i < j$, this product has a bracket $(\gamma_i - \gamma_j)$ and a bracket $(\gamma_j - \gamma_i)$. It follows that

$$N_{K/\mathbb{Q}}(f'(\gamma)) = \prod_{i<j}\left[-(\gamma_i - \gamma_j)^2\right] = (-1)^{n(n-1)/2}\Delta\left\{1, \gamma, \ldots, \gamma^{n-1}\right\}.$$

$\square$

   This can often be used to compute discriminants, especially when a primitive element γ is given explicitly; one simply takes its minimal polynomial, and evaluates the right-hand side of the proposition.

Exercise 3.12 Let $K = \mathbb{Q}(\sqrt[3]{2})$, and let $\gamma = \sqrt[3]{2}$. Use the proposition to compute $\Delta\{1, \gamma, \gamma^2\}$.

Lemma 3.32 *Suppose that $\omega_1, \ldots, \omega_n$ is a basis for K over $\mathbb{Q}$ consisting of elements of $\mathbb{Z}_K$. Then*

$$\Delta\{\omega_1, \ldots, \omega_n\}.\mathbb{Z}_K \subseteq \mathbb{Z}\omega_1 + \cdots + \mathbb{Z}\omega_n.$$

Proof Let $\alpha \in \mathbb{Z}_K$. As $\{\omega_1, \ldots, \omega_n\}$ is a basis, we can write

$$\alpha = x_1\omega_1 + \cdots + x_n\omega_n, \tag{3.2}$$

for some $x_1, \ldots, x_n \in \mathbb{Q}$. Multiply through by ω_j, and take the trace:

$$T_{K/\mathbb{Q}}(\alpha\omega_j) = \sum_{i=1}^{n} x_i T_{K/\mathbb{Q}}(\omega_i\omega_j). \tag{3.3}$$

As α and $\omega_j \in \mathbb{Z}_K$, we have $T_{K/\mathbb{Q}}(\alpha\omega_j) \in \mathbb{Z}$ by Corollary 3.17. Similarly, the traces $T_{K/\mathbb{Q}}(\omega_i\omega_j)$ on the right-hand side are also in $\mathbb{Z}$ for all i, j. So the equations (3.3) can be regarded as a set of linear equations whose solution is given by $x_1, \ldots, x_n$; Cramer's rule implies that the solutions are quotients of integers (given by suitable determinants of integers) by $\det(T_{K/\mathbb{Q}}(\omega_i\omega_j)) = \Delta\{\omega_1, \ldots, \omega_n\}$. So $\Delta\{\omega_1, \ldots, \omega_n\}x_i \in \mathbb{Z}$ for all i, and multiplying (3.2) by $\Delta\{\omega_1, \ldots, \omega_n\}$, we see that

$$\Delta\{\omega_1, \ldots, \omega_n\}.\alpha \in \mathbb{Z}\omega_1 + \cdots + \mathbb{Z}\omega_n$$

as required. □

This lemma can be very helpful in finding integral bases for a number field K. Indeed, one strategy is the following:

Step 1 Find any basis for K over $\mathbb{Q}$, and scale the basis elements so that they are in $\mathbb{Z}_K$. Let $\{\omega_1, \ldots, \omega_n\}$ be the result.

Step 2 Compute $\Delta = \Delta\{\omega_1, \ldots, \omega_n\}$. Then Lemma 3.32 shows that

$$\mathbb{Z}_K \subseteq \frac{1}{\Delta}(\mathbb{Z}\omega_1 + \cdots + \mathbb{Z}\omega_n),$$

so every integer must be of the form

$$x_1\omega_1 + \cdots + x_n\omega_n$$

for $x_i \in \mathbb{Q}$ but where the denominators divide Δ.

Step 3 For a prime $p^2 | \Delta$, check whether any element of the form $\omega = x_1\omega_1 + \cdots + x_n\omega_n$ is integral, where x_i is a rational number with denominator dividing p. If such an integral ω exists, where some x_i is not in $\mathbb{Z}$, so has denominator p, replace ω_i with ω to get a set with discriminant Δ/p^2 (by Proposition 3.22). Since the discriminant of an integral basis must be in $\mathbb{Z}$, we need only do this for primes p with $p^2 | \Delta$. Now return to Step 2. If no such element is integral, for any prime p with $p^2 | \Delta$, then we have an integral basis.

One simple consequence is the following:

Corollary 3.33 *Suppose that K is a number field and $\omega_1, \ldots, \omega_n$ are elements of $\mathbb{Z}_K$ such that $\Delta\{\omega_1, \ldots, \omega_n\}$ is squarefree. Then $\{\omega_1, \ldots, \omega_n\}$ is an integral basis.*

However, in practice, it is fairly unusual for a number field K to have squarefree discriminant, so this corollary is not as useful as one might hope.

The final result in this section will be used in Chap. 9. (The proof is, frankly, not all that interesting, and the reader is advised to skip it if possible.)

Proposition 3.34 *Suppose that $K_1 = \mathbb{Q}(\gamma_1)$ and $K_2 = \mathbb{Q}(\gamma_2)$ are two number fields of degree n_1 and n_2 respectively, such that $K = \mathbb{Q}(\gamma_1, \gamma_2)$ has degree $n_1 n_2$ over $\mathbb{Q}$. Suppose that $\{\omega_1, \ldots, \omega_{n_1}\}$ and $\{\omega'_1, \ldots, \omega'_{n_2}\}$ are integral bases for K_1 and K_2 respectively, with discriminants D_1 and D_2. If D_1 and D_2 are coprime, then $\{\omega_i \omega'_j\}$ forms an integral basis for K, of discriminant $D_1^{n_2} D_2^{n_1}$.*

Proof We first claim that $\{\omega_i \omega'_j\}$ form a basis for K over $\mathbb{Q}$. Indeed, every element of K is a polynomial expression in γ_1 and γ_2. Every power of γ_1 lies in K_1, so is a linear combination of $\{\omega_1, \ldots, \omega_{n_1}\}$, and similarly every power of γ_2 lies in K_2 and is thus a linear combination of $\{\omega'_1, \ldots, \omega'_{n_2}\}$. Thus every product $\gamma_1^a \gamma_2^b$ is a linear combination of the elements of $\{\omega_i \omega'_j\}$. Each element of K is a linear combination of these monomials, so is also a linear combination of this set. As we have $n_1 n_2$ such elements, and $[K : \mathbb{Q}] = n_1 n_2$ by hypothesis, they must be linearly independent, and are therefore a basis.

We now want to show that they form an *integral* basis.

If $\alpha \in \mathbb{Z}_K$, we can write

$$\alpha = \sum_{i=1}^{n_1} \sum_{j=1}^{n_2} x_{ij} \omega_i \omega'_j,$$

and we want to see that $x_{ij} \in \mathbb{Z}$. Then

$$\alpha = \sum_{i=1}^{n_1} \sum_{j=1}^{n_2} x_{ij} \omega_i \omega'_j = \sum_{i=1}^{n_1} \left(\sum_{j=1}^{n_2} x_{ij} \omega'_j \right) \omega_i = \sum_{i=1}^{n_1} y_i \omega_i,$$

where $y_i = \sum_{j=1}^{n_2} x_{ij} \omega'_j \in K_2$.

Since $[K : \mathbb{Q}] = n_1 n_2$ and $[K_1 : \mathbb{Q}] = n_1$, we conclude from the tower law (Theorem 3.10) that $[K : K_1] = n_2$. Since $K = K_1(\gamma_2)$, we see that there are n_2 embeddings of K into $\mathbb{C}$ which are the identity on K_1 (which we regard as a subfield of $\mathbb{C}$ using any fixed embedding). Let $\{\sigma'_1, \ldots, \sigma'_{n_2}\}$ denote these embeddings of K into $\mathbb{C}$. (Notice that these embeddings restrict to the n_2 different embeddings of K_2 into $\mathbb{C}$ if we just regard them as maps on the elements of $K_2 \subseteq K$, since they are determined by sending γ_2 to one of its conjugates.)

Then if $\mathbf{x} = \begin{pmatrix} \sigma'_1(\alpha) \\ \vdots \\ \sigma'_{n_2}(\alpha) \end{pmatrix}$, and $\mathbf{y} = \begin{pmatrix} y_1 \\ \vdots \\ y_{n_2} \end{pmatrix}$, we have $\mathbf{x} = M\mathbf{y}$, where $M_{kl} = \sigma'_k(\omega'_l)$.

By definition, $D_2 = (\det M)^2$. As in Lemma 3.32, $D_2 y_i = \sum_{j=1}^{n_2} D_2 x_{ij} \omega'_j$ has coefficients in $\mathbb{Z}$, and so $D_2 x_{ij} \in \mathbb{Z}$. In the same way (exchanging the roles of K_1 and K_2), $D_1 x_{ij} \in \mathbb{Z}$. As D_1 and D_2 are coprime, we conclude that each $x_{ij} \in \mathbb{Z}$, and so $\{\omega_i \omega'_j\}$ forms an integral basis for $\mathbb{Z}_K$.

If $\{\sigma_1, \ldots, \sigma_{n_1}\}$ denotes the embeddings of K into $\mathbb{C}$, which are the identity on K_2, then all the embeddings of K into $\mathbb{C}$ are given by $\{\sigma_i \sigma'_j\}$. (This can easily be seen by observing that an embedding is uniquely determined by its effect on γ_1

and γ_2; these in turn uniquely determine σ_i and σ'_j.) Then the discriminant of the basis $\{\omega_i \omega'_j\}$ is given by $(\det A)^2$, where A is an $n_1 n_2 \times n_1 n_2$-matrix with $A_{ki,lj} = (\sigma_k \sigma'_l)(\omega_i \omega'_j) = \sigma_k(\omega_i)\sigma'_l(\omega'_j)$. We can decompose A as $A = BC$, where B is the $n_2 \times n_2$ matrix of $n_1 \times n_1$-blocks given by

$$B = \begin{pmatrix} Q & 0 & \dots & 0 \\ 0 & Q & \dots & 0 \\ \vdots & \vdots & \ddots & \vdots \\ 0 & 0 & \dots & Q \end{pmatrix},$$

where Q is the $n_1 \times n_1$-matrix with $Q_{ki} = \sigma_k(\omega_i)$, and C is the block matrix

$$C = \begin{pmatrix} \sigma'_1(\omega'_1)I & \sigma'_2(\omega'_1)I & \dots & \sigma'_{n_2}(\omega'_1)I \\ \sigma'_1(\omega'_2)I & \sigma'_2(\omega'_2)I & \dots & \sigma'_{n_2}(\omega'_2)I \\ \vdots & \vdots & \ddots & \vdots \\ \sigma'_1(\omega'_{n_2})I & \sigma'_2(\omega'_{n_2})I & \dots & \sigma'_{n_2}(\omega'_{n_2})I \end{pmatrix},$$

where I is the $n_1 \times n_1$-identity matrix.

Clearly $\det(B) = \det(Q)^{n_2}$, so that $\det(B)^2 = ((\det Q)^2)^{n_2} = D_1^{n_2}$. Also, $\det(C) = \det(\sigma'_l(\omega'_j))^{n_1}$, so that $\det(C)^2 = D_2^{n_1}$.

As $\Delta = \det(A)^2 = \det(B)^2 \det(C)^2$, the result follows. $\square$

Exercise 3.13 If $K_1 = \mathbb{Q}(\sqrt{2})$ and $K_2 = \mathbb{Q}(\sqrt{5})$, use Proposition 3.34 to write down an integral basis for $\mathbb{Q}(\sqrt{2}, \sqrt{5})$, and its discriminant. Verify your answer directly from the definition of the discriminant.

3.6 Rings of Integers in Some Cubic and Quartic Fields

We have already computed the rings of integers for all quadratic fields, in Proposition 2.34.

In this section, we will consider several further examples in which we construct integral bases; in two of these examples, we will also show the ring of integers cannot be expressed in the form $\mathbb{Z}[\gamma]$ for any element γ.

Incidentally, fields K where $\mathbb{Z}_K = \mathbb{Z}[\gamma]$ are sometimes called *monogenic*, and the basis $\{1, \gamma, \dots, \gamma^{n-1}\}$ is sometimes called a *power basis*.

3.6.1 $K = \mathbb{Q}(\sqrt{2}, \sqrt{3})$

Given that the ring of integers of $\mathbb{Q}(\sqrt{2})$ is $\mathbb{Z}[\sqrt{2}]$ and the ring of integers of $\mathbb{Q}(\sqrt{3})$ is $\mathbb{Z}[\sqrt{3}]$, one might hope that the ring of integers of $K = \mathbb{Q}(\sqrt{2}, \sqrt{3})$ should be $\mathbb{Z}[\sqrt{2}, \sqrt{3}]$. However we have already seen that this is false, in Exercises 2.11 and 2.16. (This is not a contradiction to Proposition 3.34, since the discriminants of $\mathbb{Q}(\sqrt{2})$ and $\mathbb{Q}(\sqrt{3})$ are not coprime.)

The methods used in this section generalise to other *biquadratic* fields, that is, fields of the form $\mathbb{Q}(\sqrt{d_1}, \sqrt{d_2})$. We will deal with these fields in more generality once we have developed more theory; the result is given by Proposition 8.22.

Let $\alpha \in \mathbb{Z}_K$. Then we can write $\alpha = a + b\sqrt{2} + c\sqrt{3} + d\sqrt{6}$ for some $a, b, c, d \in \mathbb{Q}$.

As $\alpha \in \mathbb{Z}_K$, all of its conjugates

$$\alpha_2 = a - b\sqrt{2} + c\sqrt{3} - d\sqrt{6}$$

$$\alpha_3 = a + b\sqrt{2} - c\sqrt{3} - d\sqrt{6}$$

$$\alpha_4 = a - b\sqrt{2} - c\sqrt{3} + d\sqrt{6}$$

are also algebraic integers. As the set of algebraic integers is closed under addition, the following are also algebraic integers:

$$\alpha + \alpha_2 = 2a + 2c\sqrt{3}$$

$$\alpha + \alpha_3 = 2a + 2b\sqrt{2}$$

$$\alpha + \alpha_4 = 2a + 2d\sqrt{6}.$$

By Proposition 2.34, these are integral if $2a, 2b, 2c, 2d \in \mathbb{Z}$. Thus

$$\alpha = \frac{A + B\sqrt{2} + C\sqrt{3} + D\sqrt{6}}{2},$$

for $A, B, C, D \in \mathbb{Z}$, where $A = 2a$, $B = 2b$, $C = 2c$ and $D = 2d$.

Also,

$$\alpha\alpha_2 = (a + c\sqrt{3})^2 - (b\sqrt{2} + d\sqrt{6})^2$$

$$= a^2 + 2ac\sqrt{3} + 3c^2 - 2b^2 - 4bd\sqrt{3} - 6d^2$$

$$= \frac{A^2 + 3C^2 - 2B^2 - 6D^2}{4} + \frac{AC - 2BD}{2}\sqrt{3}$$

is also integral. Thus $4 | A^2 + 3C^2 - 2B^2 - 6D^2$ and $2 | AC - 2BD$. The second implies that $2 | AC$, so that at least one of A and C is even. If only one were even, then $A^2 + 3C^2 - 2B^2 - 6D^2$ would be odd, and the first requirement would fail. So both A and C are even.

Then the second divisibility is automatic, and the first reduces to $4 | 2B^2 + 6D^2$, or $2 | B^2 + D^2$, so that B and D are both even or both odd.

So integers are all of the form

$$\alpha = a + b\sqrt{2} + c\sqrt{3} + d\sqrt{6}$$

with $a, c \in \mathbb{Z}$ and b and d both integral or both halves of odd integers.

It remains to check that elements of this form are all integers. But such elements are integer linear combinations of 1, $\sqrt{2}$, $\sqrt{3}$ and $\frac{\sqrt{2}+\sqrt{6}}{2} = \frac{1+\sqrt{3}}{\sqrt{2}}$. The first three are obviously integers, and the last is integral by Exercise 2.11.

We now claim that $\mathbb{Z}_K = \mathbb{Z}[\gamma]$, where $\gamma = \frac{\sqrt{2}+\sqrt{6}}{2}$.

Indeed, $\gamma^2 = 2 + \sqrt{3}$ and $\gamma^3 = \frac{5\sqrt{2}+3\sqrt{6}}{2}$, so

$$\sqrt{2} = \gamma^3 - 3\gamma \quad \text{and} \quad \sqrt{3} = \gamma^2 - 2.$$

Then each element in the integral basis can be written as an element in $\mathbb{Z}[\gamma]$, and so $\mathbb{Z}_K \subseteq \mathbb{Z}[\gamma]$. Conversely, $\gamma \in \mathbb{Z}_K$, and so $\mathbb{Z}[\gamma] \subseteq \mathbb{Z}_K$, as $\mathbb{Z}_K$ is a ring.

Exercise 3.14 Compute the discriminant D_K using the following four methods: the original definition (Definition 3.18), the reformulation (Lemma 3.19), and the two methods which apply when $\mathbb{Z}_K$ can be written in the form $\mathbb{Z}[\gamma]$ (Example 3.21 and Proposition 3.31).

3.6.2 $K = \mathbb{Q}(\sqrt{-2}, \sqrt{-5})$

The determination of the ring of integers in this case is very similar to that of $\mathbb{Q}(\sqrt{2}, \sqrt{3})$.

Exercise 3.15 Verify that if $K = \mathbb{Q}(\sqrt{-2}, \sqrt{-5})$, then an integral basis is given by $\{1, \sqrt{-2}, \sqrt{-5}, \frac{\sqrt{-2}+\sqrt{10}}{2}\}$.

However, the reader might like to check that if $\gamma = \frac{\sqrt{-2}+\sqrt{10}}{2}$, the argument above that $\mathbb{Z}_K = \mathbb{Z}[\gamma]$ does not work in this case.

Indeed, one can show that there is no element γ such that $\mathbb{Z}_K = \mathbb{Z}[\gamma]$. Here is one way to see this. The argument is elementary, but quite intricate. (We will reinterpret it in Remark 5.47.)

Consider the elements

$$\alpha_1 = \left(1 + \sqrt{-2}\right)\left(1 + \sqrt{-5}\right)$$

$$\alpha_2 = \left(1 + \sqrt{-2}\right)\left(1 - \sqrt{-5}\right)$$

$$\alpha_3 = \left(1 - \sqrt{-2}\right)\left(1 + \sqrt{-5}\right)$$

$$\alpha_4 = \left(1 - \sqrt{-2}\right)\left(1 - \sqrt{-5}\right)$$

All lie in $\mathbb{Z}_K$, and $\alpha_1 + \alpha_2 + \alpha_3 + \alpha_4 = 4$. Notice that

$$\alpha_1\alpha_2 = \left(1 + \sqrt{-2}\right)^2 \left(1 + \sqrt{-5}\right)\left(1 - \sqrt{-5}\right) = 6\left(1 + \sqrt{-2}\right)^2,$$

and similarly $3|\alpha_i\alpha_j$ for any pair $i \neq j$.

Notice that this implies that

$$(\alpha_1 + \alpha_2 + \alpha_3 + \alpha_4)^n \equiv \alpha_1^n + \alpha_2^n + \alpha_3^n + \alpha_4^n \pmod{3};$$

expanding the left-hand side gives the right-hand side together with lots of terms involving more than one of the α_i. (Of course, the congruence actually takes place in $\mathbb{Z}_K$; it means that the two sides differ by an element of $3\mathbb{Z}_K$.)

Then

$$T_{K/\mathbb{Q}}(\alpha_1^n) = \alpha_1^n + \alpha_2^n + \alpha_3^n + \alpha_4^n \equiv (\alpha_1 + \alpha_2 + \alpha_3 + \alpha_4)^n = 4^n \equiv 1 \pmod{3}.$$

If $3|\alpha_1^n$, then it would also divide any of its conjugates, so $3|\alpha_i^n$ for all i, so that $3|T_{K/\mathbb{Q}}(\alpha_1^n)$. Thus $3 \nmid \alpha_1^n$ for any n, and similarly $3 \nmid \alpha_i^n$ for any i and any n.

Now suppose that $\mathbb{Z}_K = \mathbb{Z}[\gamma]$ for some γ, and let $f(X) \in \mathbb{Z}[X]$ be its minimal polynomial. As $\alpha_i \in \mathbb{Z}_K$, we can write $\alpha_i = f_i(\gamma)$ for some $f_i \in \mathbb{Z}[X]$.

Now $3|\alpha_i\alpha_j$ for all $i \neq j$, but $3 \nmid \alpha_i^n$ for any i and n.

Let $\overline{f}$ denote the polynomial f with its coefficients reduced modulo 3, so that $\overline{f}(X) \in \mathbb{F}_3[X]$ (recall that $\mathbb{F}_3 = \{0, 1, 2\}$, the integers modulo 3).

Exercise 3.16 Show that if $g(X) \in \mathbb{Z}[X]$, that $3|g(\gamma)$ in $\mathbb{Z}[\gamma]$ if and only if $\overline{g}$ is divisible by $\overline{f}$ in $\mathbb{F}_3[X]$.

Applying this to $f_i(X)f_j(X)$, where we know that $3|\alpha_i\alpha_j$, we conclude that $\overline{f}|\overline{f}_i\overline{f}_j$ for all $i \neq j$. But $3 \nmid \alpha_i^n$, so that $\overline{f} \nmid \overline{f}_i^n$ for any i and n.

Together, these imply that for all $i \neq j$, $\overline{f}$ has a factor dividing $\overline{f}_i$ but not $\overline{f}_j$. This means that $\overline{f}$ must have at least four different irreducible factors.

But f is a quartic, so the different factors of $\overline{f}$ must all be linear. However, there are only three different linear factors in $\mathbb{F}_3[X]$, namely, X, $X - 1$ and $X - 2$. This gives a contradiction.

3.6.3 $K = \mathbb{Q}(\sqrt[3]{2})$

For our first example of a cubic field, we will consider $K = \mathbb{Q}(\sqrt[3]{2})$. Write $\alpha = \sqrt[3]{2}$, and $\omega = e^{2\pi i/3}$. Notice that $1 + \omega + \omega^2 = 0$.

Suppose that $\theta_1 = a + b\alpha + c\alpha^2$, where $a, b, c \in \mathbb{Q}$, lies in $\mathbb{Z}_K$. Then its conjugates are also algebraic integers:

$$\theta_2 = a + ba\omega + c\alpha^2\omega^2$$
$$\theta_3 = a + ba\omega^2 + c\alpha^2\omega,$$

although they are not in K. Then

$$\theta_1 + \theta_2 + \theta_3 = 3a$$
$$\theta_1\theta_2 + \theta_2\theta_3 + \theta_3\theta_1 = 3a^2 - 6bc$$
$$\theta_1\theta_2\theta_3 = a^3 + 2b^3 + 4c^3 - 6abc$$

are all also algebraic integers. As they are also rational, they are all in $\mathbb{Z}$. Write $A = 3a$, $B = 3b$ and $C = 3c$. The first equation gives $A \in \mathbb{Z}$. Multiplying the second by 3 gives $A^2 - 2BC \equiv 0 \,(\text{mod } 3)$, and the third by 27 gives $A^3 + 2B^3 + 4C^3 - 6ABC \equiv 0 \,(\text{mod } 27)$.

Notice that the second and third give $B, C \in \mathbb{Z}$: if $A^2 - 2BC \in \mathbb{Z}$, then $2BC \in \mathbb{Z}$, and so $6ABC \in \mathbb{Z}$, so the last implies that $2B^3 + 4C^3 \in \mathbb{Z}$. But the only way that rationals can satisfy $2BC \in \mathbb{Z}$ and $2B^3 + 4C^3 \in \mathbb{Z}$ is if $B, C \in \mathbb{Z}$ (if a prime p occurs in the denominator of B, say, then as $2BC \in \mathbb{Z}$, it cannot also occur in the denominator of C, but then p will be in the denominator of $2B^3 + 4C^3$).

If $3|A$, then $2BC \equiv 0 \,(\text{mod } 3)$, so either B or C is divisible by 3, and then $3|A^3 + 2B^3 + 4C^3 - 6ABC$ implies that both must be.

If $3 \nmid A$, then the only solutions to $A^2 - 2BC \equiv 0 \,(\text{mod } 3)$ and $A^3 + 2B^3 + 4C^3 - 6ABC \equiv 0 \,(\text{mod } 3)$ are $A \equiv 1$, $B \equiv 2$ and $C \equiv 1 \,(\text{mod } 3)$ or $A \equiv 2$, $B \equiv 1$ and $C \equiv 2 \,(\text{mod } 3)$. Put $A = 1 + 3l$, $B = 2 + 3m$, $C = 1 + 3n$, and then $A^3 + 2B^3 + 4C^3 - 6ABC \equiv 9 \,(\text{mod } 27)$, whatever the choice of l, m and n. Similarly, if $A = 2+3l$, $B = 1+3m$, $C = 2+3n$, then $A^3 + 2B^3 + 4C^3 - 6ABC \equiv 18 \,(\text{mod } 27)$ for any l, m and n. This means that there are no solutions with $3 \nmid A$.

Thus $3|A$, $3|B$ and $3|C$, which implies that $a, b, c \in \mathbb{Z}$, and so the ring of integers is $\mathbb{Z}[\sqrt[3]{2}]$.

3.6.4 $K = \mathbb{Q}(\sqrt[3]{175})$

We consider a more general cubic case, where we adjoin a cube root of an integer which has a square factor.

Take $m = 175 = 5^2 \times 7$. We will compute $\mathbb{Z}_K$. Note that if $\alpha = \sqrt[3]{175}$, then $\alpha^2 = \sqrt[3]{5^4 7^2} = 5\sqrt[3]{5 \times 7^2} = 7\sqrt[3]{245}$. So $\alpha' = \sqrt[3]{245}$ is another element in K (and indeed, $K = \mathbb{Q}(\alpha')$ also). Furthermore, both α and α' are integral, as they are roots of the monic integral polynomials $X^3 - 175$ and $X^3 - 245$ respectively.

We claim that $\mathbb{Z}_K$ has integral basis $\{1, \alpha, \alpha'\}$.

This is rather like the last example, but we will also use the strategy mentioned in Sect. 3.5.

We first compute $\Delta\{1, \alpha, \alpha'\}$.

The embeddings into $\mathbb{C}$ are given by

$$\sigma_1(a + b\alpha + c\alpha') = a + b\alpha + c\alpha'$$
$$\sigma_2(a + b\alpha + c\alpha') = a + b\alpha\omega + c\alpha'\omega^2$$
$$\sigma_3(a + b\alpha + c\alpha') = a + b\alpha\omega^2 + c\alpha'\omega$$

so the discriminant is given by the square of the determinant

$$\begin{vmatrix} 1 & \alpha & \alpha' \\ 1 & \alpha\omega & \alpha'\omega^2 \\ 1 & \alpha\omega^2 & \alpha'\omega \end{vmatrix} = -3\sqrt{3}i\alpha\alpha'.$$

Note that $\alpha\alpha' = 5 \times 7 = 35$. So $\Delta\{1, \alpha, \alpha'\} = (-3\sqrt{3}i\alpha\alpha')^2 = -3^3 5^2 7^2$, which means that all integers must be of the form $\frac{a + b\alpha + c\alpha'}{d}$ where $a, b, c, d \in \mathbb{Z}$ and $d | 3 \times 5 \times 7$, using Lemma 3.32.

Let's suppose $\theta_1 = \frac{a + b\alpha + c\alpha'}{5}$ is an integer. Then so are its conjugates

$$\theta_2 = \frac{a + b\alpha\omega + c\alpha'\omega^2}{5} \text{ and } \theta_3 = \frac{a + b\alpha\omega^2 + c\alpha'\omega}{5}.$$

Then $\theta_1 + \theta_2 + \theta_3 = 3a/5$ is an integer, so $5|a$.

Now $\theta_1 = A + \frac{b\alpha + c\alpha'}{5}$, where $A \in \mathbb{Z}$, so $\frac{b\alpha + c\alpha'}{5} \in \mathbb{Z}_K$. Its norm is the product of the conjugates:

$$\frac{b^3\alpha^3 + c^3\alpha'^3}{5^3} = \frac{175b^3 + 245c^3}{125} = \frac{35b^3 + 49c^3}{25}.$$

We need this to be an integer. But if $35b^3 + 49c^3 \equiv 0 \pmod{25}$, then certainly $35b^3 + 49c^3 \equiv 0 \pmod 5$, so that $5|c^3$, and $5|c$. Then as $35b^3 + 49c^3 \equiv 0 \pmod{25}$, we see also $5|b$. Thus 5 cannot occur in the denominator of an element of $\mathbb{Z}_K$.

Exactly the same argument works for 7.

For $p = 3$, we need to consider $\theta_1 = \frac{a + b\alpha + c\alpha'}{3}$ and determine when it is an integer.

Exercise 3.17 Show, using the method of the previous example, that if $\theta_1 = \frac{a + b\alpha + c\alpha'}{3}$ is in $\mathbb{Z}_K$, then $3|a$, $3|b$ and $3|c$.

It follows that $\mathbb{Z}_K$ has $\{1, \alpha, \alpha'\}$ as an integral basis.

Suppose that there is an integral basis of the form $\{1, \gamma, \gamma^2\}$ for some γ. Writing $\gamma = a + b\alpha + c\alpha'$, then $\gamma - a = b\alpha + c\alpha'$; if $\{1, \gamma, \gamma^2\}$ is an integral basis, it is easy to see that $\{1, \gamma - a, (\gamma - a)^2\}$ is also an integral basis, so we can assume that γ is simply of the form $b\alpha + c\alpha'$. Then

$$\gamma^2 = (b\alpha + c\alpha')^2 = b^2\alpha^2 + 2bc\alpha\alpha' + c^2\alpha'^2 = 5b^2\alpha' + 70bc + 7c^2\alpha,$$

and so we have expressed the elements $\{1, \gamma, \gamma^2\}$ in terms of the basis $\{1, \alpha, \alpha'\}$. The condition that $\{1, \gamma, \gamma^2\}$ is a basis is equivalent to requiring that the change of basis matrix should have determinant ± 1; this is

$$\begin{vmatrix} 1 & 0 & 0 \\ 0 & b & c \\ 70bc & 7c^2 & 5b^2 \end{vmatrix} = 5b^3 - 7c^3.$$

But $5b^3 - 7c^3 \neq \pm 1$ for any integers b and c, as one can see by working modulo 7. The cubes modulo 7 are 0 and ± 1, so we cannot have $5b^3 \equiv \pm 1 \pmod 7$. This contradiction shows that $\mathbb{Z}_K$ has no integral basis of the form $\{1, \gamma, \gamma^2\}$.

We will state the general result in Chap. 8.

Chapter 4
Ideals

In Chap. 2, we saw that the set of integers $\mathbb{Z}_K$ in an algebraic number field K forms a *ring*. In this chapter and the next, we are going to begin the study of primes.

When we begin to study factorisation in more general settings, we quickly see that several of the results from Chap. 1 fail to hold in the generality which we would like. The major drawback is that unique factorisation no longer holds in general.

This failure was already noted by Gauss at the end of the 18th century; these examples occur in the rings of integers of some quadratic number fields.

We'll begin with a "toy example", where some of the features can be easily explained. You should notice that the set we will consider is *not* a ring!

Example 4.1 Suppose that we lived in a world where the only positive integers were $1, 4, 7, 10, \ldots$, the integers of the form $3n + 1$. In this world, we would have a definition of prime number: a prime number will be an integer which cannot be factored further. For example, the numbers 4, 7, 10, and 13 are all prime (since we only have integers of the form $3n + 1$), whereas $16 = 4 \times 4$ is not.

Now the integer 100 may be written as a product of primes in two different ways:

$$100 = 10 \times 10 = 4 \times 25.$$

All of the factors, 4, 10 and 25, are prime in this world, and the two factorisations are genuinely different.

Of course, the problem here is that we do not have enough integers; we have to enlarge our set of integers. If we also include the integers of the form $3n + 2$, then in this larger world the factors are no longer prime, as we can factorise them further:

$$4 = 2 \times 2$$
$$10 = 2 \times 5$$
$$25 = 5 \times 5.$$

F. Jarvis, *Algebraic Number Theory*, Springer Undergraduate
Mathematics Series, DOI: 10.1007/978-3-319-07545-7_4,
© Springer International Publishing Switzerland 2014

Using these factorisations, our apparent lack of unique factorisation is resolved:

$$100 = (2 \times 5) \times (2 \times 5) = (2 \times 2) \times (5 \times 5).$$

Exercise 4.1 Instead of the set in Example 4.1, consider the set of integers $1, 5, 9, \ldots$ of integers of the form $4n + 1$. Find two factorisations of 441 in this set which are genuinely different, and explain how adding the remaining odd integers resolves the problem.

4.1 Uniqueness of Factorisation Revisited

As we remarked above, in Chap. 1, we showed that $\mathbb{Z}$ has unique factorisation, but even there we cautioned that some care has to be taken in defining uniqueness. For example,

$$6 = 2 \times 3 = (-3) \times (-2)$$

should really be counted as equivalent factorisations—we've simply permuted the factors, and multiplied both by -1, and since $(-1)(-1) = 1$, this shouldn't really matter.

In general, if we have a factorisation $r = a \times b$ in some ring R (and we will be interested in the case where R is the ring of integers in some number field), and if u and v in R satisfy $uv = 1$, then we ought to consider $r = (ua) \times (vb)$ as an *equivalent* factorisation.

Let's formalise this in the following sequence of definitions:

Definition 4.2 Let R be a ring, and let $u \in R$. If there exists an element $v \in R$ with $uv = 1$, say that u is a *unit* in R.

(We have already seen this terminology in Chap. 1 in some special cases.)

Definition 4.3 Two elements $r_1, r_2 \in R$ are *associate* if there is a unit $u \in R$ such that $r_2 = ur_1$. (Note that the relation is symmetric: if $r_2 = ur_1$, then $r_1 = vr_2$ where $uv = 1$.)

Given one factorisation, we want to consider another as "equivalent" if it can be got from the first by (a) multiplying by units, and (b) rearranging the factors. This suggests the following definition.

Definition 4.4 We say that two factorisations

$$r = a_1 a_2 \ldots a_n = b_1 b_2 \ldots b_n$$

are *equivalent* if b_i is an associate of $a_{\pi(i)}$ for some permutation π of $\{1, \ldots, n\}$.

Of course, we will be particularly interested in factorisations into primes. In $\mathbb{Z}$, even prime numbers p can be factorised as $p = (-1).(-p)$, so we should restrict ourselves to factorisations which do not involve units.

There are two possible generalisations of prime numbers to more general rings:

Definition 4.5 1. Let $p \in R$. Then p is *irreducible* if

(a) p is not a unit;
(b) whenever $p = ab$, then either a or b is a unit.

2. Let $p \in R$. Then p is a *prime element* if, whenever $p|ab$ (in the sense that $ab = pr$ for some $r \in R$), then $p|a$ or $p|b$.

When $R = \mathbb{Z}$, these two are equivalent, as we saw in Chap. 1. However, we will see that they are different in general, and that this is a consequence of failure of unique factorisation.

Exercise 4.2 Using the results of Sect. 1.4, what are the irreducible elements in $\mathbb{Z}[i]$?

4.2 Non-unique Factorisation in Quadratic Number Fields

As in Chap. 1, where we looked briefly at factorisation in the Gaussian integers, we'll consider factorisation in the ring of integers of quadratic fields, giving some examples where unique factorisation fails.

Suppose that d is squarefree. For simplicity, our examples will involve cases where $d \equiv 2 \pmod 4$ or $d \equiv 3 \pmod 4$, so that the ring of integers in $\mathbb{Q}(\sqrt{d})$ is $\mathbb{Z}[\sqrt{d}]$.

Example 4.6 Gauss's examples of non-unique factorisation come from quadratic number fields. For example, when $d = 10$, one has the equalities

$$6 = 2 \times 3 = \left(4 + \sqrt{10}\right)\left(4 - \sqrt{10}\right),$$

and, just to show that the same sort of equality can hold for d negative, let's consider $d = -5$:

$$6 = 2 \times 3 = \left(1 + \sqrt{-5}\right)\left(1 - \sqrt{-5}\right),$$

It will take a little time to check that these factors are all irreducible, in the sense that they cannot be factored further, and also to show that the factorisations really are different.

To begin, we will follow the prototype of the Gaussian integers (Sect. 1.4). If $\alpha = a + b\sqrt{d} \in \mathbb{Z}[\sqrt{d}]$, we define $\overline{\alpha} = a - b\sqrt{d}$, which will play the role of complex conjugation. Next, we define the *norm* (following the definition of Sect. 3.2):

$$N\left(a + b\sqrt{d}\right) = N(\alpha) = \alpha\bar{\alpha} = \left(a + b\sqrt{d}\right)\left(a - b\sqrt{d}\right) = a^2 - db^2.$$

If $\alpha \in \mathbb{Z}[\sqrt{d}]$, then $N(\alpha) \in \mathbb{Z}$ (see Corollary 3.17).

If we are given two elements $\alpha_1 = a_1 + b_1\sqrt{d}$ and $\alpha_2 = a_2 + b_2\sqrt{d}$, we see that

$$
\begin{aligned}
N(\alpha_1\alpha_2) &= \alpha_1\alpha_2.\overline{\alpha_1\alpha_2} \\
&= \alpha_1\bar{\alpha}_1.\alpha_2\bar{\alpha}_2 \\
&= N(\alpha_1)N(\alpha_2)
\end{aligned}
$$

since $\overline{\alpha_1\alpha_2} = \overline{\alpha_1}.\overline{\alpha_2}$.

We begin with an easy lemma, which will allow us to recognise units, and therefore associates:

Lemma 4.7 *Suppose that $u \in \mathbb{Z}[\sqrt{d}]$. Then u is a unit if and only if $N(u) = \pm 1$.*

Proof If u is a unit, so that there exists v such that $uv = 1$, then

$$N(u)N(v) = N(uv) = N(1) = 1$$

so $N(u)$ and $N(v)$ must both be ± 1, as they are integers whose product is 1.

Conversely, if $N(u) = \pm 1$, then $u\bar{u} = \pm 1$. Define $v = \pm \bar{u}$—then $uv = 1$, and so u is a unit. □

With that preliminary lemma out of the way, we can check that the factorisations above are not equivalent.

Lemma 4.8 1. *In $\mathbb{Z}[\sqrt{10}]$, the two factorisations*

$$6 = 2 \times 3 = \left(4 + \sqrt{10}\right)\left(4 - \sqrt{10}\right)$$

are not equivalent.
2. *In $\mathbb{Z}[\sqrt{-5}]$, the two factorisations*

$$6 = 2 \times 3 = \left(1 + \sqrt{-5}\right)\left(1 - \sqrt{-5}\right)$$

are not equivalent.

Proof If α_1 and α_2 are associate, then there is a unit u such that $\alpha_2 = u\alpha_1$. Then

$$N(\alpha_2) = N(u\alpha_1) = N(u)N(\alpha_1) = N(\alpha_1).$$

It follows that if two factorisations are equivalent, then the norms of the factors on both sides will be the same.

However, in $\mathbb{Z}[\sqrt{10}]$,

$$N(2) = N\left(2 + 0\sqrt{10}\right) = 2^2 - 10 \times 0^2 = 4$$

$$N(3) = N\left(3 + 0\sqrt{10}\right) = 3^2 - 10 \times 0^2 = 9$$

$$N\left(4 + \sqrt{10}\right) = 4^2 - 10 \times 1^2 = 6$$

$$N\left(4 - \sqrt{10}\right) = 4^2 - 10 \times (-1)^2 = 6,$$

and so the norms on the two sides are different.

Similarly, in $\mathbb{Z}[\sqrt{-5}]$,

$$N(2) = N\left(2 + 0\sqrt{-5}\right) = 2^2 + 5 \times 0^2 = 4$$

$$N(3) = N\left(3 + 0\sqrt{-5}\right) = 3^2 + 5 \times 0^2 = 9$$

$$N\left(1 + \sqrt{-5}\right) = 1^2 + 5 \times 1^2 = 6$$

$$N\left(1 - \sqrt{-5}\right) = 1^2 + 5 \times (-1)^2 = 6,$$

and again the norms on the two sides are different. □

Now we check that all of the factors involved cannot be factorised further, so are, in a sense, prime numbers in this setting.

Lemma 4.9

1. *In $\mathbb{Z}[\sqrt{10}]$, all of the factors in the equality*

$$2 \times 3 = \left(4 + \sqrt{10}\right)\left(4 - \sqrt{10}\right)$$

   are irreducible.
2. *In $\mathbb{Z}[\sqrt{-5}]$, all of the factors in the equality*

$$2 \times 3 = \left(1 + \sqrt{-5}\right)\left(1 - \sqrt{-5}\right)$$

   are irreducible.

Proof Let's first see that there aren't any elements $\alpha \in \mathbb{Z}[\sqrt{10}]$ with $N(\alpha) = \pm 2$. If $\alpha = a + b\sqrt{10}$, then $N(\alpha) = a^2 - 10b^2 = \pm 2$. This means that either $a^2 - 10b^2 = 2$ or $a^2 - 10b^2 = -2$. Consider these equalities modulo 5: we see that we would need $a^2 \equiv 2 \pmod 5$ or $a^2 \equiv 3 \pmod 5$, but both of these are impossible. In the same way, there are no elements $\beta \in \mathbb{Z}[\sqrt{10}]$ with $N(\beta) = \pm 3$.

Suppose that 2 factorises as $\alpha\beta$ in $\mathbb{Z}[\sqrt{10}]$. Then $4 = N(2) = N(\alpha)N(\beta)$. If $N(\alpha) = \pm 1$, $N(\beta) = \pm 4$, then α is a unit; if $N(\alpha) = \pm 4$, $N(\beta) = \pm 1$, then β is a unit. So the only possibility of factorising 2 into non-units occurs if $N(\alpha) = N(\beta) = \pm 2$, and we have seen that there are no such elements.

In the same way, if 3 were to factorise as $\alpha\beta$ into non-units, then $N(\alpha) = N(\beta) = \pm 3$, and this is not possible.

Finally, the only way to factorise $4 \pm \sqrt{10}$ into non-units would be as the product of an element of norm ± 2 and an element of norm ± 3, which is impossible.

Exactly the same argument works for $\mathbb{Z}[\sqrt{-5}]$. Indeed, there are again no elements of norm ± 2, as this would require $a^2 + 5b^2 = \pm 2$, so that $a^2 + 5b^2 = 2$ (as $a^2 + 5b^2$ is necessarily positive), and there are clearly no integral solutions (again, one could also argue modulo 5). Nor are there any solutions to $a^2 + 5b^2 = 3$, so there are no elements of norm 3. The same argument as in the case of $\mathbb{Z}[\sqrt{10}]$ now applies to $\mathbb{Z}[\sqrt{-5}]$.   □

In both of Gauss's examples, therefore, we have two non-equivalent factorisations into irreducible elements, and therefore factorisation in these rings is not unique.

Let us also notice that in these cases, the factors are irreducible, but they are *not* prime elements (see Definition 4.5): $2|6$, so $2|(4 + \sqrt{10})(4 - \sqrt{10})$, but $2 \nmid 4 \pm \sqrt{10}$ as $(4 \pm \sqrt{10})/2 = 2 \pm \frac{1}{2}\sqrt{10} \notin \mathbb{Z}[\sqrt{10}]$.

Exercise 4.3 Certainly the set $\mathbb{Z}[\frac{1+\sqrt{-3}}{2}]$ is a ring, as it consists of all the algebraic integers in $\mathbb{Q}(\sqrt{-3})$. We have also seen that the ring of integers of $\mathbb{Q}(\sqrt{3})$ is $\mathbb{Z}[\sqrt{3}]$.

1. Show directly that $\mathbb{Z}[\frac{1+\sqrt{-3}}{2}]$ is a ring. (Hint: the only hard part is closure under multiplication; for this, write $\omega = \frac{1+\sqrt{-3}}{2}$, and use the fact that $\omega^2 = -(\omega + 1)$.)
2. Show that the set
$$\left\{ a + b\left(\frac{1+\sqrt{3}}{2}\right) \,\middle|\, a, b \in \mathbb{Z} \right\}$$

   is not even closed under multiplication.
3. We shall also see that $\mathbb{Z}[\frac{1+\sqrt{-3}}{2}]$ has unique factorisation. However, show that in $\mathbb{Z}[\sqrt{-3}]$, the two factorisations
$$4 = \left(1 + \sqrt{-3}\right)\left(1 - \sqrt{-3}\right) = 2 \times 2$$

   are different, so $\mathbb{Z}[\sqrt{-3}]$ fails to have unique factorisation.

Exercise 4.4 Let R denote the ring of integers of $\mathbb{Q}(\sqrt{5})$. We have
$$11 = \left(4 + \sqrt{5}\right)\left(4 - \sqrt{5}\right) = \left(2\sqrt{5} + 3\right)\left(2\sqrt{5} - 3\right).$$

Show that these two factorisations are equivalent.
[*Hint: show that each of the factors on the left-hand side is a factor of the right-hand side multiplied by a unit.*]

Exercise 4.5 In $\mathbb{Z}[i]$, explain that any element α with $N(\alpha) = p$, a prime number in $\mathbb{Z}$, is necessarily irreducible. Find an example of an irreducible $\alpha \in \mathbb{Z}[i]$ with $N(\alpha)$ not a prime.

Do the same in $\mathbb{Z}[\sqrt{-2}]$.

Exercise 4.6 Another example in $\mathbb{Z}[\sqrt{10}]$ is furnished by

$$10 = \left(\sqrt{10}\right)^2 = 2 \times 5.$$

Show that these are inequivalent factorisations into irreducible elements.

Exercise 4.7 Factor 6 in two different ways in $\mathbb{Z}[\sqrt{-6}]$. Remember to check that these really are different…

Exercise 4.8 Show that $\mathbb{Z}[\sqrt{-13}]$ does not have unique factorisation, by factoring 14 in two different ways.

Exercise 4.9 Show that in $\mathbb{Z}[\sqrt{-26}]$, one has

$$27 = 3.3.3 = \left(1 + \sqrt{-26}\right)\left(1 - \sqrt{-26}\right),$$

and that each factor is irreducible. This implies that the number of factors in two factorisations into irreducibles may differ.

4.3 Kummer's Ideal Numbers

In Example 4.1, we resolved the non-uniqueness of factorisation by enlarging our world to include "missing" integers.

Kummer tried to repair the non-uniqueness of factorisation in quadratic fields by enlarging the integers to include "*ideal numbers*" in a similar way to the example.

In the example of non-unique factorisation of $\mathbb{Z}[\sqrt{10}]$ above, Kummer's idea was to invent symbols $\mathfrak{a}_1, \mathfrak{a}_2, \mathfrak{a}_3, \mathfrak{a}_4$ such that

$$2 = \mathfrak{a}_1 \times \mathfrak{a}_2$$
$$3 = \mathfrak{a}_3 \times \mathfrak{a}_4$$
$$4 + \sqrt{10} = \mathfrak{a}_1 \times \mathfrak{a}_3$$
$$4 - \sqrt{10} = \mathfrak{a}_2 \times \mathfrak{a}_4.$$

Then the non-unique factorisation

$$2 \times 3 = \left(4 + \sqrt{10}\right)\left(4 - \sqrt{10}\right)$$

is repaired:
$$(\mathfrak{a}_1 \times \mathfrak{a}_2) \times (\mathfrak{a}_3 \times \mathfrak{a}_4) = (\mathfrak{a}_1 \times \mathfrak{a}_3) \times (\mathfrak{a}_2 \times \mathfrak{a}_4).$$

Note that these are purely invented symbols, and do not really have any meaning. Nevertheless, Kummer hoped that these symbols could be manipulated in such a way that meaningful results could be obtained.

A little later, Dedekind reformulated this idea of Kummer's in more concrete terms, which we shall now explain.

Consider again the factorisations into "ideal numbers"

$$2 = \mathfrak{a}_1 \times \mathfrak{a}_2$$

and

$$4 + \sqrt{10} = \mathfrak{a}_1 \times \mathfrak{a}_3$$

which we gave above. Then 2 would be a multiple of $\mathfrak{a}_1$, so any multiple of 2 would also be a multiple of $\mathfrak{a}_1$. Similarly, $4 + \sqrt{10}$ is again a multiple of $\mathfrak{a}_1$, so any multiple of $4 + \sqrt{10}$ is a multiple of $\mathfrak{a}_1$. Combining these, any $\mathbb{Z}[\sqrt{10}]$-linear combination of 2 and $4 + \sqrt{10}$ should be a multiple of $\mathfrak{a}_1$.

That is, if R denotes the ring of integers $\mathbb{Z}[\sqrt{10}]$ of $\mathbb{Q}(\sqrt{10})$, then the set of multiples of 2, namely $2.R$, must be contained in the set of multiples of $\mathfrak{a}_1$. Thus $2.R \subseteq \mathfrak{a}_1 R$. Similarly, $(4 + \sqrt{10})R \subseteq \mathfrak{a}_1 R$, and thus

$$2.R + \left(4 + \sqrt{10}\right) R \subseteq \mathfrak{a}_1 R.$$

A little further thought suggests that this inclusion ought to be an equality. Indeed, if we had an equality $\mathfrak{a}_1 R = R$, then $\mathfrak{a}_1$ would be invertible, and so $\mathfrak{a}_1$ would be a unit—but we don't want our factors to be units. On the other hand, a calculation gives

$$2.R + \left(4 + \sqrt{10}\right) R = \{m + n\sqrt{10} \mid m, n \in \mathbb{Z}, \ 2|m\},$$

and this has index 2 in R (informally, half of the elements of R are in this set). There is no room for anything between R and $2.R + (4 + \sqrt{10})R$, and as $\mathfrak{a}_1 R$ is strictly contained in R, and contains $2.R + (4 + \sqrt{10})R$, we must have $\mathfrak{a}_1 R = 2.R + (4 + \sqrt{10})R$.

Exercise 4.10 Expand $2.(a + b\sqrt{10}) + (4 + \sqrt{10})(c + d\sqrt{10})$ and verify that

$$2.R + \left(4 + \sqrt{10}\right) R = \{m + n\sqrt{10} \mid m, n \in \mathbb{Z}, \ 2|m\}.$$

Exercise 4.11 Suppose that $\alpha \in R$ satisfies $\alpha|2$ and $\alpha|4 + \sqrt{10}$. By taking norms, deduce that α is a unit in $\mathbb{Z}[\sqrt{10}]$. Conclude that $\mathfrak{a}_1$ really is not an element of $\mathbb{Z}[\sqrt{10}]$.

Instead of thinking of $\mathfrak{a}_1$ as an "ideal number", Dedekind's idea was to work with the set $\mathfrak{a}_1 R$. Since $\mathfrak{a}_1$ is not actually an element, writing $\mathfrak{a}_1 R$ is rather misleading, and we shall simply write $\mathfrak{a}_1$ now for the set. Then $\mathfrak{a}_1$ is the set of all linear combinations of 2 and $4 + \sqrt{10}$, i.e.,

$$\mathfrak{a}_1 = 2.R + \left(4 + \sqrt{10}\right) R.$$

In this viewpoint, even the symbol 2, which we would normally think of as a number, should be viewed as the *set* $2.R$ of all multiples of 2.

Remark 4.10 Notice that this makes perfect sense even in $\mathbb{Z}$, and we can reformulate the property that a divides b in terms of these sets. Indeed, if a divides b, then b is a multiple of a, and then any multiple of b is also a multiple of a; symbolically, $b\mathbb{Z} \subseteq a\mathbb{Z}$. Thus $a|b$ if and only if $b\mathbb{Z} \subseteq a\mathbb{Z}$.

In the example above, since $\mathfrak{a}_1$ contains all multiples of 2, one could say that $\mathfrak{a}_1$ is a *divisor* of 2. Similarly, $\mathfrak{a}_1$ is also a divisor of $4 + \sqrt{10}$, as one would hope.

Thus, although there are no *elements* of R which divide 2 and $4 + \sqrt{10}$ except units, there are certain *subsets* of R which contain $2.R$ and $(4 + \sqrt{10})R$ which are strictly contained in $1.R$.

This led Dedekind to reformulate Kummer's ideal numbers as being certain subsets of the rings of integers, which are now known as ideals.

4.4 Ideals

As motivated above, the prototype for Dedekind's sets are all the multiples of a given element of R, or, more generally (when unique factorisation fails), all the linear combinations of some set of elements. This led Dedekind to the following abstract definition:

Definition 4.11 An *ideal* I of a commutative ring R is a subset of R such that

1. $0_R \in I$;
2. if i and $i' \in I$, then $i - i' \in I$;
3. if $i \in I$ and $a \in R$, then $ai \in I$.

(Notice that the second requirement here is equivalent to I being closed under both addition and additive inverses.)

Observe that these conditions are exactly the same as those needed for I to be a *module*; the only difference is that ideals are subsets of the ring.

Let's look at some examples.

Example 4.12

1. Any ring R is an ideal in itself.
2. For any ring R, $\{0_R\}$ is an ideal in R.

3. Let R be any ring, and let $r \in R$. Let $I = rR$, all the multiples of r. Then I is an ideal in R. Indeed, writing the axioms in words, 0 is a multiple of r; the difference of any two multiples of r is again a multiple of r, and any multiple of a multiple of r is certainly a multiple of r.

The last example gives a large class of ideals, and, in some rings, all ideals are of this form:

Lemma 4.13 *In $\mathbb{Z}$, every ideal is of the form $n\mathbb{Z}$ for some integer n.*

Proof Let I be an ideal of $\mathbb{Z}$, and first suppose $I \neq \{0\}$. As I contains a non-zero integer, it will contain a positive integer (if $k \in I$, and $k < 0$, then the definition of ideal means that $(-1) \times k = -k \in I$ also).

Let n be the smallest positive integer contained in I. Clearly I then contains all multiples of n, so $I \supseteq n\mathbb{Z}$—but if $a \in I$, we can write $a = qn + r$ by the division algorithm, where $0 \leq r < n$, and as a and $n \in I$, we conclude that $r \in I$. As n was the smallest positive integer in I, we conclude that $r = 0$, so that a is a multiple of n. Thus $I = n\mathbb{Z}$.

On the other hand, if $I = \{0\}$, we can regard it as $0\mathbb{Z}$, so I is again of the required form.   □

You should be warned that, in the case of a general ring R, *not* every ideal in R is of the form rR; the reason that it holds in $\mathbb{Z}$ is because of Euclid's algorithm, but it is rare that rings have analogues of the Euclidean algorithm.

Let's now explore some of the elementary properties of ideals.

Lemma 4.14 *Let R be a ring.*

1. *If I and J are ideals of R, then so is $I \cap J$.*
2. *More generally, if $\{I_\alpha\}_{\alpha \in \Lambda}$ is any family of ideals of R, then so is their intersection $\bigcap_{\alpha \in \Lambda} I_\alpha$.*
3. *If I and J are both ideals of R, then so is*

$$IJ = \{\text{finite sums of elements of the form } ij \mid i \in I \text{ and } j \in J\}$$

   and $IJ \subseteq I \cap J$.
4. *If I and J are both ideals of R, then so is*

$$I + J = \{i + j \mid i \in I \text{ and } j \in J\}.$$

Proof These are all straightforward verifications of the axioms.

1. We check the axioms. As I and J are ideals, $0_R \in I$ and $0_R \in J$, and therefore $0_R \in I \cap J$. Next, given two elements in the intersection, we want their difference to be in the intersection. But if i and $j \in I \cap J$, then i and j each lie in both I and J. As these are ideals, $i - j \in I$ and $i - j \in J$—and thus $i - j \in I \cap J$. Finally, if $i \in I \cap J$ (so $i \in I$ and $i \in J$) and $r \in R$, then $ri \in I$ as I is an ideal, and similarly $ri \in J$, so $ri \in I \cap J$. This shows that $I \cap J$ is an ideal, as required.

2. Very similar to the first assertion.
3. As $0_R \in I$ (or J), we see that $0_R \in IJ$. Given two finite sums of terms of the form ij, their difference is clearly again a finite sum of terms of the same form, and so IJ is closed under addition. Finally, given a sum $\sum_k i_k j_k \in IJ$ and an element $r \in R$, we see that

$$r\left(\sum_k i_k j_k\right) = \sum_k (ri_k) j_k,$$

and as I is an ideal, all the bracketed terms $ri_k \in I$, so this is again a finite sum of products of elements of I with elements of J.

For the final assertion, an element of IJ is a finite sum of elements of the form ij with $i \in I$ and $j \in J$. As $J \subset R$, we have $j \in R$, and so, by definition of ideals, $ij \in IR = I$. Similarly, $I \subset R$, so $i \in R$, and $ij \in RJ = J$. It follows that all terms $ij \in I \cap J$, so that $IJ \subseteq I \cap J$.

4. As $0_R \in I$ and $0_R \in J$, we see that $0_R = 0_R + 0_R \in I + J$. Next, we take $i_1 + j_1$ and $i_2 + j_2 \in I + J$. Their difference is

$$(i_1 + j_1) - (i_2 + j_2) = (i_1 - i_2) + (j_1 - j_2) \in I + J,$$

as $i_1 - i_2 \in I$ and $j_1 - j_2 \in J$. Finally, given $i + j \in I + J$, and $r \in R$, we have $r(i + j) = ri + rj \in I + J$ as I and J are ideals. We conclude that $I + J$ is an ideal. $\square$

Note that if I and J are ideals, it is not generally true that $I \cup J$ is an ideal—if $R = \mathbb{Z}$, $I = 2\mathbb{Z}$ and $J = 3\mathbb{Z}$, then $2 \in I \subset I \cup J$, and $3 \in J \subset I \cup J$, but their sum, 5, is not in $I \cup J$. Thus $I \cup J$ is not an ideal.

The next lemma is very simple, but surprisingly useful:

Lemma 4.15 *Suppose that R is a ring, and that I is an ideal of R. If I contains a unit of R, then $I = R$.*

Proof Suppose $u \in I$ is a unit in R. Then there exists $v \in R$ such that $uv = 1_R$. Thus $1_R \in I$. Now, for all $a \in R$, $a.1_R = a$ must lie in the ideal. Thus $a \in I$, and so $R \subseteq I$. The result follows. $\square$

Along the same lines is the following:

Lemma 4.16 *Suppose that R is an integral domain (i.e., has no zero divisors). Suppose that $a, b \in R$. Then $aR = bR$ if and only if a and b are associate.*

Proof Suppose that $aR = bR$. As $a = a.1_R \in aR = bR$, we see that $a = bu$ for some element $u \in R$. Similarly, $b = av$ for some element $v \in R$. Then $a = (av)u = a(vu)$. As R is an integral domain, this only happens if $vu = 1$, i.e., u and v are units.

Conversely, if a and b are associate, then $a = bu$ for some unit u, and $b = av$ for the unit v with $uv = 1$. Thus any multiple $br \in bR$ of b can also be written avr, so lies in aR; then $bR \subseteq aR$, and the reverse inclusion is similar. $\square$

As already hinted, ideals are the right context for factorisation: in Chap. 5, we are going to prove that ideals in rings of integers of number fields factorise uniquely into "prime ideals". Using this lemma, we see that another advantage of working with ideals is that the units no longer play any role.

Lemma 4.17 *R is a field if and only if the only ideals in R are $\{0_R\}$ and R itself.*

Proof If R is a field, then every non-zero element is a unit. If I is an ideal of R, then either I only contains the zero element, or it contains a non-zero element, and therefore a unit, so $I = R$ by Lemma 4.15.

Conversely, if R is not a field, then there exists some non-zero element r which is not a unit. Then the collection $rR = \{ra \mid a \in R\}$ is an ideal in R; it is non-zero as it contains $r = r.1_R \neq 0$, but nor is it all of R, as there is no $a \in R$ such that $ra = 1_R$, as r is not a unit.   □

4.5 Generating Sets for Ideals

Now we consider ideals generated by sets. In practice, this will be the simplest way to specify ideals, and we will use this notation a lot.

Definition 4.18 Let X be a (possibly infinite) subset of R. Then the intersection of all ideals containing X is an ideal of R (by Lemma 4.14(2)), and is clearly contained in all ideals containing X. This ideal is denoted by $\langle X \rangle$, and called the *ideal generated by X*.

Proposition 4.19 *Let X be a subset of R. Then*

$$\langle X \rangle = \{all\ finite\ sums\ of\ elements\ of\ the\ form\ rx,\ with\ r \in R, x \in X\}.$$

Proof Let $I = \{$all finite sums of elements of the form rx, with $r \in R$, $x \in X\}$. We want to show that $I = \langle X \rangle$. One inclusion is clear from the definition; as I is an example of an ideal containing X, the intersection $\langle X \rangle$ of all such ideals must be a subset of I. We need to check the converse, that $I \subseteq \langle X \rangle$.

Let J be any ideal containing all $x \in X$. For any $r \in R$, as $x \in J$ and J is an ideal, $rx \in J$. Further, J is closed under addition as well, and as we have just shown that all elements $r_1x_1, \ldots, r_nx_n$ with $r_i \in R$ and $x_i \in X$ lie in J, so does their sum $r_1x_1 + \cdots + r_nx_n$. But any element of I is of this form, so each element of I lies in J. This shows that if J is any ideal containing all $x \in X$, then $J \supseteq I$.

However, $\langle X \rangle$ is an ideal containing every element of X, so $\langle X \rangle \supseteq I$.   □

So the typical element of $\langle X \rangle$ is $r_1x_1 + r_2x_2 + \cdots + r_kx_k$ for some $k \in \mathbb{N}$.

In particular, if $X = \{x_1, \ldots, x_n\}$ is a finite set, then the ideal $\langle X \rangle$, which in this case we also denote by $\langle x_1, \ldots, x_n \rangle$, consists of all sums of the form $\sum_{i=1}^{n} r_i x_i$ with $r_i \in R$. In other words,

$$\langle x_1, \ldots, x_n \rangle = x_1 R + \cdots + x_n R.$$

This should remind you strongly of the notion of the span of a set of vectors in a vector space.

Example 4.20 Consider the ideal $\langle 2, 3 \rangle$ in $\mathbb{Z}$. This consists of every integer n which can be written as $2a + 3b$ for integers a, b. But every integer may be written in this way ($n = 2.(-n) + 3.n$), so $\mathbb{Z} = \langle 2, 3 \rangle = \langle 1 \rangle$. Note that $\langle 2 \rangle$ and $\langle 3 \rangle$ are both proper subsets of $\mathbb{Z}$, so this shows that $\{2, 3\}$ is a minimal set of generators, in the sense that no proper subset generates the whole ideal. As both $\{1\}$ and $\{2, 3\}$ are minimal generating sets, we see that ideals may have minimal generating sets of different sizes (in contrast to vector spaces).

Exercise 4.12 Show that another minimal generating set is $\{6, 10, 15\}$.

Exercise 4.13 In the ring $\mathbb{Z}$, write each of the following ideals in the form $\langle n \rangle$ for some integer n:

$$\langle 12 \rangle \cap \langle 20 \rangle; \quad \langle 12 \rangle \langle 20 \rangle; \quad \langle 12 \rangle + \langle 20 \rangle; \quad \langle 12, 20 \rangle.$$

Exercise 4.14 Let $I = \langle a_1, \ldots, a_m \rangle$ and $J = \langle b_1, \ldots, b_n \rangle$ be two ideals of a ring R. Show that

1. $I + J = \langle a_1, \ldots, a_m, b_1, \ldots, b_n \rangle$.
2. $IJ = \langle a_1 b_1, a_1 b_2, \ldots, a_1 b_n, a_2 b_1, \ldots, a_m b_n \rangle$.

Definition 4.21 Ideals of the form $\langle r \rangle$, with one generator, are called *principal*.

Now let's remark that rings exist where not every ideal is of the form $\langle r \rangle = rR$ for some $r \in R$. Here is an example of a non-principal ideal:

Example 4.22 In $\mathbb{Z}[X]$, the set I of polynomials whose constant term is divisible by 2 is an ideal (treat this as an exercise!), and is not of the form $r\mathbb{Z}[X]$ for any r.

Indeed, both 2 and X would have to be multiples of r, which means that r would have to be ± 1. But ± 1 does not belong to I, so this is also not possible. In fact, $I = \langle 2, X \rangle$: given a polynomial

$$f(X) = \sum_{n=0}^{d} a_n X^n \in I,$$

we can write it as

$$f(X) = a_0 + X. \sum_{n=1}^{d} a_n X^{n-1},$$

and $a_0 \in 2\mathbb{Z} \subseteq 2\mathbb{Z}[X]$, while clearly $X. \sum_{n=1}^{d} a_n X^{n-1} \in X.\mathbb{Z}[X]$. It follows that every polynomial in I can be written as the sum of something in $2\mathbb{Z}[X]$ and something in $X\mathbb{Z}[X]$. It follows that $I \subseteq \langle 2, X \rangle$. The opposite inclusion is clear.

Rings of integers of quadratic fields with non-unique factorisation also have this property. Here is an example:

Example 4.23 Suppose that $R = \mathbb{Z}[\sqrt{10}]$. Then the set $\mathfrak{a}_1 = 2.R + (4 + \sqrt{10})R = \langle 2, 4 + \sqrt{10} \rangle$ is an ideal in R, and it is not possible to write $\mathfrak{a}_1$ as $\langle \alpha \rangle$ for any $\alpha \in R$. If it were, then every element of $\mathfrak{a}_1$ would be a multiple of α. In particular, we would have $2 = \alpha\beta$ and $4 + \sqrt{10} = \alpha\gamma$. Take norms:

$$4 = N(2) = N(\alpha)N(\beta)$$
$$6 = N\left(4 + \sqrt{10}\right) = N(\alpha)N(\gamma),$$

and as we observed above, this means that $N(\alpha) = 1$ or $N(\alpha) = 2$. We have already seen that there are no elements in $\mathbb{Z}[\sqrt{10}]$ with norm 2, but if $N(\alpha) = 1$, α would be a unit, so $\langle \alpha \rangle$ would equal R. However, $1 \notin \mathfrak{a}_1$; we saw that every element in $\mathfrak{a}_1$ has an even number as a coefficient of 1.

Incidentally, it is traditional to use Gothic letters $\mathfrak{a}$, $\mathfrak{b}$, etc., for ideals in rings of integers of number fields. However, we will use I, J, etc., for ideals in more general rings.

Exercise 4.15 Let $R = \mathbb{Z}[\sqrt{-5}]$, and consider the ideal $\mathfrak{a} = \langle 2, 1 + \sqrt{-5} \rangle$.

1. By calculating the general element $2\beta + (1 + \sqrt{-5})\gamma$ in $\mathfrak{a}$ (where $\beta, \gamma \in R$), show that every element $a + b\sqrt{-5} \in \mathfrak{a}$ with $a, b \in \mathbb{Z}$ has the property that $a \equiv b \pmod{2}$. Deduce that $1 \notin \mathfrak{a}$.
2. Show that if $\mathfrak{a} = \langle \alpha \rangle$, then $N(\alpha)|2$, and deduce that $\mathfrak{a}$ is not principal.
3. Show that we can also write $\mathfrak{a}$ as $\langle 2, 1 - \sqrt{5} \rangle$.

I hope that you see from these examples that there seems to be a close relation between the fact that R does not have unique factorisation, and the fact that we have an ideal in R which is not principal. We shall explore this relationship more closely later.

An ideal of R which has a finite generating set is called *finitely generated*. Rings in which every ideal is finitely generated are called *Noetherian*, and are of special interest for us, since we shall see that rings of integers of number fields have this property; while there are rings in which some ideals are not finitely generated, we shall not consider this situation.

Remark 4.24 In Lemma 4.14(3), we claimed that $IJ \subseteq I \cap J$. Sometimes we can have equality here: if $R = \mathbb{Z}$, $I = 2\mathbb{Z}$, $J = 3\mathbb{Z}$, then

$$IJ = \langle 2 \rangle.\langle 3 \rangle = \langle 6 \rangle,$$

and also $I \cap J$ consists of all integers in $I \cap J$, which are those integers simultaneously divisible by 2 (so lie in I) and by 3 (so lie in J), and are therefore all multiples of 6. So $I \cap J = \langle 6 \rangle = IJ$.

But it is easy to give examples where $IJ \neq I \cap J$. If $R = \mathbb{Z}$, $I = J = 2\mathbb{Z}$, then $IJ = \langle 2 \rangle . \langle 2 \rangle = \langle 4 \rangle$, whereas $I \cap J = \langle 2 \rangle$. From these examples, and others you should invent for yourselves, you should get the impression that the equality of IJ and $I \cap J$ should be related to whether they are "coprime" in a certain sense.

Definition 4.25 As already remarked (see Remark 4.10), the notation $\alpha | \beta$ means that β is a multiple of α. In particular, any multiple of β is a multiple of α, and so $\langle \beta \rangle \subseteq \langle \alpha \rangle$. We extend the notation to ideals by writing $\mathfrak{a} | \mathfrak{b}$ to mean $\mathfrak{b} \subseteq \mathfrak{a}$, and we may use either notation interchangeably.

In Chap. 5, we will see that if $\mathfrak{a}$ and $\mathfrak{b}$ are ideals in the ring of integers $\mathbb{Z}_K$ of a number field K, then $\mathfrak{b} \subseteq \mathfrak{a}$ if and only if there is some ideal $\mathfrak{c}$ of $\mathbb{Z}_K$ such that $\mathfrak{b} = \mathfrak{a}\mathfrak{c}$, which is another definition of division one might have come up with.

4.6 Ideals in Quadratic Fields

We know that $\mathbb{Q}(\sqrt{d})$ does not always have unique factorisation. For example, if $d = -5$, the ring of integers is $\mathbb{Z}[\sqrt{-5}]$, and then

$$6 = 2 \times 3 = \left(1 + \sqrt{-5}\right)\left(1 - \sqrt{-5}\right).$$

In terms of ideals, we can consider the ideal $\langle 6 \rangle \subset \mathbb{Z}[\sqrt{-5}]$; then the above factorisations correspond to factorisations of ideals:

$$\langle 6 \rangle = \langle 2 \rangle \langle 3 \rangle = \left\langle 1 + \sqrt{-5} \right\rangle \left\langle 1 - \sqrt{-5} \right\rangle,$$

where $\langle a \rangle$ denotes the principal ideal generated by a, namely $a\mathbb{Z}[\sqrt{-5}]$. However, although 2 and 3 are irreducible in $\mathbb{Z}[\sqrt{-5}]$ as we saw above, which means that the ideal $\langle 3 \rangle$, say, cannot be written as the product of principal ideals (if $3 = \alpha \cdot \beta$, then $\langle 3 \rangle = \langle \alpha \rangle \langle \beta \rangle$), the principal ideal $\langle 3 \rangle$ may nonetheless be factored as a product of non-principal ideals. The obstruction to unique factorisation is coming from the fact that not every ideal in $\mathbb{Z}[\sqrt{-5}]$ is principal. Indeed, consider the two ideals

$$\mathfrak{a}_1 = \langle 3, 1 + \sqrt{-5} \rangle, \quad \mathfrak{a}_2 = \langle 3, 1 - \sqrt{-5} \rangle.$$

Let's work out the product $\mathfrak{a}_1 \mathfrak{a}_2$. Then

$$\mathfrak{a}_1 \mathfrak{a}_2 = \left\langle 3 \cdot 3, 3\left(1 - \sqrt{-5}\right), 3\left(1 + \sqrt{-5}\right), \left(1 + \sqrt{-5}\right)\left(1 - \sqrt{-5}\right) \right\rangle$$
$$= \left\langle 9, 3 - 3\sqrt{-5}, 3 + 3\sqrt{-5}, 6 \right\rangle$$

That is, every element of the product $\mathfrak{a}_1\mathfrak{a}_2$ is of the form

$$9\alpha + \left(3 - 3\sqrt{-5}\right)\beta + \left(3 + 3\sqrt{-5}\right)\gamma + 6\delta$$

for some α, β, γ and $\delta \in \mathbb{Z}[\sqrt{-5}]$. It is clear that the collection of such elements must be contained in the set $\{A + B\sqrt{-5} \mid 3|A, 3|B\}$. Conversely, any element $3A + 3B\sqrt{-5}$ lies in $\mathfrak{a}_1\mathfrak{a}_2$ on taking $\alpha = A + B\sqrt{-5}$, $\beta = \gamma = 0$, $\delta = -\alpha$ (and there are lots of other ways of doing this). It follows that

$$\mathfrak{a}_1\mathfrak{a}_2 = \left\{A + B\sqrt{-5} \mid 3|A, 3|B\right\},$$

i.e., all multiples of 3. Thus $\mathfrak{a}_1\mathfrak{a}_2 = \langle 3 \rangle$.

In the same way, let

$$\mathfrak{b} = \left\langle 2, 1 + \sqrt{-5}\right\rangle.$$

Then a typical element of $\mathfrak{b}^2$ is given by

$$4\alpha + \left(2 + 2\sqrt{-5}\right)\beta + \left(1 + \sqrt{-5}\right)^2\gamma,$$

and an easy check shows that $\mathfrak{b}^2 = \langle 2 \rangle$.

Further, we leave it as an exercise to verify that the ideal $\langle 1 + \sqrt{-5} \rangle$ (resp. $\langle 1 - \sqrt{-5} \rangle$) is given by $\mathfrak{a}_1\mathfrak{b}$ (resp. $\mathfrak{a}_2\mathfrak{b}$).

Later we will explain that these ideals are "prime". However, you can already see that the two distinct factorisations we gave,

$$6 = 2 \times 3 = \left(1 + \sqrt{-5}\right)\left(1 - \sqrt{-5}\right),$$

actually become the same factorisation in terms of the *ideals*:

$$\langle 6 \rangle = \mathfrak{b}^2.\mathfrak{a}_1\mathfrak{a}_2 = (\mathfrak{a}_1\mathfrak{b}).(\mathfrak{a}_2\mathfrak{b}).$$

By introducing ideals, we have repaired the non-uniqueness of factorisation in this case.

Exercise 4.16 Again, let $R = \mathbb{Z}[\sqrt{-5}]$, and consider the ideals $\mathfrak{a}_1 = \langle 3, 1 + \sqrt{-5} \rangle$ and $\mathfrak{a}_2 = \langle 3, 1 - \sqrt{-5} \rangle$.

1. Show that every element $a + b\sqrt{-5} \in \mathfrak{a}_2$ has the property that $3|a + b$, and similarly every element $c + d\sqrt{-5} \in \mathfrak{a}_1$ satisfies $3|c - d$.
2. Show that $\mathfrak{a}_1$ and $\mathfrak{a}_2$ are not principal.

Exercise 4.17 Now let R denote the ring $\mathbb{Z}[\sqrt{-6}]$. We have

$$10 = 2.5 = \left(2 + \sqrt{-6}\right)\left(2 - \sqrt{-6}\right).$$

1. Show that R has no elements of norm 2 or 5, and deduce that the two factorisations of 10 are distinct factorisations into irreducibles.
2. Let $\mathfrak{a}$ denote the ideal $\langle 2, \sqrt{-6} \rangle$ of R. Show that $\mathfrak{a}$ is not principal.
   [*Hint: to show that $\mathfrak{a}$ is proper, show that $\mathfrak{a}$ is the set of elements $a + b\sqrt{-6} \in R$ such that a is even.*]
3. Show that $\mathfrak{a}^2 = \langle 2 \rangle$.
4. Let $\mathfrak{b}_1$ and $\mathfrak{b}_2$ denote the ideals $\langle 5, 2 + \sqrt{-6} \rangle$ and $\langle 5, 2 - \sqrt{-6} \rangle$ respectively. Compute $\mathfrak{b}_1 \mathfrak{b}_2$, and write $\langle 10 \rangle$ as a product of ideals of R.
5. Explain that the non-uniqueness of factorisation has been resolved by using ideals.

In Chap. 5, we will consider what it means for ideals to be "prime", and we will show that ideals in rings of integers of number fields may be factorised *uniquely* into prime ideals.

However, we saw an example at the end of Chap. 1 where it was important that the rings of integers themselves have unique factorisation, and before we think about factorisation of ideals into "prime ideals", we will now study unique factorisation in rings from a more abstract, ring-theoretical point of view.

4.7 Unique Factorisation Domains and Principal Ideal Domains

What do we mean by unique factorisation? Clearly we should regard two factorisations as the same if the factors merely occur in a different order, so that 2×3 and 3×2 are the same factorisation. But we should also say that 2×3 is the same as $(-2) \times (-3)$. We are always going to have problems with units in any ring; if u is a unit, then $ab = (au)(bu^{-1})$, and we should treat these as the same. This is a general ring-theoretic version of Definitions 4.3 and 4.4:

Definition 4.26 A ring R is a *unique factorisation domain* (UFD) if it is an integral domain in which every non-zero $a \in R$ may be written

$$a = up_1 \ldots p_n,$$

where u is a unit and each p_i is irreducible (i.e., factorisation into irreducibles exists); further, if $a = vq_1 \ldots q_m$ is another such factorisation, then $n = m$ and p_i is an associate of $q_{\pi(i)}$ for some permutation of $\{1, \ldots, n\}$ (i.e., factorisation into irreducibles is unique).

We have seen that $\mathbb{Z}$ and $\mathbb{Z}[i]$ both have unique factorisation, and are therefore UFDs. However, we have seen above that neither $\mathbb{Z}[\sqrt{10}]$ nor $\mathbb{Z}[\sqrt{-5}]$ are UFDs.

We will explain next that there is a class of rings for which one can easily prove unique factorisation.

Recall that a principal ideal is one of the form aR. For some rings, such as $\mathbb{Z}$, these are the only ideals (Lemma 4.13).

Definition 4.27 Let R be an integral domain. Then R is a *principal ideal domain* (abbreviated *PID*) if every ideal of R is principal.

We have already seen examples of integral domains which are not PIDs, such as $\mathbb{Z}[X]$, which has an ideal $\langle 2, X \rangle$ which was not principal, and similarly $\mathbb{Z}[\sqrt{10}]$, which has an ideal $\langle 2, \sqrt{10} \rangle$ which is not principal.

We are going to explain that every PID is also a UFD.

Of course, every field K is a PID, because its only ideals are $\langle 0_K \rangle$ and K itself, which may also be written as $\langle 1_K \rangle$. Also, we have already seen that $\mathbb{Z}$ is a PID, as in Lemma 4.13. The proof that $\mathbb{Z}$ is a PID relies on Euclid's algorithm. Any ring with some kind of "Euclidean algorithm" is going to be a PID, using exactly the same proof. This motivates the following definition:

Definition 4.28 An integral domain R is a *Euclidean domain* if there is a function

$$\phi : R - \{0_R\} \longrightarrow \mathbb{Z}_{>0}$$

such that

1. $a|b \Rightarrow \phi(a) \leq \phi(b)$,
2. if $a \in R, b \in R - \{0_R\}$, then there exist q and r in R such that $a = bq + r$ and either $r = 0$ or $\phi(r) < \phi(b)$.

ϕ is called a *Euclidean function* on R.

So $\mathbb{Z}$ is a Euclidean domain (define $\phi(n) = |n|$) and so is $K[X]$ for any field K (define $\phi(f) = \deg f$). We saw earlier that there is also a Euclidean algorithm in $\mathbb{Z}[i]$ (define $\phi(a + ib) = a^2 + b^2$).

The argument of Lemma 4.13 shows the following

Proposition 4.29 *Every Euclidean domain is a principal ideal domain.*

Proof Let I be an ideal of the Euclidean domain R, and suppose $I \neq \{0\}$. Consider the set $D = \{\phi(i) \mid i \in I, i \neq 0\} \subseteq \mathbb{Z}_{>0}$ of all values taken by the Euclidean function ϕ on the nonzero elements of the ideal I. Choose $b \in I$ such that $\phi(b)$ is the minimal value in D. As $b \in I$, certainly I then contains all multiples of b; that is, $I \supseteq \langle b \rangle$.

Conversely, take $a \in I$. We can write $a = qb + r$ by Definition 4.28, where either $r = 0$ or $\phi(r) < \phi(b)$. As a and $b \in I$, we conclude that $r = a - qb \in I$. But if $\phi(r) < \phi(b)$, then this contradicts the choice of b as an element of I with the least possible value of the Euclidean function. So $r = 0$, and therefore $a = qb$; thus every element of I is a multiple of b, and $I = \langle b \rangle$ is principal. $\qquad\square$

The converse to this is false, but it is not so easy to write down an example. In fact, if $\rho = \frac{1+\sqrt{-19}}{2}$, then it is known that $\mathbb{Z}[\rho]$ is a PID, but not Euclidean. This is probably the easiest example of such a ring. We will prove this in Chap. 6.

In Chap. 1, we saw that the Euclidean algorithm was a key ingredient in our proof of unique factorisation in $\mathbb{Z}$ (and also in $\mathbb{Z}[i]$). The Euclidean algorithm gives us a way to compute the highest common factor of two elements in a Euclidean domain. But this notion exists more generally, in principal ideal domains, and we shall show the general result that every PID is a UFD.

Definition 4.30 Let R be a PID, and let a and b be in R. The ideal $\langle a, b \rangle = aR + bR$ is principal, so can be written $\langle d \rangle = dR$ for some element $d \in R$. Then d is a *highest common factor* of a and b. Note that highest common factors are unique up to multiplication by a unit.

This agrees with the usual notion in $\mathbb{Z}$. The difference between PIDs and Euclidean domains is *not* in the essential point that highest common factors exist, but rather that there is a good way to compute them in Euclidean domains, namely the Euclidean algorithm, which may be absent in more general PIDs.

We already saw that the existence of the Euclidean algorithm implied that unique factorisation exists. In fact, it is enough that we have the weaker concept of highest common factor.

Theorem 4.31 *Every PID is a UFD.*

Proof Suppose first that there exists an element a without any factorisation. Call such elements 'bad', and other elements 'good'. Then a is not a unit, nor an irreducible, so we must have $a = a_1 b_1$ for some a_1, b_1; at least one of a_1 and b_1 must be bad (otherwise the product of the factorisations for a_1 and b_1 gives a factorisation of a). Suppose a_1 is bad. Then, in the same way, $a_1 = a_2 b_2$ with a_2 bad. Continuing in this way, we get a sequence of bad elements $a_1, a_2, \ldots$. Further, as a_i is a multiple of a_{i+1}, we see that $\langle a_{i+1} \rangle \supset \langle a_i \rangle$ (and these are different as no b_{i+1} is a unit). Let $I = \bigcup_{i=1}^{\infty} \langle a_i \rangle$. It is easy to check that this is an ideal, and therefore $I = \langle c \rangle$ for some $c \in R$. Thus $c \in I$, so c lies in some $\langle a_n \rangle$. Then $I = \langle c \rangle \subseteq \langle a_n \rangle \subset \langle a_{n+1} \rangle \subseteq I$. This is a contradiction—so no bad elements exist, and everything has some factorisation.

Now suppose that there is an element with two factorisations. We first show that every irreducible element $p \in R$ satisfies the following:

$$p|ab \Rightarrow p|a \text{ or } p|b.$$

Let p be an irreducible element, therefore, and suppose $p|ab$. If $p \nmid a$, we show that $p|b$ (here, the notion of divisibility is that of Definition 1.2).

Consider the ideal $\langle p, a \rangle = pR + aR$. Then, as R is a PID, $\langle p, a \rangle = \langle d \rangle$. Then $d|p$ and $d|a$. As p is irreducible, either d is a unit or d is an associate of p, but this latter is impossible as $p \nmid a$. Thus $\langle p, a \rangle$ is generated by a unit d, so $\langle p, a \rangle = R$. Thus we can find r and $s \in R$ such that $pr + as = 1_R$. Multiply by b to get

$$p(br) + (ab)s = b,$$

and we see that b is a multiple of p as $p|ab$.

Now we show unique factorisation. Suppose we had an element n with two factorisations:

$$n = up_1 \ldots p_r = vq_1 \ldots q_s$$

(where u and v are units, and the p_i, q_j are irreducible). Then p_1 divides n and therefore the right-hand side. Using the above claim, we see that p_1 divides one of the q_i, q_1 say (permute the q_i if not). As both p_1 and q_1 are irreducible, we must have $q_1 = u_1 p_1$ where u_1 is a unit. We can cancel p_1 and q_1 from the factorisations (as R is an integral domain). We can continue in this way until all prime factors on the left-hand side are paired off with factors on the right-hand side, and only units are left.   □

Note that the proof that every element has some factorisation (i.e., there are no bad elements) would work given only the weaker statement that every ideal (in particular, the ideal I) has a *finite* generating set, so that $I = \langle d_1, \ldots, d_k \rangle$; this is the defining property of a *Noetherian ring*, and we shall remark in the next section that rings of integers of number fields are always Noetherian.

For rings of integers in number fields, the converse to this theorem is true: such a ring is a PID if and only if it is a UFD. We will prove this in Chap. 5. (It turns out to be false in a general ring—it is known that if R is a UFD, then so is the polynomial ring $R[X]$; this shows that $\mathbb{Z}[X]$ is a UFD, but we have alreay seen that it is not a PID, as the ideal $\langle 2, X \rangle$ is not principal.)

4.8 The Noetherian Property

We've remarked already that the property of unique factorisation in a number field is equivalent to the ring of integers having the property that every ideal is principally generated, i.e., has 1 generator.

We've also seen that many number fields do *not* have unique factorisation, and therefore do not have this property. However, there is a weaker property that they all satisfy which will be very important for us in Chap. 5. It is named after Emmy Noether, who introduced it:

Definition 4.32 A *Noetherian ring* is a ring R in which every ideal is finitely generated.

(Note that there are other equivalent formulations you might see in the literature.)

Recall from Theorem 3.27 that $\mathbb{Z}_K$ has an integral basis, and that this is equivalent to the property that $\mathbb{Z}_K$ is a free abelian group of rank $[K : \mathbb{Q}]$. We will explain that this implies that $\mathbb{Z}_K$ is Noetherian.

We need one more result about free abelian groups.

Proposition 4.33 *Suppose H is a subgroup of a free abelian group G of rank n. Then H is also a free abelian group of rank at most n.*

Proof We will prove the claim by induction. For $n = 1$, $G \cong \mathbb{Z}$, and the Euclidean algorithm shows that $H = k\mathbb{Z}$ for some k (this argument is exactly that of Lemma 4.13); if $k = 0$, then H has rank 0, and otherwise H has rank 1 and is of finite index.

Now let G have rank n, and suppose the result is true for free abelian groups of rank $n - 1$.

We can write $G = \mathbb{Z}\omega_1 + \cdots + \mathbb{Z}\omega_n$. Let $\pi : G \longrightarrow \mathbb{Z}$ map $a_1\omega_1 + \cdots + a_n\omega_n$ to a_1, and let $K = \ker \pi = \mathbb{Z}\omega_2 + \cdots + \mathbb{Z}\omega_n$, a free abelian group of rank $n - 1$. Then $\pi(H) \subseteq \mathbb{Z}$, and by the rank 1 case, $\pi(H) = \{0\}$ or $\pi(H)$ is infinite cyclic.

If $\pi(H) = \{0\}$, then $H \subset \mathbb{Z}\omega_2 + \cdots + \mathbb{Z}\omega_n$, so is a subgroup of a free abelian group of rank $n - 1$, and the result follows by the inductive hypothesis.

If $\pi(H)$ is infinite cyclic, choose $h_1 \in H$ such that $\pi(h_1)$ generates $\pi(H)$. It is easy to prove (exercise, but use the method of the First Isomorphism Theorem 5.5 if you get stuck) that $H = \mathbb{Z}h_1 \oplus (H \cap K)$, and $H \cap K$ is contained in K, a free abelian group of rank $n - 1$. Then the inductive hypothesis shows that $H \cap K$ is a free abelian group, $\mathbb{Z}h_2 + \cdots + \mathbb{Z}h_r$, say, and again the claim follows. $\square$

An ideal is an (additive) subgroup of $\mathbb{Z}_K$, and so we can apply this result to conclude that every ideal is also a free abelian group of finite rank. Equivalently, it is finitely generated as a $\mathbb{Z}$-module. Thus $I = \mathbb{Z}\omega_1 + \cdots + \mathbb{Z}\omega_r$ for some elements $\omega_1, \ldots, \omega_r \in I$. (In fact, if $I \neq \{0\}$, it is a consequence of the results of Chap. 5—see Lemma 5.20, for example—that I has finite index in $\mathbb{Z}_K$, and this implies that $r = n$.)

That isn't quite what we want to prove. We'd really like to see that it is finitely generated as an ideal, so that there is a finite set $\alpha_1, \ldots, \alpha_k$ such that $I = \langle \alpha_1, \ldots, \alpha_k \rangle$.

Fortunately, this is nearly immediate. Take a set $\{\omega_1, \ldots, \omega_r\}$ which generates I as a $\mathbb{Z}$-module. Notice that we must have $\omega_i \in I$. As I is an ideal, $\mathbb{Z}_K\omega_i \subseteq I$. Clearly $\mathbb{Z}_K\omega_i \supseteq \mathbb{Z}\omega_i$, and so

$$\mathbb{Z}_K\omega_1 + \cdots + \mathbb{Z}_K\omega_r \supseteq \mathbb{Z}\omega_1 + \cdots + \mathbb{Z}\omega_r.$$

On the other hand, the right-hand side is exactly I, as we chose $\{\omega_1, \ldots, \omega_r\}$ to be a generating set for I as a $\mathbb{Z}$-module. But the left-hand side is contained in I, as each $\mathbb{Z}_K\omega_i \subseteq I$. So

$$I \supseteq \mathbb{Z}_K\omega_1 + \cdots + \mathbb{Z}_K\omega_r \supseteq \mathbb{Z}\omega_1 + \cdots + \mathbb{Z}\omega_r = I,$$

and all the inclusions are equalities. In particular, we see that

$$I = \mathbb{Z}_K\omega_1 + \cdots + \mathbb{Z}_K\omega_r,$$

and so I is finitely generated as an ideal.

We have therefore shown:

Theorem 4.34 *If K is a number field, then $\mathbb{Z}_K$ is Noetherian.*

Exercise 4.18 If $K = \mathbb{Q}(\sqrt{2})$, this shows that $\mathbb{Z}_K$ is Noetherian. Is it a PID or UFD?

Later we will need one further result about Noetherian rings.

Definition 4.35 A ring R is said to *satisfy the ascending chain condition* (or *ACC*) if for every chain of ideals

$$I_1 \subseteq I_2 \subseteq I_3 \subseteq \cdots$$

of ideals of R, there exists some positive integer n such that $I_n = I_{n+1} = I_{n+2} = \cdots$.

Proposition 4.36 *A ring R is Noetherian if and only if it satisfies the ACC.*

Proof Assume that R is Noetherian, and let $I_1 \subseteq I_2 \subseteq \cdots$ be an ascending chain of ideals. Let $I = \bigcup_{i=1}^{\infty} I_i$; it is easy to check that this is an ideal. As R is Noetherian, I has a finite generating set, $\{r_1, \ldots, r_n\}$. Each element r_j of the generating set must occur in some I_{n_j}; if $n = \max(n_j)$, the largest of these numbers, then each element of the generating set is already contained in I_n. Thus $I_n = I_{n+1} = \cdots = I$, and so the chain becomes stationary.

Conversely, if the ACC is satisfied, then every ideal must be finitely generated. If not, there is an ideal I which has no finite generating set. Pick $r_1 \in I$. Then $I \neq \langle r_1 \rangle$, as otherwise $\{r_1\}$ would generate I, so we may pick $r_2 \in I - \langle r_1 \rangle$. Again, $I \neq \langle r_1, r_2 \rangle$, so we may pick $r_3 \in I - \langle r_1, r_2 \rangle$. In this way, we find an infinite sequence of elements $r_1, r_2, \ldots$, and an infinite strictly ascending chain $\langle r_1 \rangle \subseteq \langle r_1, r_2 \rangle \subseteq \langle r_1, r_2, r_3 \rangle \subseteq \cdots$, contradicting the ACC. $\square$

Remark 4.37 There is also the notion of a *descending chain condition*, in which every descending chain must eventually become stationary. Rings satisfying the DCC are said to be *Artinian*. Note that $\mathbb{Z}$ is Noetherian (as it is a PID), but that it is not Artinian, as the descending chain

$$\langle 2 \rangle \supset \langle 4 \rangle \supset \langle 8 \rangle \supset \cdots$$

never becomes stationary. Rings of integers in number fields are never Artinian, and we will not refer to Artinian rings again in this book.

The Ascending Chain Condition can be used to prove results like the following (similar examples will be seen in Chap. 5).

Lemma 4.38 *Suppose that I is a proper ideal in a Noetherian ring R. Then I is contained in a maximal ideal.*

Proof Either I is maximal itself, or it is strictly contained in a larger proper ideal I_1. If I_1 is maximal, the result follows; otherwise it is strictly contained in a larger ideal I_2. Repeat—since there cannot be arbitrarily long chains $I \subset I_1 \subset I_2 \subset \cdots$, at some point, one of the ideals must be maximal, and the result follows. $\square$

Chapter 5
Prime Ideals and Unique Factorisation

We have already studied unique factorisation in $\mathbb{Z}$, and seen how it fails in certain rings of integers of number fields. We have also seen the suggestion that non-uniqueness of factorisation may be remedied by working with ideals. In order to show that this procedure will work generally, we will need to have some concept of what it means for an ideal to be prime.

We have two equivalent ways to define prime numbers in $\mathbb{Z}$. In Chap. 1, we defined a prime number as one which has no divisor other than 1 and itself. Later in that chapter (Lemma 1.15), we pointed out that in $\mathbb{Z}$, prime numbers have the property that if $p|ab$, then $p|a$ or $p|b$, and it is easy to see that this property characterises prime numbers in $\mathbb{Z}$.

We have already generalised these two properties (Definition 4.5) to give two possible generalisations of prime numbers in rings. We shall now generalise these two notions to ideals, to get ideals called *maximal* and *prime* ideals respectively.

But first we'll develop some ring theory. If you are familiar with ring homomorphisms and quotient rings, feel free to skip this section!

5.1 Some Ring Theory

The notion of a ring homomorphism is very similar to that of a group homomorphism; it is a map from one ring to another which preserves all the algebraic ring structure.

Definition 5.1 Let R and S be rings. Then $\phi : R \longrightarrow S$ is a *ring homomorphism* if it preserves the additive and multiplicative structures and the multiplicative identity:

1. for all $a, b \in R$, we have $\phi(a + b) = \phi(a) + \phi(b)$,
2. for all $a, b \in R$, we have $\phi(ab) = \phi(a)\phi(b)$.
3. $\phi(1_R) = 1_S$.

F. Jarvis, *Algebraic Number Theory*, Springer Undergraduate
Mathematics Series, DOI: 10.1007/978-3-319-07545-7_5,
© Springer International Publishing Switzerland 2014

If also ϕ is bijective, then ϕ is an *isomorphism*. The *kernel of* ϕ is defined as the set $\{r \in R \mid \phi(r) = 0_S\}$, and is written ker ϕ; and similarly the set $\{s \in S \mid s = \phi(r)$ for some $r \in R\}$ is called the *image* of ϕ, and is written im ϕ.

Note that some books do not insist on the third of the requirements when defining a homomorphism.

It is easy to check that if $\phi : R \longrightarrow S$ is a ring homomorphism, then

1. $\phi(0_R) = 0_S$,
2. if $a \in R$, then $\phi(-a) = -\phi(a)$.

Our first lemma will have a useful corollary:

Lemma 5.2 *Let $\phi : R \longrightarrow S$ be a ring homomorphism. Let I be an ideal of S, and let $\phi^{-1}(I) = \{a \in R \mid \phi(a) \in I\}$. Then $\phi^{-1}(I)$ is an ideal of R.*

Proof We have to check that

1. $0_R \in \phi^{-1}(I)$,
2. if a and $b \in \phi^{-1}(I)$, then so is $a - b$,
3. if $i \in \phi^{-1}(I)$ and $a \in R$, then $ai \in \phi^{-1}(I)$.

These are all very easy. For the first, as $\phi(0_R) = 0_S \in I$, we see that $0_R \in \phi^{-1}(I)$. The second follows because $\phi(a - b) = \phi(a) - \phi(b)$, so if $\phi(a)$ and $\phi(b) \in I$, so is $\phi(a - b)$. Equivalently, if a and $b \in \phi^{-1}(I)$, then $a - b \in \phi^{-1}(I)$. Finally, let $i \in \phi^{-1}(I)$ and $a \in R$; then $\phi(ai) = \phi(a)\phi(i)$, so as $\phi(i) \in I$, $\phi(ai) \in I$, so $ai \in \phi^{-1}(I)$. □

As a corollary, we see that if $\phi : R \longrightarrow S$ is a ring homomorphism, then ker $\phi = \{a \in R \mid \phi(a) = 0_S\}$ is an ideal in R, simply because it can be rewritten as $\phi^{-1}(0_S)$, and $\{0_S\}$ is an ideal in S. (It is also easy to verify this directly.) It is also straightforward to check that im ϕ is a subring of S.

Lemma 5.3 *A ring homomorphism $\phi : R \longrightarrow S$ is injective if and only if its kernel just consists of the zero element.*

Proof Suppose that ker $\phi = \{0_R\}$. Then

$$\phi(r_1) = \phi(r_2) \Rightarrow \phi(r_1 - r_2) = 0_S$$
$$\Rightarrow r_1 - r_2 \in \ker \phi$$
$$\Rightarrow r_1 - r_2 = 0_R$$
$$\Rightarrow r_1 = r_2$$

so that ϕ is injective.

Conversely, if ϕ is injective, then its kernel is trivial, for if $\phi(r) = 0_S$, we have $\phi(r) = \phi(0_R)$, so that $r = 0_R$ as ϕ is injective. □

Ideals in rings may be viewed as analogous to normal subgroups of groups. For example, the kernel of a group homomorphism is a normal subgroup, and the kernel of a ring homomorphism is an ideal.

Exercise 5.1 Suppose that I is an ideal in a ring R, and that $r \in R$. Show that $r \in I$ if and only if $r + I = I$, where

$$r + I = \{r + i \mid i \in I\}$$

is the *coset* of I.

We should emphasise the following consequence of this:

$$r + I = r' + I \Longleftrightarrow r - r' \in I.$$

We shall see next that one can also take a quotient of a ring by an ideal in the same way that one can form the quotient of a group by a normal subgroup. There is an analogue of the First Isomorphism Theorem for the quotient of a ring by the kernel of a homomorphism.

Proposition 5.4 *Let I be an ideal in the ring R. If $a \in R$, let*

$$a + I = \{a + i \mid i \in I\}$$

be the coset of I. Then the collection of cosets,

$$R/I = \{a + I \mid a \in R\},$$

may be given the structure of a ring, called the quotient ring.

Proof We define
$$(a + I) + (b + I) = (a + b) + I$$

and
$$(a + I)(b + I) = (ab) + I.$$

We first verify that this is well-defined, in that choosing a different coset representative gives the same coset as the answer.

If $a + I = a' + I$, then $a - a' \in I$. Let $i = a - a'$. As $i \in I, i + I = I$. Then

$$(a+I)+(b+I) = (a+b)+I = (a'+i+b)+I = (a'+b)+I = (a'+I)+(b+I)$$

and

$$(a+I)(b+I) = ab+I = (a'+i)b+I = a'b+ib+I = a'b+I = (a'+I)(b+I),$$

the penultimate equality holding as $ib \in I$ (because $i \in I$ and I is an ideal).

Now we check the ring axioms. These are all straightforward, as they are inherited from R. For example, the additive identity is $I = 0_R + I$, because

$$(a + I) + I = (a + I) + (0_R + I) = (a + 0_R) + I = a + I.$$

In the same way, the multiplicative identity is $1_R + I$. Let's just check one of the axioms—commutativity of addition, say:

$$\begin{aligned}(a + I) + (b + I) &= (a + b) + I \\ &= (b + a) + I \\ &= (b + I) + (a + I)\end{aligned}$$

where the middle equality holds because of commutativity of addition in R, and the others from our definition of addition of cosets. Checking the other axioms is similar; all properties are inherited from the corresponding properties of R.   □

The proposition should remind you of the construction of the integers modulo n, which we now recall.

Fix the positive integer $n \geq 2$. For each integer a, let

$$\bar{a} = \{\ldots, a - 2n, a - n, a, a + n, a + 2n, \ldots\}$$

consist of all integers congruent to $a \pmod{n}$. Note that if $a \equiv b \pmod{n}$, then $\bar{a} = \bar{b}$. Then the integers modulo n are given by

$$\{\bar{0}, \bar{1}, \ldots, \overline{n - 1}\},$$

with the addition and multiplication defined using arithmetic modulo n (which is well-defined): $\bar{a} + \bar{b} = \overline{a + b}$ and $\bar{a}.\bar{b} = \overline{ab}$. To check any given ring axiom, write down the corresponding axiom for $\mathbb{Z}$, and reduce it modulo n. Thus all the axioms are inherited from those for $\mathbb{Z}$.

Note that $\bar{a} = a + n\mathbb{Z}$, so that the integers mod n are given by $\{a + n\mathbb{Z} \mid a \in \mathbb{Z}\}$. It follows they can be viewed as the quotient of the ring $\mathbb{Z}$ by the ideal $n\mathbb{Z}$ of all integers divisible by n, i.e., as $\mathbb{Z}/n\mathbb{Z}$.

We will henceforth use the notation $\mathbb{Z}/n\mathbb{Z}$ to denote the integers modulo n. We will also omit the bars on top of the numbers, so that we view $\mathbb{Z}/n\mathbb{Z}$ as the set $\{0, \ldots, n - 1\}$, with addition and multiplication taken modulo n.

Sometimes in books, you will see the integers mod n written as $\mathbb{Z}_n$. However, if p is a prime number, number theorists use $\mathbb{Z}_p$ for a completely different set than $\mathbb{Z}/p\mathbb{Z}$. For this reason, number theorists never use $\mathbb{Z}_n$ for the integers modulo n, and always use $\mathbb{Z}/n\mathbb{Z}$. We will mention the construction of $\mathbb{Z}_p$ in Sect. 10.6; it is a central concept in modern research (both in number theory and in other areas).

If I is an ideal in the ring R, then there is a naturally-defined *quotient map*

$$R \longrightarrow R/I$$
$$r \mapsto r + I$$

which is always a homomorphism. In the case $R = \mathbb{Z}$, $I = n\mathbb{Z}$, then this is exactly the map $\mathbb{Z} \longrightarrow \mathbb{Z}/n\mathbb{Z}$ which takes m to m (mod n) given above.

Note the following two special cases:

1. if $I = R$, then R/I is the trivial ring, with just one element. For this, take any element $a + R \in R/R$. As $a \in R$, we have $a + R = R$. Thus the only element of R/R is $0_R + R = R$.
2. if $I = (0_R)$, then R/I is isomorphic to R. Every element of R/I is of the form $a + (0_R)$ for some $a \in R$, but these are all distinct: $a + (0_R) = b + (0_R)$ implies that $a = b$.

Consider the case $R = \mathbb{Z}[X]$ and $I = \langle X^2 \rangle$ consisting of all multiples of X^2. Then a typical element of R/I may be written $f(X) + \langle X^2 \rangle$, where $f(X)$ is a polynomial with integer coefficients. If $f(X)$ is the polynomial $a_0 + a_1 X + \cdots + a_{d-1} X^{d-1} + a_d X^d$, then f may be written $a_0 + a_1 X + g(X)X^2$ for some polynomial $g(X)$. It follows that $f + \langle X^2 \rangle = a_0 + a_1 X + \langle X^2 \rangle$ because $g(X)X^2$ is in the ideal $\langle X^2 \rangle$. So elements of $\mathbb{Z}[X]/\langle X^2 \rangle$ are parametrised only by their constant and linear terms. The coset corresponding to $a + bX$ is the collection of all polynomials whose constant term is a and whose linear term is b. We can add two elements:

$$(a + bX + \langle X^2 \rangle) + (c + dX + \langle X^2 \rangle) = (a + c) + (b + d)X + \langle X^2 \rangle$$

and multiply them:

$$(a + bX + \langle X^2 \rangle)(c + dX + \langle X^2 \rangle) = ac + (ad + bc)X + \langle X^2 \rangle$$

simply by adding and multiplying in the usual way, and ignoring all terms X^2 and above (there should be a term bdX^2, of course, but that belongs to the ideal $\langle X^2 \rangle$, and so $bdX^2 + \langle X^2 \rangle = \langle X^2 \rangle$).

A more instructive example, of a kind we will consider again later, is obtained by letting $I = \langle X^2 - 2 \rangle$, say, the ideal of all multiples of $X^2 - 2$, in the ring $R = \mathbb{Z}[X]$ again. Then each $f \in \mathbb{Z}[X]$ can be written as $q(X)(X^2 - 2) + r(X)$, where $q(X)$ and $r(X)$ are the quotient and remainder after dividing f by $X^2 - 2$. The degree of $r(X)$ is at most 1, so $r(X) = b_0 + b_1 X$ for some $b_0, b_1 \in \mathbb{Z}$. Thus $f + I = r + I$, and so every coset is parametrised by a linear polynomial as before. The addition rule is the same as before:

$$(a + bX + \langle X^2 - 2 \rangle) + (c + dX + \langle X^2 - 2 \rangle) = (a + c) + (b + d)X + \langle X^2 - 2 \rangle;$$

however, the multiplication rule looks rather different:

$$(a + bX + \langle X^2 - 2 \rangle)(c + dX + \langle X^2 - 2 \rangle)$$
$$= ac + (ad + bc)X + bdX^2 + \langle X^2 - 2 \rangle$$
$$= ac + (ad + bc)X + bd(2 + (X^2 - 2)) + \langle X^2 - 2 \rangle$$
$$= ac + (ad + bc)X + 2bd + bd(X^2 - 2) + \langle X^2 - 2 \rangle$$
$$= (ac + 2bd) + (ad + bc)X + \langle X^2 - 2 \rangle.$$

Note that if $\alpha = a + b\sqrt{2} \in \mathbb{Z}[\sqrt{2}]$ and $\beta = c + d\sqrt{2} \in \mathbb{Z}[\sqrt{2}]$, then

$$\alpha + \beta = (a + c) + (b + d)\sqrt{2}$$
$$\alpha\beta = (ac + 2bd) + (ad + bc)\sqrt{2}$$

which very closely resembles the addition and multiplication law above for the quotient ring $\mathbb{Z}[X]/\langle X^2 - 2 \rangle$. In fact, the map

$$\mathbb{Z}[X]/\langle X^2 - 2 \rangle \longrightarrow \mathbb{Z}[\sqrt{2}]$$
$$a + bX + \langle X^2 - 2 \rangle \mapsto a + b\sqrt{2}$$

is an isomorphism of rings. That it is a homomorphism is verified explicitly using the above calculations; it is left as an easy exercise to check that it is a bijection.

The First Isomorphism Theorem will provide many similar examples, as we explain next. You can probably guess what the statement of the First Isomorphism Theorem for Rings should be, if you remember the statement for groups.

Theorem 5.5 (First Isomorphism Theorem) *Let $\phi : R \longrightarrow S$ be a ring homomorphism. Then there is an isomorphism*

$$R/\ker\phi \cong \operatorname{im}\phi.$$

Proof Define a map $\widetilde{\phi} : R/\ker\phi \longrightarrow \operatorname{im}\phi$ by $\widetilde{\phi}(r + \ker\phi) = \phi(r)$.

We should be a little careful here! After all, there may be many ways to write a coset in $R/\ker\phi$ as a set $r + \ker\phi$, and we should check that if there are two different ways to write the coset, say, as both $r + \ker\phi$ and $r' + \ker\phi$, then our definition above gives the same answer. (That is, we need to check that $\widetilde{\phi}$ is *well-defined*.)

So suppose that the coset $r + \ker\phi$ may also be written as $r' + \ker\phi$. With one definition of $\widetilde{\phi}$ we get $\phi(r)$, and with the other we get $\phi(r')$. We must check that these are the same.

As $r + \ker\phi = r' + \ker\phi$, it follows that the element $r - r' \in \ker\phi$. In other words, there exists $k \in \ker\phi$ such that $r = r' + k$. But now

$$\phi(r) = \phi(r' + k) = \phi(r') + \phi(k) = \phi(r') + 0_S = \phi(r').$$

Now we know that the map $\widetilde{\phi}$ exists and makes sense. Clearly, it is also valued in $\operatorname{im}\phi$, since applying $\widetilde{\phi}$ to any coset gives an element which is in the image of ϕ.

Now we need to check that $\widetilde{\phi}$ is a homomorphism, and is bijective. Then $\widetilde{\phi}$ will be an isomorphism, as required.

To see that $\widetilde{\phi}$ is a homomorphism is easy; this follows easily from our definition of addition and multiplication of cosets; let's just do addition, and multiplication is similar:

$$\widetilde{\phi}((a + \ker \phi) + (b + \ker \phi)) = \widetilde{\phi}((a + b) + \ker \phi)$$
$$= \phi(a + b)$$
$$= \phi(a) + \phi(b)$$
$$= \widetilde{\phi}(a + \ker \phi) + \widetilde{\phi}(b + \ker \phi).$$

Next, let's check that $\widetilde{\phi}$ is injective. Suppose we have an element $r + \ker \phi$ in the kernel. Then

$$\phi(r) = \widetilde{\phi}(r + \ker \phi) = 0_S,$$

so certainly $r \in \ker \phi$, and so $r + \ker \phi = \ker \phi$, the zero element of the quotient ring $R/\ker \phi$. Thus the kernel just consists of the zero element of the quotient ring, and so $\widetilde{\phi}$ is injective.

It is clear that the image of $\widetilde{\phi}$ is exactly the same as the image of ϕ, so $\widetilde{\phi}$ is surjective onto im ϕ.

This shows that $\widetilde{\phi}$ is an isomorphism, and completes the proof. $\qquad\square$

Remark 5.6 First Isomorphism Theorems occur all over algebra. Probably the first time you met them was with the rank-nullity theorem for vector spaces over fields, which is a relationship between the dimensions of the image of a linear map between vector spaces and of the kernel, but the proof essentially works in the same way. We will also use a First Isomorphism Theorem between modules of a ring: if $\phi : M \to N$ is a homomorphism of modules over a ring R (i.e., a map ϕ such that $\phi(m + m') = \phi(m) + \phi(m')$ and $\phi(rm) = r\phi(m)$), then the collection $M/\ker \phi$ of cosets $m + \ker \phi$ is isomorphic to im ϕ. The idea of the proof is very similar to Theorem 5.5 (or to the rank-nullity theorem), and is left as an exercise.

Consider the following situation:

Lemma 5.7 *Let K be a field, and suppose that γ is algebraic over K, i.e., satisfies a polynomial equation with coefficients in K. Suppose that $f \in K[X]$ is the minimal polynomial of γ. Then there is an isomorphism*

$$K[X]/\langle f \rangle \cong K(\gamma)$$

got by mapping X to γ.

Proof Consider the homomorphism (you should check that it is one!)

$$\phi_\gamma : K[X] \longrightarrow K(\gamma)$$
$$g(X) \mapsto g(\gamma)$$

The kernel of this homomorphism consists of all polynomials which have γ as a root, and this is precisely the set of all multiples of f, namely $\langle f \rangle$.

Clearly ϕ_γ is also surjective: every element of $K(\gamma)$ is just a polynomial $a_d \gamma^d + \cdots + a_0$, with each $a_i \in K$, but this is the image under ϕ_γ of the polynomial $a_d X^d + \cdots + a_0 \in K[X]$. Thus ϕ_γ gives an isomorphism

$$\widetilde{\phi}_\gamma : K[X]/\langle f \rangle \cong K(\gamma)$$

by the First Isomorphism Theorem 5.5.   □

For another example, take $K = \mathbb{Q}$ and $\gamma = \sqrt{2}$. Then γ has minimal polynomial $X^2 - 2$. By the lemma, there is an isomorphism $\mathbb{Q}[X]/\langle X^2 - 2 \rangle \cong \mathbb{Q}(\sqrt{2})$ given by sending X to $\sqrt{2}$, i.e., sending a coset $a + bX + \langle X^2 - 2 \rangle$ to $a + b\sqrt{2}$, just as we saw above for $\mathbb{Z}$.

5.2 Maximal Ideals

Recall that we want to generalise the notion of prime number to ideals. We have two equivalent definitions of prime number in $\mathbb{Z}$, and they will generalise naturally in different ways.

The first definition of a prime number is that it is a natural number p with no divisor other than 1 and itself.

It is easy to reformulate this in terms of ideals in $\mathbb{Z}$. Indeed, if a natural number a exists with $a|p$, then $\langle p \rangle \subset \langle a \rangle$. If a differs from p, then the inclusion $\langle p \rangle \subset \langle a \rangle$ must be *strict*. Furthermore, if $a \neq 1$, then $\langle a \rangle \subset \mathbb{Z}$ is also a strict inclusion, and so we have two strict inclusions $\langle p \rangle \subset \langle a \rangle \subset \mathbb{Z}$.

But if p is prime, there is no natural number a such that we have strict inclusions $\langle p \rangle \subset \langle a \rangle \subset \mathbb{Z}$; in other words, there is no proper ideal which is strictly bigger than $\langle p \rangle$.

This motivates the following definition for a more general integral domain:

Definition 5.8 Let R be an integral domain. An ideal I of R is said to be *maximal* if

1. $I \neq R$,
2. there is no ideal $J \neq R$ which strictly contains I.

To confirm we are on the right track, let's verify that maximal principal ideals are generated by irreducible elements (in the sense of Definition 4.5).

Lemma 5.9 *Let R be an integral domain, and let $p \in R$. If $\langle p \rangle$ is maximal, then p is irreducible.*

Proof If p is not irreducible, either p would be a unit, when we would have $\langle p \rangle = R$, so $\langle p \rangle$ is not maximal, or we could write $p = ab$ for two non-unit elements a and b. In the latter case, p is a multiple of a, so $\langle p \rangle \subseteq \langle a \rangle$; on the other hand, a is not a multiple of p since b is not a unit, and so $a \notin \langle p \rangle$, showing that $\langle a \rangle$ strictly contains $\langle p \rangle$, so that $\langle p \rangle$ is not maximal. $\qquad\square$

Exercise 5.2 If R is a principal ideal domain, show that the converse of this lemma is also true.

In particular, maximal ideals exactly correspond to irreducible elements in PIDs (two associate irreducible elements will give the same maximal ideal), suggesting that the notion of a maximal ideal might be a suitable generalisation to ideals of the notion of an irreducible element.

There is a useful characterisation of maximal ideals in terms of quotient rings:

Lemma 5.10 *I is a maximal ideal of R if and only if R/I is a field.*

Proof First suppose that I is a maximal ideal of R. Let $a \in R$, but $a \notin I$. Then the set $\langle a, I \rangle = aR + I$ is an ideal of R which is strictly larger than I as it contains a. Thus we must have $aR + I = R$. In particular, $1_R \in aR + I$, so there exists $b \in R$ such that $1_R \in ab + I$. It follows that $1_R + I = ab + I = (a + I)(b + I)$, and so $b + I$ is a multiplicative inverse for $a + I$ in R/I. Thus every non-zero coset is invertible, so R/I is a field.

Conversely, if R/I is a field, then every non-zero coset is invertible. Suppose that J is an ideal of R strictly containing I, and let $a \in J - I$. Then there exists $b \in R$ such that $(a + I)(b + I) = ab + I = 1_R + I$, and, as $J \supset I$ and $ab \in J$, we must have $1_R \in J$. But then it follows that $J = R$ (any ideal containing a unit must be the whole ring by Lemma 4.15). $\qquad\square$

Our first example will concern the simplest ring of all, namely $\mathbb{Z}$.

Example 5.11 The maximal ideals of $\mathbb{Z}$ are precisely $\langle p \rangle = p\mathbb{Z}$ where p is prime.

Proof The ideals of $\mathbb{Z}$ are $\langle 0 \rangle$ and $\langle n \rangle = n\mathbb{Z}$ where n is a positive integer (we saw this in Lemma 4.13). $\langle 0 \rangle$ is not maximal because it is contained in any proper ideal, $\langle 2 \rangle$, for example. (Alternatively, $\mathbb{Z}/\langle 0 \rangle \cong \mathbb{Z}$, which is not a field.) If n is not prime, it has a divisor d greater than 1 – then $n\mathbb{Z}$ is not maximal as it is contained in $d\mathbb{Z}$ (or alternatively because d is not invertible in $\mathbb{Z}/n\mathbb{Z}$). However, if $n = p$ is prime, then we know that $\mathbb{Z}/p\mathbb{Z}$ is a field, because any non-zero element $a \in \mathbb{Z}/p\mathbb{Z}$ has an inverse; find elements b and s by the Euclidean algorithm such that $ab + ps = 1$, and then $ab \equiv 1 \pmod{p}$. $\qquad\square$

Thus maximal ideals in $\mathbb{Z}$ match up nicely with the prime numbers.

Exercise 5.3 Similarly verify that the maximal ideals in $\mathbb{Z}[i]$ correspond with the irreducible elements (see Exercise 4.2).

5.3 Prime Ideals

The other possible way to generalise the idea of a prime number to ideals is to recall the property that p is prime if, whenever $p|ab$, then $p|a$ or $p|b$ (see Definition 4.5 again).

As above, we start by translating this property into ideals. The property says that, if ab is a multiple of p, then either a is a multiple of p or b is a multiple of p, i.e., if $ab \in \langle p \rangle$, then either $a \in \langle p \rangle$ or $b \in \langle p \rangle$. This suggests the generalisation.

Definition 5.12 Let R be an integral domain. An ideal I of R is said to be *prime* if

1. $I \neq R$,
2. if $xy \in I$, then $x \in I$ or $y \in I$.

We could reformulate this further; the original property of prime numbers could be written: $\langle a \rangle \langle b \rangle \subseteq \langle p \rangle$ implies that $\langle a \rangle \subseteq \langle p \rangle$ or $\langle b \rangle \subseteq \langle p \rangle$. This suggests an alternative generalisation to ideals; unsurprisingly, these are equivalent.

Lemma 5.13 *I is a prime ideal of R if and only if whenever J_1 and J_2 are ideals of R such that $J_1 J_2 \subseteq I$, either $J_1 \subseteq I$ or $J_2 \subseteq I$.*

Proof ($\Rightarrow$) Suppose $J_1 \nsubseteq I$, $J_2 \nsubseteq I$, but $J_1 J_2 \subseteq I$. Then there is some $a_1 \in J_1 - I$ and some $a_2 \in J_2 - I$. But $J_1 J_2 \subseteq I$, so $a_1 a_2 \in I$. As I is prime, either a_1 or a_2 must lie in I, contradicting our choices.

($\Leftarrow$) Conversely, if I is not prime, there are elements a_1 and a_2 not in I but with $a_1 a_2 \in I$. Let $J_1 = \langle a_1 \rangle$, $J_2 = \langle a_2 \rangle$; note that neither J_1 nor J_2 is contained in I, but $J_1 J_2 = \langle a_1 a_2 \rangle \subseteq I$. □

We will need this second formulation later, when we think about factorisation of ideals in rings of integers of number fields into prime ideals.

Again, it really is true that principal prime ideals correspond to prime elements. The situation here is even better than Lemma 5.9 as both directions of the implication hold for any integral domain:

Lemma 5.14 *Let R be an integral domain, and let $p \in R$. Then $\langle p \rangle$ is a prime ideal in R if and only if p is a prime element.*

Proof Suppose that $\langle p \rangle$ is a prime ideal, and let's show that p is a prime element. Suppose then that $p = ab$ in R. Then $ab \in \langle p \rangle$, so that either $a \in \langle p \rangle$ or $b \in \langle p \rangle$. But this means that $p|a$ or $p|b$ respectively. So p is a prime element.

Conversely, suppose that p is a prime element, and let's show that $\langle p \rangle$ is a prime ideal. Take a and b with $ab \in \langle p \rangle$. Then $ab = cp$ for some $c \in R$. So $p|ab$, and by Definition 4.5, $p|a$ or $p|b$. But this means that $a \in \langle p \rangle$ or $b \in \langle p \rangle$, showing that $\langle p \rangle$ is a prime ideal. □

In particular, if R were a ring of integers of some number field, where every ideal was principal, the prime ideals would correspond exactly to prime elements (of course, two associate prime elements give the same prime ideal).

Therefore the notion of a prime ideal is a suitable generalisation to ideals of the definition of a prime element, and therefore of prime numbers.

We have a similar result to Lemma 5.10.

Lemma 5.15 *Let R be a ring and I an ideal of R. Then I is a prime ideal if and only if R/I is an integral domain.*

Proof Suppose that I is a prime ideal of R. We have to check that R/I has no zero divisors. Suppose that

$$(a + I)(b + I) = 0_R + I = I.$$

But $(a + I)(b + I) = ab + I$, and $ab + I = I$ implies that $ab \in I$. As I is prime, either $a \in I$, which implies that $a + I = I$, or $b \in I$, which implies that $b + I = I$, so that one of $a + I$ and $b + I$ is the zero element $0_R + I$. It follows that there are no zero divisors in R/I.

Conversely, if R/I has no zero divisors, then if a and b are elements of R such that $a \notin I$, and $ab \in I$, then $a + I$ is a non-zero coset such that

$$(a + I)(b + I) = 0_R + I.$$

As there are no zero-divisors, we must have $b + I = 0_R + I$, so that $b \in I$. Thus I is prime. □

Corollary 5.16 *Maximal ideals are prime.*

Proof If I is a maximal ideal, then R/I is a field. Every field is an integral domain, so R/I is an integral domain, and therefore I is prime. □

To see that the converse is not quite true, let's work out the prime ideals of $\mathbb{Z}$.

Example 5.17 The prime ideals of $\mathbb{Z}$ are precisely $\langle p \rangle = p\mathbb{Z}$ where p is prime, and also $\langle 0 \rangle$.

Proof This time $\langle 0 \rangle$ is prime, because $\mathbb{Z}/\langle 0 \rangle \cong \mathbb{Z}$, which is an integral domain.

The other ideals are all of the form $n\mathbb{Z}$ for some positive integer n; if n is not prime, then $\mathbb{Z}/n\mathbb{Z}$ has zero divisors, so $\langle n \rangle$ is not prime, and if $n = p$ is prime, then we saw in Example 5.11 that $\mathbb{Z}/p\mathbb{Z}$ is a field, so certainly an integral domain. □

Therefore $\langle 0 \rangle$ is a prime ideal of $\mathbb{Z}$ which is not maximal, but all other prime ideals are also maximal.

Note 5.18 Notice that the prime ideal $\langle 0 \rangle$ of $\mathbb{Z}$ is contained inside all the other prime ideals of $\mathbb{Z}$; it might seem odd at first to have one prime ideal contained inside another. Of course, the zero ideal $\langle 0 \rangle$ is in some sense a rather exceptional prime ideal.

However, perhaps surprisingly, one can find many examples of rings R in which one prime ideal can contain another, non-trivial, prime ideal. For example, if $R = K[X, Y]$, then $P_1 = \langle X, Y \rangle$ and $P_2 = \langle X \rangle$ are both prime, and $P_1 \supset P_2$.

We shall see next, however, that this sort of example does not occur for rings of integers of number fields, and that every non-zero prime ideal in such rings is also maximal.

The fact that every maximal ideal is prime followed from the assertion that every field is an integral domain. The converse is not true (e.g., $\mathbb{Z}$), but finite integral domains are always fields.

Lemma 5.19 *If R is a finite integral domain, then R is a field.*

Proof We just need to check that every non-zero $r \in R$ is invertible. Consider the map (not a homomorphism) $\phi : R \longrightarrow R$ given by $\phi(s) = rs$. It is injective: if $\phi(s_1) = \phi(s_2)$, then $rs_1 = rs_2$, and so $r(s_1 - s_2) = 0$, so that $s_1 - s_2 = 0$ as R is an integral domain and has no zero divisors.

But an injective map from a finite set to itself is also surjective. So there is some s with $\phi(s) = 1$, so that $rs = 1$. $\qquad\qquad\square$

We can use this to show that non-zero prime ideals in rings of integers of number fields are always maximal. We need a lemma:

Lemma 5.20 *Let K be a number field. If $\mathfrak{p}$ is a non-zero prime ideal in $\mathbb{Z}_K$, then $\mathbb{Z}_K/\mathfrak{p}$ is finite.*

Proof Let $\mathfrak{p}$ be a non-zero prime ideal in $\mathbb{Z}_K$. Then there is a non-zero element $\alpha \in \mathfrak{p}$. Its norm $N = N_{K/\mathbb{Q}}(\alpha)$ lies in $\mathbb{Z}$ (by Corollary 3.17), and is the product of $\alpha \in \mathfrak{p}$ with all its conjugates. So $N \in \mathfrak{p}$.

Now $\mathbb{Z}_K$ has an integral basis, by the results of Chap. 3; so we can write $\mathbb{Z}_K = \mathbb{Z}\omega_1 + \cdots + \mathbb{Z}\omega_n$. Notice that as $N \in \mathfrak{p}$, we have $N\omega_i \in \mathfrak{p}$ for each i (just by the defining rule of ideals). It follows that every element $a_1\omega_1 + \cdots + a_n\omega_n$ is congruent modulo $\mathfrak{p}$ to some element of the form $b_1\omega_1 + \cdots + b_n\omega_n$ with $0 \le b_i < N$. There are finitely many such elements, so $\mathbb{Z}_K/\mathfrak{p}$ is finite. $\qquad\square$

It is easy to see that this proof is valid for any non-zero ideal, not just prime ideals.

Proposition 5.21 *Let K be a number field. Then every non-zero prime ideal $\mathfrak{p}$ in $\mathbb{Z}_K$ is maximal.*

Proof By Lemma 5.20, $\mathbb{Z}_K/\mathfrak{p}$ is finite., As $\mathbb{Z}_K/\mathfrak{p}$ is a finite integral domain, it is also a field by Lemma 5.19. But then $\mathfrak{p}$ must be maximal, using Lemma 5.10. $\qquad\square$

Exercise 5.4 Using Exercise 4.2, what are the prime ideals in the Gaussian integers $\mathbb{Z}[i]$?

Exercise 5.5 Let $K = \mathbb{Q}(\sqrt{2})$, and let $\mathfrak{a} = \langle 3 \rangle = 3\mathbb{Z}_K$. Show that $\mathbb{Z}_K/\mathfrak{a}$ is a finite field with 9 elements, which can be identified with $\{a + b\sqrt{2} \mid a, b \in \mathbb{Z}/3\mathbb{Z}\}$. Conclude that $\mathfrak{a}$ is a prime ideal of $\mathbb{Z}_K$.

Exercise 5.6 Again let $K = \mathbb{Q}(\sqrt{2})$, and consider the ideal $\mathfrak{a} = \langle 2 \rangle = 2\mathbb{Z}_K$. Is $\mathbb{Z}_K/\mathfrak{a}$ a field?

If $\mathfrak{b} = \langle 7 \rangle = 7\mathbb{Z}_K$, show that $\mathbb{Z}_K/\mathfrak{b}$ is not a field.

5.4 Unique Factorisation into Prime Ideals

If K is a number field, we have already seen in Theorem 4.34 that $\mathbb{Z}_K$ is Noetherian.

It will be useful to define the notion of a fractional ideal in a number field, and to show that the collection of all fractional ideals form a group at the same time as showing unique factorisation into prime ideals.

Definition 5.22 A *fractional ideal* of $\mathbb{Z}_K$ is a subset of K which is of the form $\frac{1}{\gamma}\mathfrak{c}$, where $\mathfrak{c}$ is an ideal of $\mathbb{Z}_K$ and γ is a non-zero element of $\mathbb{Z}_K$. We say that the fractional ideal is *principal* if $\mathfrak{c}$ is principal.

Notice that fractional ideals are subsets of K, not just of $\mathbb{Z}_K$, and so (despite the name) are not generally ideals of $\mathbb{Z}_K$.

Remark 5.23 Since the product of two ideals is again an ideal, we can see that the product of two fractional ideals is again a fractional ideal.

We are going to prove two crucial results about ideals in $\mathbb{Z}_K$; firstly, that the non-zero fractional ideals of $\mathbb{Z}_K$ form an abelian group under multiplication, and secondly (and even more important!), that every non-zero ideal of $\mathbb{Z}_K$ can be written uniquely as a product of prime ideals. This second statement about unique factorisation is what we have been aiming for throughout the book so far.

We prove these together, in a series of lemmas.

Lemma 5.24 *Let* $\mathfrak{a}$ *be a non-zero ideal of* $\mathbb{Z}_K$. *Then there exist prime ideals* $\mathfrak{p}_1, \ldots, \mathfrak{p}_r$ *such that* $\mathfrak{p}_1 \ldots \mathfrak{p}_r \subseteq \mathfrak{a}$.

Proof If not, then we can choose $\mathfrak{a}$ as large as possible subject to the condition that the statement is false (so we choose $\mathfrak{a}$ so that any larger ideal does have prime ideals $\mathfrak{p}_1, \ldots, \mathfrak{p}_r$ as in the statement).

This is one point where we use the fact that $\mathbb{Z}_K$ is Noetherian: the Noetherian condition is equivalent to the Ascending Chain Condition (see Proposition 4.36), and so if we consider the set of all ideals such that the statement fails, and choose one, $\mathfrak{a}_1$ say, either $\mathfrak{a}_1$ is as large as possible in this sense, or there is a bigger ideal $\mathfrak{a}_2$ contradicting the statement of the Lemma. We repeat this process, but the ACC (Definition 4.35) guarantees that this process must eventually produce an ideal which is as large as possible with this property.

Clearly $\mathfrak{a}$ is not prime (otherwise take $\mathfrak{p}_1 = \mathfrak{a}$!), so there exist ideals $\mathfrak{a}_1$ and $\mathfrak{a}_2$ of $\mathbb{Z}_K$ with $\mathfrak{a}_1\mathfrak{a}_2 \subseteq \mathfrak{a}$, $\mathfrak{a}_1 \not\subseteq \mathfrak{a}$, $\mathfrak{a}_2 \not\subseteq \mathfrak{a}$. Write

$$\mathfrak{b}_1 = \mathfrak{a} + \mathfrak{a}_1, \quad \mathfrak{b}_2 = \mathfrak{a} + \mathfrak{a}_2.$$

Then $\mathfrak{b}_1\mathfrak{b}_2 \subseteq \mathfrak{a}$, but $\mathfrak{b}_1$ and $\mathfrak{b}_2$ both strictly contain $\mathfrak{a}$. By maximality of $\mathfrak{a}$, there exist prime ideals $\mathfrak{p}_i$ such that

$$\mathfrak{p}_1 \ldots \mathfrak{p}_s \subseteq \mathfrak{b}_1$$
$$\mathfrak{p}_{s+1} \ldots \mathfrak{p}_t \subseteq \mathfrak{b}_2$$

Then $\mathfrak{p}_1 \ldots \mathfrak{p}_t \subseteq \mathfrak{b}_1 \mathfrak{b}_2 \subseteq \mathfrak{a}$, contradicting the choice of $\mathfrak{a}$.   □

As remarked above, we are going to prove that the set of fractional ideals of $\mathbb{Z}_K$ forms an abelian group. The next lemma introduces what will turn out to be the inverse of an ideal $\mathfrak{a}$, but we won't be able to prove that just yet.

Lemma 5.25 *If $\mathfrak{a}$ is an ideal of $\mathbb{Z}_K$, define*

$$\mathfrak{a}^{-1} = \{\alpha \in K \mid \alpha\mathfrak{a} \subseteq \mathbb{Z}_K\}.$$

Then $\mathfrak{a}^{-1}$ is a fractional ideal.

Proof Take any $\gamma \in \mathfrak{a}$, and put $\mathfrak{c} = \gamma\mathfrak{a}^{-1}$. We claim that $\mathfrak{c}$ is an ideal of $\mathbb{Z}_K$. Clearly $0 \in \mathfrak{c}$. If $i, i' \in \mathfrak{c}$, so that $i = \gamma\beta$ and $i' = \gamma\beta'$, with $\beta, \beta' \in \mathfrak{a}^{-1}$, we need $i + i' \in \mathfrak{c}$. But $i + i' = \gamma(\beta + \beta')$, so we need $\beta + \beta' \in \mathfrak{a}^{-1}$; this follows easily since $(\beta + \beta')\mathfrak{a} = \beta\mathfrak{a} + \beta'\mathfrak{a} \subseteq (\mathbb{Z}_K + \mathbb{Z}_K) = \mathbb{Z}_K$, as required. Finally, if $i = \gamma\beta \in \mathfrak{c}$, so $\beta \in \mathfrak{a}^{-1}$, and $r \in \mathbb{Z}_K$, we need $ri \in \mathfrak{c}$, which would follow from $r\beta \in \mathfrak{a}^{-1}$. Again this is easy, since $(r\beta)\mathfrak{a} = r(\beta\mathfrak{a}) \subseteq r\mathbb{Z}_K \subseteq \mathbb{Z}_K$, as $r \in \mathbb{Z}_K$. We have now explained that $\mathfrak{c} = \gamma\mathfrak{a}^{-1}$ is an ideal, and so $\mathfrak{a}^{-1} = \frac{1}{\gamma}\mathfrak{c}$ is a fractional ideal, as required.   □

Lemma 5.26 *If $\mathfrak{a}$ is a proper ideal of $\mathbb{Z}_K$, then $\mathfrak{a}^{-1}$ strictly contains $\mathbb{Z}_K$.*

Proof Clearly $\mathfrak{a}^{-1}$ contains $\mathbb{Z}_K$; we need to check that the inclusion is strict.

It is easy to see that if $\mathfrak{a} \subseteq \mathfrak{b}$, then $\mathfrak{b}^{-1} \subseteq \mathfrak{a}^{-1}$. As $\mathfrak{a}$ is contained in a maximal ideal $\mathfrak{p}$ (Lemma 4.38), it suffices to show that $\mathfrak{p}^{-1}$ strictly contains $\mathbb{Z}_K$. Clearly $\mathfrak{p}^{-1} \supseteq \mathbb{Z}_K$, but we must find a non-integer in $\mathfrak{p}^{-1}$.

Choose any non-zero $\alpha \in \mathfrak{p}$, so $\langle\alpha\rangle \subseteq \mathfrak{p}$. Choose the smallest r such that there exist prime ideals $\mathfrak{p}_1, \ldots, \mathfrak{p}_r$ with

$$\mathfrak{p}_1 \ldots \mathfrak{p}_r \subseteq \langle\alpha\rangle \subseteq \mathfrak{p}.$$

It exists by Lemma 5.24. As $\mathfrak{p}$ is prime, some $\mathfrak{p}_i \subseteq \mathfrak{p}$; after re-ordering, we may suppose it to be $\mathfrak{p}_1$. As non-zero prime ideals are maximal (Proposition 5.21), and maximal ideals cannot be properly contained in one another, we have $\mathfrak{p}_1 = \mathfrak{p}$. As r is minimal, $\mathfrak{p}_2 \ldots \mathfrak{p}_r \not\subseteq \langle\alpha\rangle$. So there is some $\beta \in \mathfrak{p}_2 \ldots \mathfrak{p}_r$ not in $\langle\alpha\rangle$. Then $\beta\mathfrak{p} \subseteq \langle\alpha\rangle$, so $\beta\alpha^{-1}\mathfrak{p} \subseteq \mathbb{Z}_K$ and $\beta\alpha^{-1} \in \mathfrak{p}^{-1}$. As $\beta\alpha^{-1} \notin \mathbb{Z}_K$ (as $\beta \notin \alpha\mathbb{Z}_K$), the result follows.   □

The method of proof of the next lemma will be familiar from Chap. 2 (see Proposition 2.33, for example).

Lemma 5.27 *If $\mathfrak{a}$ is a non-zero ideal of $\mathbb{Z}_K$, and $\theta \in K$ satisfies $\mathfrak{a}\theta \subseteq \mathfrak{a}$, then $\theta \in \mathbb{Z}_K$.*

Proof As $\mathbb{Z}_K$ is Noetherian, $\mathfrak{a}$ is finitely generated; $\mathfrak{a} = \langle\omega_1, \ldots, \omega_m\rangle$. Then

$$\omega_1 \theta = a_{11}\omega_1 + \cdots + a_{1m}\omega_m$$
$$\vdots \qquad \vdots$$
$$\omega_m \theta = a_{m1}\omega_1 + \cdots + a_{mm}\omega_m$$

with $a_{ij} \in \mathbb{Z}$. This is exactly (2.3); as in the proof of Proposition 2.25, θ is an eigenvalue of $A = (a_{ij})$, and is therefore a root of the characteristic polynomial of a matrix of integers, and so θ is an algebraic integer. As $\theta \in K$, it follows that $\theta \in \mathbb{Z}_K$, as required. □

Next we can prove that the fractional ideals defined in Lemma 5.25 are genuinely inverses, first for maximal ideals, and then in general.

Lemma 5.28 *If $\mathfrak{p}$ is a maximal ideal of $\mathbb{Z}_K$, then $\mathfrak{p}\mathfrak{p}^{-1} = \mathbb{Z}_K$.*

Proof Since $\mathfrak{p}^{-1}$ is a fractional ideal, and $\mathfrak{p}$ is an ideal (and therefore also a fractional ideal), the product $\mathfrak{p}\mathfrak{p}^{-1}$ is a fractional ideal. However, by definition of $\mathfrak{p}^{-1}$, $\mathfrak{p}\mathfrak{p}^{-1} \subseteq \mathbb{Z}_K$, so the product is an ideal of $\mathbb{Z}_K$. Certainly $\mathfrak{p}\mathfrak{p}^{-1} \supseteq \mathfrak{p}$ as $\mathfrak{p}^{-1} \supseteq \mathbb{Z}_K$. As $\mathfrak{p}$ is maximal, either $\mathfrak{p}\mathfrak{p}^{-1} = \mathfrak{p}$ or $\mathfrak{p}\mathfrak{p}^{-1} = \mathbb{Z}_K$. The first is impossible, as $\mathfrak{p}$ contains a non-integer element θ, and by Lemma 5.27, $\mathfrak{p}\theta \nsubseteq \mathfrak{p}$. The claim follows. □

Lemma 5.29 *If $\mathfrak{a}$ is any non-zero ideal of $\mathbb{Z}_K$, then $\mathfrak{a}\mathfrak{a}^{-1} = \mathbb{Z}_K$.*

Proof If not, choose $\mathfrak{a}$ to be an ideal such that $\mathfrak{a}\mathfrak{a}^{-1} \neq \mathbb{Z}_K$ which is as large as possible (as in Lemma 5.24). Let $\mathfrak{p}$ be a maximal ideal containing $\mathfrak{a}$, and consider $\mathfrak{a}\mathfrak{p}^{-1}$. As $\mathbb{Z}_K \subseteq \mathfrak{p}^{-1} \subseteq \mathfrak{a}^{-1}$, we see that

$$\mathfrak{a} \subseteq \mathfrak{a}\mathfrak{p}^{-1} \subseteq \mathfrak{a}\mathfrak{a}^{-1} \subseteq \mathbb{Z}_K.$$

So $\mathfrak{a}\mathfrak{p}^{-1} \subseteq \mathbb{Z}_K$, and so $\mathfrak{a}\mathfrak{p}^{-1}$ is genuinely an ideal (not just a fractional ideal) of $\mathbb{Z}_K$. But we cannot have $\mathfrak{a}\mathfrak{p}^{-1} = \mathfrak{a}$, because $\mathfrak{p}^{-1}$ contains some non-integral θ by Lemma 5.26, leading to a contradiction with Lemma 5.27. So $\mathfrak{a}$ is strictly contained in $\mathfrak{a}\mathfrak{p}^{-1}$, and by our choice of $\mathfrak{a}$ as being as large as possible subject to the condition that the statement is false, we have

$$\mathfrak{a}\mathfrak{p}^{-1}(\mathfrak{a}\mathfrak{p}^{-1})^{-1} = \mathbb{Z}_K.$$

Thus

$$\mathfrak{p}^{-1}(\mathfrak{a}\mathfrak{p}^{-1})^{-1} \subseteq \mathfrak{a}^{-1},$$

and so

$$\mathbb{Z}_K = \mathfrak{a}\mathfrak{p}^{-1}(\mathfrak{a}\mathfrak{p}^{-1})^{-1} \subseteq \mathfrak{a}\mathfrak{a}^{-1} \subseteq \mathbb{Z}_K,$$

and the result follows. □

Now we prove the first of the two main results.

Theorem 5.30 *The set of fractional ideals form an abelian group.*

Proof We already know how to multiply ideals (and thus fractional ideals), and this is clearly associative and commutative. The whole ring $\mathbb{Z}_K$ forms the identity. The only thing left to check is that we can define an inverse for any given fractional ideal. But Lemma 5.29 gives us the inverse for any ideal, and any fractional ideal is of the form $\mathfrak{b} = \gamma^{-1}\mathfrak{c}$ for some ideal of $\mathbb{Z}_K$ and some non-zero $\gamma \in K$. Then its inverse $\mathfrak{b}^{-1}$ will be $\gamma\mathfrak{c}^{-1}$—note that

$$\mathfrak{b}\mathfrak{b}^{-1} = \gamma^{-1}\mathfrak{c}.\gamma\mathfrak{c}^{-1} = \mathfrak{c}\mathfrak{c}^{-1} = \mathbb{Z}_K,$$

as required.   □

Lemma 5.31 *Every non-zero ideal $\mathfrak{a}$ is a product of prime ideals.*

Proof If not, let $\mathfrak{a}$ be maximal subject to the condition that it is not a product of prime ideals (again, as in Lemma 5.24). Then $\mathfrak{a}$ is contained in some maximal ideal $\mathfrak{p}$, and because $\mathfrak{a}$ is strictly contained in $\mathfrak{a}\mathfrak{p}^{-1}$, we can write

$$\mathfrak{a}\mathfrak{p}^{-1} = \mathfrak{p}_1 \ldots \mathfrak{p}_r,$$

a product of prime ideals. Now $\mathfrak{a} = \mathfrak{p}\mathfrak{p}_1, \ldots \mathfrak{p}_r$.   □

Finally, we can prove uniqueness of factorisation of ideals in $\mathbb{Z}_K$ into prime ideals.

Theorem 5.32 *Factorisation of ideals into prime ideals is unique.*

Proof Lemma 5.31 gives a factorisation into ideals; we just need to see that this decomposition is unique.

Let r be minimal such that there is an ideal $\mathfrak{a}$ with two different factorisations

$$\mathfrak{a} = \mathfrak{p}_1 \ldots \mathfrak{p}_r = \mathfrak{q}_1 \ldots \mathfrak{q}_s$$

into prime ideals. Then $\mathfrak{p}_1 \supseteq \mathfrak{q}_1 \ldots \mathfrak{q}_s$. As $\mathfrak{p}_1$ is a prime ideal, $\mathfrak{p}_1 \supseteq \mathfrak{q}_i$ for some i. But both are maximal ideals, so $\mathfrak{p}_1 = \mathfrak{q}_i$. Cancel these (i.e., multiply by $\mathfrak{p}_1^{-1}$), and we get two different factorisations of an ideal $\mathfrak{a}\mathfrak{p}_1^{-1}$ where at least one expression is of shorter length than r, contradicting our choice of r. This proves the result.   □

Exercise 5.7 In Exercise 5.6, the two ideals $\mathfrak{a} = \langle 2 \rangle$ and $\mathfrak{b} = \langle 7 \rangle$ were not prime ideals in $\mathbb{Z}_K$, where $K = \mathbb{Q}(\sqrt{2})$. Factor $\mathfrak{a}$ and $\mathfrak{b}$ into prime ideals in $\mathbb{Z}_K$.

5.5 Coprimality

In $\mathbb{Z}$, two integers are coprime if their highest common factor is 1, and unique factorisation shows that this is equivalent to the statement that no prime number divides both. Now that we have unique factorisation in $\mathbb{Z}_K$, a similar statement is available.

Given two ideals $\mathfrak{a}$ and $\mathfrak{b}$ of $\mathbb{Z}_K$, there is a natural notion of coprimality. We say that $\mathfrak{a}$ and $\mathfrak{b}$ are *coprime* if $\mathfrak{a} + \mathfrak{b} = \mathbb{Z}_K$, that is, if the ideal generated by both $\mathfrak{a}$ and $\mathfrak{b}$ is the whole ring.

Alternatively, we could factor the two ideals into primes, and then the ideals will be coprime when no prime ideal occurs in both factorisations.

To see that these are equivalent, suppose first that $\mathfrak{a}$ and $\mathfrak{b}$ both have a prime ideal $\mathfrak{p}$ in their factorisation. Then $\mathfrak{a} + \mathfrak{b} \subseteq \mathfrak{p}$, and so they will not generate the whole ring. Conversely, the proof above shows that if $\mathfrak{a} + \mathfrak{b}$ is strictly contained in $\mathbb{Z}_K$, then it has a prime ideal $\mathfrak{p}$ in its factorisation. Clearly then $\mathfrak{a} \subseteq \mathfrak{p}$ and $\mathfrak{b} \subseteq \mathfrak{p}$, and then the proof shows that $\mathfrak{a}$ and $\mathfrak{b}$ both have $\mathfrak{p}$ in their factorisations.

More generally, the *highest common factor* of two ideals $\mathfrak{a}$ and $\mathfrak{b}$ is the ideal $\mathfrak{h}$ such that

1. $\mathfrak{h}|\mathfrak{a}$ and $\mathfrak{h}|\mathfrak{b}$;
2. if $\mathfrak{c}|\mathfrak{a}$ and $\mathfrak{c}|\mathfrak{b}$, then $\mathfrak{c}|\mathfrak{h}$.

(Compare this definition with Definition 1.5.)

Exercise 5.8 Given two ideals $\mathfrak{a} = \prod_{\mathfrak{p}} \mathfrak{p}^{a_{\mathfrak{p}}}$ and $\mathfrak{b} = \prod_{\mathfrak{p}} \mathfrak{p}^{b_{\mathfrak{p}}}$, show that the highest common factor is

$$\mathfrak{a} + \mathfrak{b} = \prod_{\mathfrak{p}} \mathfrak{p}^{\min(a_{\mathfrak{p}}, b_{\mathfrak{p}})}.$$

Theorem 5.33 (Chinese Remainder Theorem) *Suppose that K is a number field. Suppose that $\mathfrak{a}_1, \ldots, \mathfrak{a}_n$ are ideals in $\mathbb{Z}_K$, which are coprime in the sense that $\mathfrak{a}_i + \mathfrak{a}_j = \mathbb{Z}_K$ for all $i \neq j$. Then*

$$\mathbb{Z}_K/(\mathfrak{a}_1 \cap \ldots \cap \mathfrak{a}_n) \cong \mathbb{Z}_K/\mathfrak{a}_1 \oplus \cdots \oplus \mathbb{Z}_K/\mathfrak{a}_n.$$

Proof The result is clear for $n = 1$, so we assume that $n \geq 2$.

There is a homomorphism

$$\theta : \mathbb{Z}_K \longrightarrow \mathbb{Z}_K/\mathfrak{a}_1 \oplus \cdots \oplus \mathbb{Z}_K/\mathfrak{a}_n$$
$$\alpha \mapsto (\alpha \,(\mathrm{mod}\ \mathfrak{a}_i), \ldots, \alpha \,(\mathrm{mod}\ \mathfrak{a}_i))$$

and the kernel consists of $\alpha \in \mathbb{Z}_K$ mapping to $(0, \ldots, 0)$, i.e., those α such that $\alpha \in \mathfrak{a}_i$ for each i, which is the intersection $\mathfrak{a}_1 \cap \ldots \cap \mathfrak{a}_n$. We need to see that the map is surjective.

We can write $1 = \alpha_i + \beta_i$ where $\alpha_i \in \mathfrak{a}_i$, and $\beta_i \in \mathfrak{a}_j$ for all $j \neq i$.

Indeed, in the case $i = 1$, $\mathfrak{a}_1 + \mathfrak{a}_i = \mathbb{Z}_K$ for all $i \neq 1$, so we can write $1 = x_i + y_i$ for $x_i \in \mathfrak{a}_1$, $y_i \in \mathfrak{a}_i$. Then

$$y_2 y_3 \cdots y_n = (1 - x_2)(1 - x_3) \cdots (1 - x_n);$$

write $\beta_1 = y_2 y_3 \cdots y_n \in \mathfrak{a}_i$ for each $i = 2, \ldots, n$; expanding the right-hand side gives an expression $1 - \alpha_1$ where all the terms defining α_1 are divisible by some $x_i \in \mathfrak{a}_1$, so that $\alpha_1 \in \mathfrak{a}_1$.

Then θ is surjective; given $(x_1, \ldots, x_n) \in \mathbb{Z}_K/\mathfrak{a}_1 \oplus \cdots \oplus \mathbb{Z}_K/\mathfrak{a}_n$, we have $\theta(x_1 \beta_1 + \cdots + x_n \beta_n) = (x_1, \ldots, x_n)$. $\qquad\square$

You should notice that we haven't used any properties of number fields here; this result is valid for any commutative ring, using the definition of coprimality of the statement.

5.6 Norms of Ideals

Before we study unique factorisation into prime ideals, we will introduce the notion of *norm* of an ideal. As we shall see in a moment, this could be regarded as a generalisation of the concept of norm of an element.

Definition 5.34 The *norm* $N_{K/\mathbb{Q}}(\mathfrak{a})$ of a non-zero ideal $\mathfrak{a}$ in $\mathbb{Z}_K$ is the cardinality $|\mathbb{Z}_K/\mathfrak{a}|$. It is finite by Lemma 5.20.

We will use this definition frequently later in the book.

We now have two notions of norm, one for elements (Sect. 3.2) and the one for ideals just given. If an ideal is principal, generated by a single element, then there is a nice relationship between the two.

Lemma 5.35 *Let $\alpha \in \mathbb{Z}_K$ be non-zero. Then $N_{K/\mathbb{Q}}(\langle \alpha \rangle) = |N_{K/\mathbb{Q}}(\alpha)|$.*

Proof Let $\mathbb{Z}_K = \mathbb{Z}\omega_1 + \cdots + \mathbb{Z}\omega_n$. Then $\langle \alpha \rangle = \mathbb{Z}\alpha\omega_1 + \cdots + \mathbb{Z}\alpha\omega_n$. Then $N_{K/\mathbb{Q}}(\langle \alpha \rangle) = |\mathbb{Z}_K/\langle \alpha \rangle|$; if we write $\alpha\omega_i = \sum_{j=1}^{n} a_{ji}\omega_j$, then the index of $\langle \alpha \rangle$ in $\mathbb{Z}_K$ is just $|\det(a_{ij})|$. But we know that $N_{K/\mathbb{Q}}(\alpha) = \det(a_{ij})$, and so we see that $N_{K/\mathbb{Q}}(\langle \alpha \rangle) = |N_{K/\mathbb{Q}}(\alpha)|$.

Just like the notion of norm of an element, the ideal norm is multiplicative in the sense that $N_{K/\mathbb{Q}}(\mathfrak{a}\mathfrak{b}) = N_{K/\mathbb{Q}}(\mathfrak{a})N_{K/\mathbb{Q}}(\mathfrak{b})$. We need a lemma.

Lemma 5.36 *Suppose that $\mathfrak{a}$ is a non-zero ideal of $\mathbb{Z}_K$ and that $\mathfrak{p}$ is a non-zero prime ideal of $\mathbb{Z}_K$. Then $|\mathbb{Z}_K/\mathfrak{p}| = |\mathfrak{a}/\mathfrak{a}\mathfrak{p}|$.*

Proof If $\mathfrak{a} \supseteq \mathfrak{b} \supseteq \mathfrak{a}\mathfrak{p}$, then, multiplying through by $\mathfrak{a}^{-1}$ gives $\mathbb{Z}_K \supseteq \mathfrak{a}^{-1}\mathfrak{b} \supseteq \mathfrak{p}$. As $\mathfrak{p}$ is a non-zero prime ideal, it is maximal, and so either $\mathfrak{a}^{-1}\mathfrak{b} = \mathbb{Z}_K$ or $\mathfrak{a}^{-1}\mathfrak{b} = \mathfrak{p}$. This shows that $\mathfrak{b} = \mathfrak{a}$ or $\mathfrak{a}\mathfrak{p}$.

Fix $\alpha \in \mathfrak{a}$, but not in $\mathfrak{a}\mathfrak{p}$. Then the ideal generated by α and $\mathfrak{a}\mathfrak{p}$ is clearly contained in $\mathfrak{a}$, but is strictly bigger than $\mathfrak{a}\mathfrak{p}$, so must equal $\mathfrak{a}$. Then define the map

$$\phi : \mathbb{Z}_K \to \mathfrak{a}/\mathfrak{a}\mathfrak{p};$$

$$x \mapsto \alpha x + \mathfrak{a}\mathfrak{p}$$

ϕ is a homomorphism of $\mathbb{Z}_K$-modules, and is surjective by the above remark. The kernel clearly contains $\mathfrak{p}$, since if $x \in \mathfrak{p}$, then $\alpha x \in \mathfrak{ap}$. But $1 \notin \ker \phi$, as $\phi(1) = \alpha + \mathfrak{ap}$, and we chose $\alpha \notin \mathfrak{ap}$. As $\mathfrak{p}$ is maximal, we see that $\ker \phi = \mathfrak{p}$. Then the First Isomorphism Theorem for modules (see Remark 5.6) gives $\mathbb{Z}_K/\mathfrak{p} \cong \mathfrak{a}/\mathfrak{p}$, which in turn gives the lemma. □

With this lemma, it is easy to deduce the multiplicativity of the norm.

Theorem 5.37 *Suppose that $\mathfrak{a}$ and $\mathfrak{b}$ are two ideals of $\mathbb{Z}_K$. Then*

$$N_{K/\mathbb{Q}}(\mathfrak{ab}) = N_{K/\mathbb{Q}}(\mathfrak{a})N_{K/\mathbb{Q}}(\mathfrak{b}).$$

Proof By factorising $\mathfrak{b}$ into prime ideals, it suffices to deal with the case that $\mathfrak{b} = \mathfrak{p}$, a prime ideal, and to show that $N_{K/\mathbb{Q}}(\mathfrak{ap}) = N_{K/\mathbb{Q}}(\mathfrak{a})N_{K/\mathbb{Q}}(\mathfrak{p})$. By applying the First Isomorphism Theorem (Theorem 5.5) to the homomorphism

$$\phi : \mathbb{Z}_K/\mathfrak{ap} \to \mathbb{Z}_K/\mathfrak{a},$$
$$\alpha + \mathfrak{ap} \mapsto \alpha + \mathfrak{a}$$

which is clearly surjective, and whose kernel is easily seen to be the set $\mathfrak{a}/\mathfrak{ap} = \{\alpha + \mathfrak{ap} \mid \alpha \in \mathfrak{a}\}$, we see that

$$\left| \frac{\mathbb{Z}_K/\mathfrak{ap}}{\mathfrak{a}/\mathfrak{ap}} \right| = |\mathbb{Z}_K/\mathfrak{a}|.$$

Thus

$$|\mathbb{Z}_K/\mathfrak{ap}| = |\mathbb{Z}_K/\mathfrak{a}|.|\mathfrak{a}/\mathfrak{ap}|,$$

and then the previous lemma gives

$$|\mathbb{Z}_K/\mathfrak{ap}| = |\mathbb{Z}_K/\mathfrak{a}|.|\mathbb{Z}_K/\mathfrak{p}|.$$

Now the definition of the ideal norm gives $N_{K/\mathbb{Q}}(\mathfrak{ap}) = N_{K/\mathbb{Q}}(\mathfrak{a})N_{K/\mathbb{Q}}(\mathfrak{p})$, as required. □

5.7 The Class Group

We now know several important results:

- Chapter 4: Elements in rings of integers of number fields do not generally factorise uniquely into irreducible elements.
- Theorem 4.31: Every domain in which all ideals are principal (a principal ideal domain) is one where we do have unique factorisation of elements (a unique factorisation domain).

- Theorem 5.32: Ideals in rings of integers of number fields always factorise uniquely into prime ideals.

As a consequence, if unique factorisation fails, some ideals are not principal. This serves as a test for uniqueness or non-uniqueness of factorisation.

Using the fact (Theorem 5.30) that the fractional ideals form a group, we can construct a group which measures the success or failure of uniqueness of factorisation.

Suppose that K is a number field, with ring of integers $\mathbb{Z}_K$. Let's form the collection of all ideals:

$$\mathfrak{I}_K = \{\text{ideals in } \mathbb{Z}_K\}.$$

Every element $\alpha \in \mathbb{Z}_K$ generates a principal ideal, $\langle \alpha \rangle = \alpha \mathbb{Z}_K$. So we can form the collection

$$\mathfrak{P}_K = \{\text{principal ideals in } \mathbb{Z}_K\} \subseteq \mathfrak{I}_K,$$

and then unique factorisation would follow from the equality $\mathfrak{P}_K = \mathfrak{I}_K$. If this does not happen, it can be useful to quantify the extent to which it fails, and to estimate what proportion of ideals are principal.

We do this by shifting from ideals to fractional ideals, which have a good group structure. Write

$$\mathfrak{F}_K = \{\text{fractional ideals of } \mathbb{Z}_K\},$$

and

$$\mathfrak{P}\mathfrak{F}_K = \{\text{principal fractional ideals of } \mathbb{Z}_K\}.$$

Then $\mathfrak{F}_K$ forms a group, as already noted; $\mathfrak{P}\mathfrak{F}_K$ is also a group—after all, its elements are simply $\alpha \mathbb{Z}_K$ for $\alpha \in K$. Since $\mathfrak{F}_K$ is abelian, every subgroup is normal, and the quotient

$$C_K = \frac{\mathfrak{F}_K}{\mathfrak{P}\mathfrak{F}_K}$$

is again a group, called the *class group* of K.

Notice that if C_K is the trivial group, then $\mathfrak{F}_K = \mathfrak{P}\mathfrak{F}_K$, and intersecting with the collection of genuine ideals of $\mathbb{Z}_K$ gives $\mathfrak{I}_K = \mathfrak{P}_K$, which implies unique factorisation.

Later we will prove that the group C_K is always finite. The number of elements h_K in C_K measures the proportion of ideals which are principal, and is known as the *class number*.

When the class number is 1, this means that C_K is trivial, so that every ideal is principal, and we have unique factorisation. More generally, if the class number is h_K, we see that the proportion of ideals which are principal is $1/h_K$.

The group structure also provides further information.

There is clearly a surjective group homomorphism

$$\mathfrak{F}_K \longrightarrow C_K,$$

sending a fractional ideal $\mathfrak{f}$ to its class $[\mathfrak{f}] \in C_K$. We will usually just use this in the case where $\mathfrak{f}$ is a genuine ideal.

Since this is a homomorphism, $[\mathfrak{a}][\mathfrak{b}] = [\mathfrak{ab}]$ for all (fractional) ideals $\mathfrak{a}$ and $\mathfrak{b}$.

By Lagrange's Theorem, the order of every element in a group divides the order of the group. If $\mathfrak{a}$ is any (fractional) ideal in $\mathbb{Z}_K$, it belongs to a class $[\mathfrak{a}] \in C_K$, and $[\mathfrak{a}^{h_K}] = [\mathfrak{a}]^{h_K}$ is trivial. Thus $\mathfrak{a}^{h_K}$ is in the identity class, which consists of all principal fractional ideals. So $\mathfrak{a}^{h_K}$ is principal for any $\mathfrak{a}$.

Exercise 5.9 Conversely, explain that if m is coprime to h_K, and if $\mathfrak{a}$ is an ideal in $\mathbb{Z}_K$ such that $\mathfrak{a}^m$ is principal, then $\mathfrak{a}$ is itself principal.

5.8 Splitting of Primes

Suppose that K is a number field, and that $\mathfrak{p}$ is a non-zero prime ideal in $\mathbb{Z}_K$. Then $\mathbb{Z}_K/\mathfrak{p}$ is a finite field, by Lemma 5.20.

It is well-known that finite fields must have cardinality p^f for some prime number p and some exponent f. Indeed, if k is a finite field, we can consider the sequence $1, 1 + 1, 1 + 1 + 1, \ldots$, and eventually the sequence must repeat, as k has only finitely many elements. Subtracting the shorter expression from the longer gives a sum $1 + \cdots + 1 = 0$. That is, for some number n, we must have $n = 0$ in the field. If we suppose that n is the smallest positive integer with this property, it is easy to see that n must be prime; if $n = rs$, and $n = 0$, then either r or s must be 0 as fields have no non-trivial zero-divisors; by the minimality of n, we cannot have $1 < r < n$ or $1 < s < n$. Then k contains a copy of $\mathbb{F}_p$, the finite field of integers modulo p (sometimes denoted $\mathbb{Z}_p$ or $\mathbb{Z}/p\mathbb{Z}$); k can be regarded as a field extension of $\mathbb{F}_p$, and therefore as a vector space over $\mathbb{F}_p$. As $\mathbb{F}_p$ has p elements, any vector space over it has p^f elements, where f denotes $[k : \mathbb{F}_p]$.

If $\mathfrak{p}$ is a prime ideal of $\mathbb{Z}_K$, we have already seen (by Lemma 5.20) that $\mathbb{Z}_K/\mathfrak{p}$ is a finite field, so $\mathfrak{p}$ is associated to some prime number $p \in \mathbb{Z}$, and its norm is p^f for some f. We sometimes say that $\mathfrak{p}$ *lies above* p, or that p *lies below* $\mathfrak{p}$.

Conversely, we can find prime ideals in $\mathbb{Z}_K$ by trying to factor primes $p \in \mathbb{Z}$ in $\mathbb{Z}_K$. For example, if $K = \mathbb{Q}(i)$, when $\mathbb{Z}_K = \mathbb{Z}[i]$, we can factor the first few primes as follows:

$$2 = (1 + i)(1 - i), \quad 3 = 3, \quad 5 = (2 + i)(2 - i), \quad 7 = 7, \ldots$$

and we notice that some primes can be factorised and some can't. A little thought should convince the reader that if $p = (a + bi)(c + di)$, then in order that there should be no imaginary part in the product, we need $c + di = a - bi$, and then $p = a^2 + b^2$, so that the primes which are the sums of squares (which we know to be $p = 2$ and $p \equiv 1 \pmod 4$) will factor, and those which are not (the primes $p \equiv 3 \pmod 4$) will not (see Theorem 1.19).

These representations are not unique. We also have $5 = (1 + 2i)(1 - 2i)$, but this is easily seen to be an equivalent factorisation to the one already given; the factors differ by units. We know that we can avoid this by working with ideals instead. We note that $\langle 1 + i \rangle = \langle 1 - i \rangle$ (as $1 + i = i(1 - i)$, so $1 + i$ and $1 - i$ are associates), and the factorisations so far become

$$\langle 2 \rangle = \mathfrak{p}_2^2, \quad \langle 3 \rangle = \mathfrak{p}_3, \quad \langle 5 \rangle = \mathfrak{p}_5 \mathfrak{p}_5', \quad \langle 7 \rangle = \mathfrak{p}_7$$

where $\mathfrak{p}_2 = \langle 1 + i \rangle$ has norm 2, $\mathfrak{p}_3 = \langle 3 \rangle$ is a prime ideal in $\mathbb{Z}[i]$ of norm 9, $\mathfrak{p}_5 = \langle 2 + i \rangle$ and $\mathfrak{p}_5' = \langle 2 - i \rangle$ are prime ideals of norm 5.

Indeed, these three primes demonstrate the three different sorts of factorisation possible in $K = \mathbb{Q}(i)$, or indeed in any quadratic field.

In a quadratic field, the following things can happen.

Definition 5.38 Let p a prime, and suppose that K is a quadratic field.

- We say that p *splits* in K if $p\mathbb{Z}_K = \mathfrak{p}\mathfrak{p}'$, for two ideals $\mathfrak{p} \neq \mathfrak{p}'$ of norm p.
- We say that p is *inert* in K if $p\mathbb{Z}_K$ is a prime ideal in $\mathbb{Z}_K$, necessarily of norm p^2.
- We say that p is *ramified* in K if $p\mathbb{Z}_K = \mathfrak{p}^2$ for some prime ideal $\mathfrak{p}$ of norm p.

Remark 5.39 Of course there are similar definitions for any number field K; but for arbitrary number fields, some combination of these may occur. For example, it may be that in some higher degree number field, $p\mathbb{Z}_K = \mathfrak{p}^2\mathfrak{p}'$, which shows aspects of ramification (because of the exponent of $\mathfrak{p}$), and of splitting (as there is more than one distinct prime ideal appearing), and if the norms of $\mathfrak{p}$ or $\mathfrak{p}'$ were greater than p, there would also be aspects of inert behaviour shown.

Now let's develop some notation for working in more general number fields. If p is a prime number in $\mathbb{Z}$, consider $\langle p \rangle = p\mathbb{Z}_K$. This is an ideal in $\mathbb{Z}_K$, and therefore it should factorise uniquely as a product $\mathfrak{p}_1^{e_1} \ldots \mathfrak{p}_r^{e_r}$ of prime ideals in $\mathbb{Z}_K$. In the equality $p\mathbb{Z}_K = \mathfrak{p}_1^{e_1} \ldots \mathfrak{p}_r^{e_r}$, the exponents e_i are called the *ramification indices*.

As $\mathfrak{p}_i$ is a prime ideal in $\mathbb{Z}_K$, the quotient $\mathbb{Z}_K/\mathfrak{p}_i$ is a finite field for each i, and $\mathbb{Z}_K/\mathfrak{p}_i$ is a field extension of $\mathbb{Z}/p\mathbb{Z} = \mathbb{F}_p$, so both have the same characteristic, and we can define $f_i = [\mathbb{Z}_K/\mathfrak{p}_i : \mathbb{F}_p]$ to be the *inertia degree*. Note that $N_{K/\mathbb{Q}}(\mathfrak{p}_i) = |\mathbb{Z}_K/\mathfrak{p}_i| = p^{f_i}$.

Example 5.40 In the case of quadratic fields, a prime p splits if $\langle p \rangle = \mathfrak{p}\mathfrak{p}'$, where $\mathfrak{p}$ and $\mathfrak{p}'$ both have ramification index and inertia degree equal to 1. A prime p is inert if $\langle p \rangle$ is a prime ideal with ramification index 1, and inertia degree 2. A prime p is ramified if $\langle p \rangle = \mathfrak{p}^2$ where $\mathfrak{p}$ has ramification index 2, and inertia degree 1.

This is a special case of the following theorem for general number fields.

Theorem 5.41 *Let K be a number field of degree n, and suppose that $p\mathbb{Z}_K = \mathfrak{p}_1^{e_1} \ldots \mathfrak{p}_r^{e_r}$, and that $f_i = [\mathbb{Z}_K/\mathfrak{p}_i : \mathbb{F}_p]$. Then $n = \sum_{i=1}^{r} e_i f_i$.*

Proof By the Chinese Remainder Theorem (Theorem 5.33), we have

$$\mathbb{Z}_K / p\mathbb{Z}_K \cong \bigoplus_{i=1}^{r} \mathbb{Z}_K / \mathfrak{p}_i^{e_i}.$$

All these are vector spaces over $\mathbb{F}_p$; we will show that $\dim_{\mathbb{F}_p} \mathbb{Z}_K / p\mathbb{Z}_K = n$, and $\dim_{\mathbb{F}_p} \mathbb{Z}_K / \mathfrak{p}_i^{e_i} = e_i f_i$. Indeed,

$$|\mathbb{Z}_K / p\mathbb{Z}_K| = p^{[K:\mathbb{Q}]} = p^n,$$

as $p \in \mathbb{Z}$, which gives the first claim, and

$$|\mathbb{Z}_K / \mathfrak{p}_i^{e_i}| = N_{K/\mathbb{Q}}(\mathfrak{p}_i^{e_i}) = N_{K/\mathbb{Q}}(\mathfrak{p}_i)^{e_i} = (p^{f_i})^{e_i},$$

using Theorem 5.37, which gives the second. □

There remains the general problem of computing how a prime number factorises in a number field K. It turns out that, for all but finitely many primes, there is a simple way to do this.

Suppose that K is a number field. Then we know that K is generated over $\mathbb{Q}$ by a single element γ (Theorem 2.17).

For the rest of the section, we make the simplifying assumption that $\gamma \in \mathbb{Z}_K$ has the property that $\mathbb{Z}_K = \mathbb{Z}[\gamma]$. The results we shall give can be proven under much weaker hypotheses; we have already seen in Chap. 3 that this does not always occur.

Proposition 5.42 *Suppose that K is a number field, and that $\mathbb{Z}_K = \mathbb{Z}[\gamma]$. Write $g(X) \in \mathbb{Z}[X]$ for its minimal polynomial.*

Let p be a prime in $\mathbb{Z}$, and let

$$\overline{g}(X) = \overline{g}_1(X)^{e_1} \cdots \overline{g}_r(X)^{e_r}$$

be the factorisation of the minimal polynomial g modulo p of γ into irreducibles. Then

$$p\mathbb{Z}_K = \mathfrak{p}_1^{e_1} \ldots \mathfrak{p}_r^{e_r},$$

for certain distinct ideals $\mathfrak{p}_i$ of $\mathbb{Z}_K$; the inertia degree of $\mathfrak{p}_i$ is simply given by the degree of $\overline{g}_i(X)$.

Proof Let $g_i(X)$ denote any polynomial whose reduction modulo p is $\overline{g}_i(X)$.

Define the ideal $\mathfrak{p}_i = \langle p, g_i(\gamma) \rangle$. Then

$$\mathbb{Z}_K / \mathfrak{p}_i = \mathbb{Z}[\gamma] / \langle p, g_i(\gamma) \rangle.$$

The map $\mathbb{Z}[X] \longrightarrow \mathbb{Z}[\gamma]$ induced by $X \mapsto \gamma$ has kernel $\langle g(X) \rangle$, and induces an isomorphism $\mathbb{Z}[X] / \langle g(X) \rangle \cong \mathbb{Z}[\gamma]$. Thus

$$\mathbb{Z}[\gamma]/\langle p, g_i(\gamma)\rangle \cong \mathbb{Z}[X]/\langle g(X), p, g_i(X)\rangle.$$

On the other hand, the homomorphism $\mathbb{Z}[X] \longrightarrow \mathbb{F}_p[X]$ got by reducing the coefficients mod p gives an isomorphism

$$\mathbb{Z}[X]/\langle g(X), p, g_i(X)\rangle \cong \mathbb{F}_p[X]/\langle \overline{g}(X), \overline{g}_i(X)\rangle.$$

As $\overline{g}_i(X)$ divides $\overline{g}(X)$, we see that this last quotient is just $\mathbb{F}_p[X]/\langle \overline{g}_i(X)\rangle$. Combining all the isomorphisms above, we get

$$\mathbb{Z}_K/\mathfrak{p}_i \cong \mathbb{F}_p[X]/\langle \overline{g}_i(X)\rangle.$$

As $\overline{g}_i(X)$ is irreducible, the right-hand side is a field, and so $\mathfrak{p}_i$ is a prime ideal.

Similarly, there are isomorphisms

$$\mathbb{Z}_K/p\mathbb{Z}_K \cong \mathbb{Z}[\gamma]/p\mathbb{Z}[\gamma] \cong \mathbb{Z}[X]/\langle p, g(X)\rangle \cong \mathbb{F}_p[X]/\langle \overline{g}(X)\rangle.$$

The Chinese Remainder Theorem (Theorem 5.33) implies that

$$\mathbb{F}_p[X]/\langle \overline{g}(X)\rangle \cong \mathbb{F}_p[X]/\langle \overline{g}_1(X)^{e_1}\rangle \times \cdots \times \mathbb{F}_p[X]/\langle \overline{g}_r(X)^{e_r}\rangle.$$

The map $\mathbb{Z}_K \longrightarrow \mathbb{Z}_K/p\mathbb{Z}_K$ has kernel $p\mathbb{Z}_K$. Using the above isomorphism, we can view this as a map

$$\mathbb{Z}_K \longrightarrow \mathbb{F}_p[X]/\langle \overline{g}_1(X)^{e_1}\rangle \times \cdots \times \mathbb{F}_p[X]/\langle \overline{g}_r(X)^{e_r}\rangle.$$

Unravelling the maps above, the map is given by $\gamma \mapsto (X, \ldots, X)$, and so the kernel is

$$\langle p, g_1(\gamma)^{e_1}\rangle \cap \ldots \cap \langle p, g_r(\gamma)^{e_r}\rangle.$$

Next, note that $\mathfrak{p}_i^{e_i} \subseteq \langle p, g_i(\gamma)^{e_i}\rangle$; to see this, observe that the generators of $\mathfrak{p}_i^{e_i} = \langle p, g_i(\gamma)\rangle^{e_i}$ are all divisible by p except for $g_i(\gamma)^{e_i}$ itself.

Combining everything, we have

$$p\mathbb{Z}_K = \langle p, g_1(\gamma)^{e_1}\rangle \cap \ldots \cap \langle p, g_r(\gamma)^{e_r}\rangle \supseteq \mathfrak{p}_1^{e_1} \cdots \mathfrak{p}_r^{e_r};$$

the norm of the left-hand side is p^n; the norm of the right-hand side is $(p^{f_1})^{e_1}$ $\ldots (p^{f_r})^{e_r}$, and these two are the same, by Theorem 5.41. It follows that the inclusion is an equality, so that

$$p\mathbb{Z}_K = \mathfrak{p}_1^{e_1} \cdots \mathfrak{p}_r^{e_r}. \qquad \square$$

Remark 5.43 The proof shows that we can take $\mathfrak{p}_i = \langle p, g_i(\gamma)\rangle = p\mathbb{Z}_K + g_i(\gamma)\mathbb{Z}_K$.

Exercise 5.10 How do the ideals $\langle 5\rangle$, $\langle 7\rangle$ and $\langle 31\rangle$ factor into prime ideals in $\mathbb{Q}(\sqrt[3]{2})$?

We say that a prime ideal $\mathfrak{p}_i$ of $\mathbb{Z}_K$ above a prime p is *unramified* if its exponent in the decomposition

$$p\mathbb{Z}_K = \mathfrak{p}_1^{e_1} \ldots \mathfrak{p}_r^{e_r}$$

is $e_i = 1$. If $e_i > 1$, we say that $\mathfrak{p}_i$ is *ramified*. We say that p is *unramified* if $e_1 = \cdots = e_r = 1$, and *ramified* otherwise.

Proposition 5.44 *If K is a number field, then there are only finitely many primes p which are ramified in K. Indeed, p is ramified in K if and only if p divides D_K.*

Proof By Proposition 5.42,

$$p\mathbb{Z}_K = \langle p, f_1(\gamma)\rangle^{e_1} \times \cdots \times \langle p, f_r(\gamma)\rangle^{e_r}.$$

By definition, p ramifies in K if and only if some $e_i > 1$. Thus the polynomial $\overline{f}(X)$ does not have distinct roots modulo p. But these primes are the ones that divide the discriminant of $f(X)$. Under the assumption that $\mathbb{Z}_K = \mathbb{Z}[\gamma]$, the discriminant of $f(X)$ is equal to D_K, by Example 3.21.   □

Exercise 5.11 Let $K = \mathbb{Q}(\sqrt{-2})$. Using Proposition 5.42, which prime numbers are ramified, split and inert in K?

Repeat for other quadratic fields of your choice.

Remark 5.45 We have seen (in Sect. 3.6) that it is not always possible to find elements γ such that $\mathbb{Z}_K = \mathbb{Z}[\gamma]$, although they exist when K is a quadratic field, and we shall see further examples ("cyclotomic fields") in Chap. 9 where such elements exist.

More generally, we can pick any element $\gamma \in \mathbb{Z}_K$ such that $K = \mathbb{Q}(\gamma)$. With a small amount of extra work, one can show that Proposition 5.42 holds more generally in this setting for the primes p not dividing $|\mathbb{Z}_K/\mathbb{Z}[\gamma]|$.

However, Proposition 5.44 remains valid; p ramifies in the number field K if and only if p divides the discriminant of K. The proof in the general case is a little harder.

Remark 5.46 Much of the content of this section is valid also in a more general situation of an extension L/K of number fields. The proofs generalise with little difficulty, although sometimes slight alterations are required to the results.

Remark 5.47 In chap. 3, we gave a rather complicated proof that $K = \mathbb{Q}(\sqrt{-2}, \sqrt{-5})$ does not have an integral basis of the form $\{1, \gamma, \gamma^2, \gamma^3\}$ (i.e., K is not *monogenic* in the terminology of Sect. 3.6). The proof we gave there was a rather obscured version of the following sketch proof.

Suppose that $\mathbb{Z}_K = \mathbb{Z}[\gamma]$. Then Proposition 5.42 tells us that the factorisation of $\langle 3 \rangle = 3\mathbb{Z}_K$ corresponds to the factorisation of the minimal polynomial f of γ modulo 3. We chose the field K so that $\langle 3 \rangle$ factors as the product of 4 distinct prime ideals in $\mathbb{Z}_K$; the proposition implies that f must factor into 4 distinct linear factors modulo 3, but there are only 3 irreducible linear polynomials modulo 3, which gives a contradiction.

Exercise 5.12 In the same way as Remark 5.47, show that if K is a cubic field, and $\langle 2 \rangle = 2\mathbb{Z}_K$ factors as the product of 3 distinct prime ideals $\langle 2 \rangle = \mathfrak{p}_1 \mathfrak{p}_2 \mathfrak{p}_3$, then $\mathbb{Z}_K$ does not have a basis of the form $\{1, \gamma, \gamma^2\}$.

[*Hint: Deduce that the minimal polynomial of γ would factor into 3 distinct linear factors modulo 2, but there are only 2 possible distinct factors.*]

5.9 Primes in Quadratic Fields

As an extended example of the results of the last section, we consider the case of a quadratic field $K = \mathbb{Q}(\sqrt{d})$, where d is a squarefree integer. Quadratic fields are monogenic, so do have the property that $\mathbb{Z}_K = \mathbb{Z}[\gamma]$ for some element $\gamma \in \mathbb{Z}_K$, and so all the results of the previous section are valid.

By Proposition 5.44, the primes p which ramify in K are those which divide the discriminant D_K. Recall that $D_K = d$ if $d \equiv 1 \pmod 4$ and $D_K = 4d$ otherwise.

We can see how $p\mathbb{Z}_K$ factorises into prime ideals using Proposition 5.42, at least for p odd. As already noted, p ramifies in K when $p | D_K$

For $d \equiv 2, 3 \pmod 4$, we know $\mathbb{Z}_K = \mathbb{Z}[\sqrt{d}]$, and the minimal polynomial of $\sqrt{d}$ is just $X^2 - d$ (note that this quadratic has discriminant $4d$). Then a prime p factorises in $\mathbb{Z}_K$ in the same way that $X^2 - d$ factorises modulo p.

- p is split in $\mathbb{Z}[\sqrt{d}]$ if and only if $X^2 - d$ factors into two linear factors modulo p. That is, if and only if $X^2 - d$ has two (distinct) roots modulo p, i.e., if $(\frac{d}{p}) = 1$.
- p is inert in $\mathbb{Z}[\sqrt{d}]$ if and only if $X^2 - d$ has no root modulo p, i.e., if $(\frac{d}{p}) = -1$.

For $d \equiv 1 \pmod 4$, $\mathbb{Z}_K = \mathbb{Z}[\frac{1+\sqrt{d}}{2}]$, and the minimal polynomial of $\frac{1+\sqrt{d}}{2}$ is $X^2 - X + (\frac{1-d}{4})$ (a polynomial with discriminant d). The results are identical: a prime p not dividing D_K is split if and only if $(\frac{d}{p}) = 1$, and inert if and only if $(\frac{d}{p}) = -1$.

Using Quadratic Reciprocity, it is not hard to see that these conditions are characterised by congruence conditions modulo D_K.

For example, in the case $K = \mathbb{Q}(i)$, with $d = -1$, we have $D_K = -4$, and a prime p is split if and only if $(\frac{-1}{p}) = 1$, and it is well-known that this is equivalent to $p \equiv 1 \pmod 4$. A prime p is inert if and only if $(\frac{-1}{p}) = -1$, and this is equivalent to $p \equiv 3 \pmod 4$.

For $K = \mathbb{Q}(\sqrt{-3})$, with $D_K = -3$, a prime p is split if and only if $(\frac{-3}{p}) = 1$, which is equivalent to $p \equiv 1 \pmod 3$, and is inert if and only if $(\frac{-3}{p}) = -1$, which is equivalent to $p \equiv 2 \pmod 3$.

Exercise 5.13 Classify the splitting or ramification behaviour of a prime number p in $\mathbb{Q}(\sqrt{5})$ in terms of the congruence class of p. Try to verify this with some explicit examples.

Repeat for other quadratic fields of your choice.

Chapter 6
Imaginary Quadratic Fields

It won't be a surprise that fields of low degree over $\mathbb{Q}$ are going to be the easiest cases to understand. Quadratic fields prove a good place to start, but it turns out to be convenient to split this case into imaginary quadratic fields and real quadratic fields, since the two cases turn out to be very different. We will deal with some aspects of the theory of real quadratic fields in Chap. 8.

The case of imaginary quadratic fields, which we treat in this chapter, will form the most complete example of the general theory that we consider in this book.

Throughout this chapter, we will be considering a field $K = \mathbb{Q}(\sqrt{d})$ where d is a negative, squarefree integer, that is, it is not divisible by the square of any prime. Every imaginary quadratic field can be written in this way for a unique choice of d (as we noted in Example 2.14(3)).

We recall from Proposition 2.34 that the ring of integers is $\mathbb{Z}[\frac{1+\sqrt{d}}{2}]$ if $d \equiv 1$ (mod 4), and is $\mathbb{Z}[\sqrt{d}]$ if not. As d is squarefree, d is not divisible by the square of any prime, so the case $d \equiv 0$ (mod 4) is not permitted, so this latter case arises when $d \equiv 2$ (mod 4) or $d \equiv 3$ (mod 4).

What we would really like to be able to do is to determine the fields with unique factorisation, and to understand the failure of unique factorisation in the other cases. It will turn out that there are very few imaginary quadratic fields with unique factorisation.

There are many numerical calculations in this chapter. There are comparatively few exercises—but the reader is strongly invited to make up exercises of your own, by making up examples in different quadratic fields. (Indeed, it's probably the best way to master the material!)

6.1 Units

One difference between the case of imaginary quadratic fields and that of real quadratic fields concerns the group of units. As we shall see, real quadratic fields have infinitely many units. Imaginary quadratic fields will have only finitely many

F. Jarvis, *Algebraic Number Theory*, Springer Undergraduate
Mathematics Series, DOI: 10.1007/978-3-319-07545-7_6,
© Springer International Publishing Switzerland 2014

units, all roots of unity, and it is easy to determine these units. (In fact, we shall see later that imaginary quadratic fields are the only fields other than $\mathbb{Q}$ for which the ring of integers has a finite group of units.)

Recall from Lemma 4.7 that $\alpha \in \mathbb{Z}_K$ is a unit precisely when $N(\alpha) = 1$. Since there are two possibilities for $\mathbb{Z}_K$, depending on d (mod 4), we will divide the calculation into two cases.

6.1.1 $d \equiv 2, 3$ *(mod 4)*

In this case, we have $\mathbb{Z}_K = \mathbb{Z}[\sqrt{d}]$. Then a typical element is $\alpha = a + b\sqrt{d}$, where a and b are in $\mathbb{Z}$, and the norm of α is

$$N_{K/\mathbb{Q}}(\alpha) = (a + b\sqrt{d})(a - b\sqrt{d}) = a^2 - db^2.$$

We need to solve $N_{K/\mathbb{Q}}(\alpha) = 1$. But notice that a^2 is a non-negative integer, and $-db^2$ is also a non-negative integer (as $d < 0$). In order that two non-negative integers add to 1, we need that one of them is 0, and the other is 1.

If $a^2 = 1$ and $-db^2 = 0$, then $a = \pm 1$ and $b = 0$ (as $d \neq 0$). This tells us that ± 1 is always a unit (and indeed, they are obviously invertible in $\mathbb{Z}_K$).

The other case is where $a^2 = 0$ and $-db^2 = 1$. If $d < -1$, then there is clearly no solution to $-db^2 = 1$. However, if $d = -1$, then $b = \pm 1$ is also possible. So in the field $\mathbb{Q}(\sqrt{-1})$, we also have units $0 \pm \sqrt{-1}$; in other words, $\pm i$ are units.

6.1.2 $d \equiv 1$ *(mod 4)*

The analysis here is very similar, if slightly more complicated by the fact that the ring of integers is now $\mathbb{Z}[\frac{1+\sqrt{d}}{2}]$. A typical element is therefore $\alpha = a + b(\frac{1+\sqrt{d}}{2}) = (2a + b + b\sqrt{d})/2$. Again, we must compute the norm of α:

$$N(\alpha) = \alpha\bar{\alpha} = \frac{(2a + b + b\sqrt{d})(2a + b - b\sqrt{d})}{4} = \frac{(2a + b)^2 - db^2}{4}.$$

Thus the equality $N(\alpha) = 1$ is equivalent to finding integral solutions to $(2a+b)^2 - db^2 = 4$.

We first consider the case where $d < -3$, in which case $d \leq -7$ (remember that $d \equiv 1$ (mod 4)). If $b \neq 0$, then $-db^2 \geq 7$, and as $(2a + b)^2 \geq 0$, there are no solutions. So $b = 0$. In this case, our equation becomes $(2a + 0)^2 = 4$, so that $a = \pm 1$. It follows that ± 1 are the only units in this case.

Now let's consider the case $d = -3$. If $|b|$ were to be at least 2, then $-db^2$ would be at least 12, and so there would be no solutions. The only possible solutions occur when $b = -1, 0$ or 1.

- When $b = -1$, we must solve $(2a - 1)^2 + 3 = 4$, giving $2a - 1 = \pm 1$, and so $a = 0$ or 1.
- When $b = 0$, we must solve $(2a)^2 = 4$, giving $a = \pm 1$.
- When $b = 1$, we must solve $(2a + 1)^2 + 3 = 4$, giving $2a + 1 = \pm 1$, and so $a = -1$ or 0.

We therefore have six solutions: $(a, b) = (0, -1), (1, -1), (-1, 0), (1, 0), (-1, 1)$ and $(0, 1)$. The corresponding units $\alpha = a + b\frac{1+\sqrt{-3}}{2}$ are given by:

$$\frac{-1 - \sqrt{-3}}{2}, \quad \frac{1 - \sqrt{-3}}{2}, \quad -1, \quad 1, \quad \frac{-1 + \sqrt{-3}}{2}, \quad \text{and} \quad \frac{1 + \sqrt{-3}}{2}.$$

These numbers, $\frac{\pm 1 \pm \sqrt{-3}}{2}$ and ± 1, are the sixth roots of unity.

Earlier, we used ω to denote the primitive cube root of unity $\frac{-1+\sqrt{-3}}{2} = e^{\frac{2\pi i}{3}}$. Notice that the ring of integers of $\mathbb{Q}(\sqrt{-3})$ is given by $\mathbb{Z}[\omega]$. Then $\omega^2 = e^{\frac{4\pi i}{3}} = \frac{-1-\sqrt{-3}}{2}$, and so the units are given by $\{\pm 1, \pm \omega, \pm \omega^2\}$.

6.1.3 Summary

We have shown that the only units in the imaginary quadratic field $\mathbb{Q}(\sqrt{d})$ are the elements of $\{\pm 1\}$, except in two cases. The first is when $d = -1$; the units in the Gaussian integers $\mathbb{Z}[i]$ are $\{\pm 1, \pm i\}$. The other exceptional case is when $d = -3$; the units in the ring of integers $\mathbb{Z}[\omega]$ of $\mathbb{Q}(\sqrt{-3})$ are $\{\pm 1, \pm \omega, \pm \omega^2\}$.

Notice that these units are all roots of unity (fourth roots, in the case of $\mathbb{Q}(i)$, and sixth roots in the case of $\mathbb{Q}(\sqrt{-3})$). So the units in every imaginary quadratic field are the roots of unity.

Conversely, it is easy to see that every root of unity is a unit. If λ is a root of unity in $\mathbb{Z}_K$, then $\lambda^n = 1$ for some n. Then $\lambda \times \lambda^{n-1} = 1$, and so λ is invertible, with inverse λ^{n-1} (this lies in $\mathbb{Z}_K$ as $\lambda \in \mathbb{Z}_K$, and $\mathbb{Z}_K$ is a ring).

Write μ_k for the set of kth roots of unity in $\mathbb{C}$; then we have proven the following result:

Theorem 6.1 *Let $K = \mathbb{Q}(\sqrt{d})$, with $d \in \mathbb{Z}_{<0}$ squarefree. Then λ is a unit in $\mathbb{Z}_K$ if and only if λ is a root of unity, and the units in $\mathbb{Z}_K$ are:*

$$U(\mathbb{Z}_K) = \mathbb{Z}_K^\times = \begin{cases} \mu_4 = \{\pm 1, \pm i\}, & \text{if } d = -1, \\ \mu_6 = \{\pm 1, \pm \omega, \pm \omega^2\}, & \text{if } d = -3, \\ \mu_2 = \{\pm 1\}, & \text{otherwise.} \end{cases}$$

6.2 Euclidean Imaginary Quadratic Fields

We have already explained in Chap. 1 that the Gaussian integers $\mathbb{Z}[i]$ possess unique factorisation, and this was critical when answering the question of which prime numbers can be written as the sum of two squares. Our proof involved showing that the norm function is Euclidean, and therefore that $\mathbb{Z}[i]$ is a Euclidean domain using this norm function.

In this section we will work out which imaginary quadratic fields can be shown to have unique factorisation in the same way; that is, when the ring of integers is a Euclidean domain.

We shall see later that there are imaginary quadratic fields with unique factorisation which are not Euclidean in this sense, and this provides examples of UFDs which are not Euclidean.

First, we will try to generalise the argument that worked for the Gaussian integers to more general imaginary quadratic fields K. We therefore need to know when the norm function is Euclidean.

Choose any α and β in $\mathbb{Z}_K$. We must be able to find a quotient $\kappa \in \mathbb{Z}_K$ and a remainder $\rho \in \mathbb{Z}_K$ such that $\alpha = \kappa\beta + \rho$, and $N(\rho) < N(\beta)$. As in Chap. 1, the method is to consider the quotient α/β, and to define κ to be the integer "closest" to it. Then we define $\rho = \alpha - \kappa\beta$.

It follows that

$$\rho = \beta \left(\frac{\alpha}{\beta} - \kappa \right),$$

and $N(\rho) = N(\beta)N(\alpha/\beta - \kappa)$, so $N(\rho) < N(\beta)$ as long as $N(\alpha/\beta - \kappa) < 1$. In particular, $\mathbb{Z}_K$ is Euclidean if, for any $\alpha/\beta \in \mathbb{Q}(\sqrt{d})$, there is $\kappa \in \mathbb{Z}_K$ with $N(\alpha/\beta - \kappa) < 1$.

Again, the two different forms of $\mathbb{Z}_K$ mean that we will treat this in two cases.

6.2.1 $d \equiv 2, 3 \pmod 4$

Now $\mathbb{Z}_K = \mathbb{Z}[\sqrt{d}]$. Suppose that $\alpha/\beta = a + b\sqrt{d}$, with a and b in $\mathbb{Q}$. We choose κ to be the "nearest" integer $m + n\sqrt{d} \in \mathbb{Z}_K$.

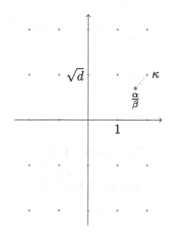

κ is chosen as the closest point of $\mathbb{Z}[\sqrt{d}]$ to $\frac{\alpha}{\beta}$

This means choosing $|m - a| \leq \frac{1}{2}$ and $|n - b| \leq \frac{1}{2}$. Then

$$N(\alpha/\beta - \kappa) = N((a + b\sqrt{d}) - (m + n\sqrt{d}))$$
$$= N((a - m) + \sqrt{d}(b - n))$$
$$= (a - m)^2 - d(b - n)^2$$
$$\leq (\tfrac{1}{2})^2 - d(\tfrac{1}{2})^2.$$

So if $d = -1$ or -2, this will definitely be strictly less than 1, as required.

On the other hand, if $d \leq -5$, there are certainly quotients α/β where there is no integer κ satisfying $N(\alpha/\beta - \kappa) < 1$; for example, just take $\alpha = 1 + \sqrt{d}$, $\beta = 2$.

Thus the only imaginary quadratic fields $\mathbb{Q}(\sqrt{d})$ with $d \equiv 2, 3 \pmod 4$ which are Euclidean with respect to the norm function are $\mathbb{Q}(\sqrt{-1})$ and $\mathbb{Q}(\sqrt{-2})$.

6.2.2 $d \equiv 1 \pmod 4$

Now $\mathbb{Z}_K = \mathbb{Z}[\frac{1+\sqrt{d}}{2}]$. Suppose that $\alpha/\beta = a + b\sqrt{d}$, with a and b in $\mathbb{Q}$, and again choose κ to be the "nearest" integer $m + n\sqrt{d} \in \mathbb{Z}_K$. However, $\mathbb{Z}_K$ looks a little different:

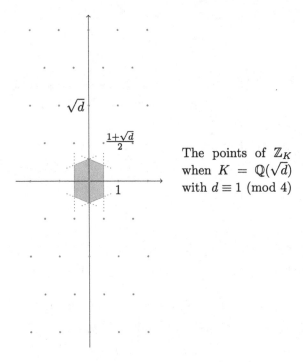

The points of $\mathbb{Z}_K$
when $K = \mathbb{Q}(\sqrt{d})$
with $d \equiv 1 \pmod 4$

The collection of points nearest to the origin lie in the shaded hexagon.

As in the first case, it will do to show that every point in $\mathbb{Q}(\sqrt{d})$ lies at a distance strictly less than 1 from some point in $\mathbb{Z}_K$. That is, it would suffice that the hexagon above lie inside the unit circle. Since the edges are given by the bisectors of the lines joining the origin and the points ± 1 and $\frac{\pm 1 \pm \sqrt{d}}{2}$, computing the vertices of the hexagon is elementary (though slightly painful!), and is left as an exercise.

Exercise 6.1 View the hexagon above as plotted in the (x, y)-plane, and bounded by the bisectors of the lines joining $(0, 0)$ to $(\pm 1, 0)$ and to $(\pm \frac{1}{2}, \pm \frac{\sqrt{|d|}}{2})$ with $|d| > 1$. Verify that the vertices of the hexagon above are at $\left(0, \pm(\frac{|d|+1}{4\sqrt{|d|}})\right)$ and $\left(\pm \frac{1}{2}, \pm(\frac{|d|-1}{4\sqrt{|d|}})\right)$.

It is then easy to check that if $d = -3$, $d = -7$ or $d = -11$, the hexagon lies entirely within the unit circle, and the corresponding field then has a Euclidean algorithm. However, if $d \leq -19$, then $\sqrt{|d|} > 4$, and the hexagon contains points of $\mathbb{Q}(\sqrt{d})$ at a distance of more than 1 from any point of $\mathbb{Z}_K$ (e.g., $\frac{1}{4}\sqrt{d}$), so that there can be no Euclidean algorithm using the norm function in these cases.

Exercise 6.2 Suppose that $K = \mathbb{Q}(\sqrt{-19})$. Find examples of α and β in $\mathbb{Z}_K$ such that there is no quotient κ and remainder ρ with $\alpha = \kappa\beta + \rho$ with $N_{K/\mathbb{Q}}(\rho) < N_{K/\mathbb{Q}}(\beta)$.

So we conclude that there are only five imaginary quadratic fields, $\mathbb{Q}(\sqrt{-1})$, $\mathbb{Q}(\sqrt{-2})$, $\mathbb{Q}(\sqrt{-3})$, $\mathbb{Q}(\sqrt{-7})$ and $\mathbb{Q}(\sqrt{-11})$ which are Euclidean with respect to their norm function.

It follows from Proposition 4.29 and Theorem 4.31 that these five fields all have unique factorisation.

We have now seen that there are exactly five imaginary quadratic fields whose rings of integers are Euclidean domains with respect to their norm function.

But one might wonder whether there might be other functions, different from the norm function, which make other rings of integers into Euclidean domains. It turns out that this is false, but the proof is comparatively recent; it was proven by Motzkin in 1949. We'll give the complete argument below in the more interesting case where $d \equiv 1 \pmod 4$, and leave the other case as an exercise.

We first need a general algebraic result.

Theorem 6.2 *Suppose that R is a Euclidean domain with respect to a Euclidean function ϕ, but that R is not a field. Then there is a non-zero element u of R, which is not a unit, such that for all $x \in R$, either $u|x$, or $u|x - v$ for some unit $v \in R$.*

Proof Let S denote the set of non-zero elements of R which are not units. As R is not a field, S is not empty. Consider the set

$$\phi(S) = \{\phi(s) \mid s \in S\}.$$

As $\phi(r)$ is a positive integer for all $r \in R$ (see Definition 4.28), we can choose $u \in S$ with $\phi(u)$ minimal amongst all the values in $\phi(S)$.

Given $x \in R$, the Euclidean property means that $x = qu + r$, where either $r = 0$, or $\phi(r) < \phi(u)$. When $r = 0$, we clearly have $x = qu$, so that $u|x$. In the second case, where $\phi(r) < \phi(u)$, we can't have $r \in S$, as $\phi(u)$ was the smallest value in $\phi(S)$. Since $r \notin S$ and $r \neq 0$, we must have that r is a unit in R. Let's write v for this unit, and then $qu = x - v$ shows that $u|x - v$ as required.   □

(The element u is often called a *universal side divisor* in the literature).

We can restate this:

Corollary 6.3 *Suppose that R is an integral domain that is not a field. If there are no elements u as in the theorem, then R is not Euclidean.*

Let's use this to deduce our main result:

Theorem 6.4 *Suppose that $K = \mathbb{Q}(\sqrt{d})$ with d squarefree and negative. Suppose that $d \equiv 1 \pmod 4$, and that $d < -11$. Then $\mathbb{Z}_K$ is not Euclidean.*

Proof By Corollary 6.3, we simply have to show that there are no elements u as in Theorem 6.2. In order to get a contradiction, we'll suppose that u does exist. We know that $\mathbb{Z}_K = \mathbb{Z}[\frac{1+\sqrt{d}}{2}]$, and, from Theorem 6.1, we know that the units in $\mathbb{Z}_K$ are simply $\{\pm 1\}$, and so the property implies that for all $\alpha \in \mathbb{Z}_K$, we need $u|\alpha$, or $u|\alpha \pm 1$.

Let's apply this with $\alpha = 2$. So we need $u|2$, or $u|2 \pm 1$. That is, u divides 1, 2 or 3. But u cannot divide 1, since u is not a unit. So u is a divisor of either 2 or 3. Let's see that 2 and 3 are irreducible: if not, there would be some element β of norm 2 or 3. But if $\beta = a + b\left(\frac{1+\sqrt{d}}{2}\right)$, then $N_{K/\mathbb{Q}}(\beta) = a^2 + ab + b^2\left(\frac{1-d}{4}\right)$, and we note that as $d < -11$, we have $k = \frac{1-d}{4} \geq 4$. It is easy to see that $a^2 + ab + kb^2 = 2$ and $a^2 + ab + kb^2 = 3$ have no solution for $k \geq 4$; once $b \neq 0$, the left-hand side is too large, and then there is clearly no solution for a. So both 2 and 3 are irreducible, and as u divides either 2 or 3, we must have $u = 2, -2, 3$ or -3.

But now we can take $\alpha = \frac{1+\sqrt{d}}{2}$ instead. Again, we should have $u|\alpha$ or $u|\alpha \pm 1$. However, none of these elements have ± 2 or ± 3 as divisors.

So there can be no element u, and so $\mathbb{Z}_K$ is not Euclidean. $\square$

A very similar argument applies in the remaining cases.

Exercise 6.3 Suppose $K = \mathbb{Q}(\sqrt{d})$ with d squarefree chosen such that $\mathbb{Z}_K = \mathbb{Z}[\sqrt{d}]$, and where $d < -2$. Recall that in this case, the units in $\mathbb{Z}_K$ are $\{\pm 1\}$; deduce that any element u as in Theorem 6.2 must be a divisor of 1, 2 or 3. Conclude as above that $u = \pm 2$ or ± 3, but as these do not divide α or $\alpha \pm 1$, where $\alpha = \sqrt{d}$, no element u can exist, and $\mathbb{Z}_K$ is not Euclidean.

We have therefore proven the following theorem:

Theorem 6.5 *Suppose that $K = \mathbb{Q}(\sqrt{d})$ with $d < 0$ squarefree. Then $\mathbb{Z}_K$ is Euclidean if and only if $d = -1, -2, -3, -7$ or -11.*

6.3 Quadratic Forms

In the rest of the chapter, we will relate uniqueness of factorisation to the theory of quadratic forms.

As already remarked, the class group may be viewed as the obstruction to unique factorisation: if the class group is trivial, then the number field has unique factorisation, but otherwise unique factorisation fails.

We would like a way to calculate the class group to be able to ascertain whether or not the field has unique factorisation. While some of the ideas that we will introduce in the remainder of this chapter have analogues for other number fields, they are most easily introduced in the context of imaginary quadratic fields.

We will show that the class number can be computed by counting a certain set of "binary quadratic forms", which we now introduce.

Definition 6.6 A *quadratic form in n variables* is a homogeneous polynomial of degree 2, and is therefore of the form

$$\sum_{i=1}^{n}\sum_{j=1}^{n} a_{ij}x_i x_j.$$

Notice that if we write $\mathbf{v} = (x_1 \cdots x_n)^t$, and A for the matrix (a_{ij}), then we can write the form as $\mathbf{v}^t A \mathbf{v}$.

We will only consider the situation where $a_{ij} \in \mathbb{Z}$.

We will focus on the case where $n = 2$; such forms are known as *binary* quadratic forms.

Definition 6.7 A *binary quadratic form* is a quadratic form in 2 variables, and is therefore of the form

$$f(x, y) = ax^2 + bxy + cy^2,$$

for some $a, b, c \in \mathbb{Z}$. The *discriminant* of this form is $b^2 - 4ac$.
We may abbreviate the form $ax^2 + bxy + cy^2$ by (a, b, c).

For example, the discriminant of $x^2 + y^2$ is -4 and the discriminant of $x^2 + (-d)y^2$ is $4d$.

Definition 6.8 Say that a quadratic form $f(x, y)$ is *positive definite* if

1. $f(x, y) \geq 0$ for all $x, y \in \mathbb{R}$;
2. $f(x, y) = 0$ means that we must have $(x, y) = (0, 0)$.

For completeness, although we will not need these in this book:

Definition 6.9 A quadratic form is *positive semi-definite* if $f(x, y) \geq 0$ for all $x, y \in \mathbb{R}$. Forms which take both positive and negative values are known as *indefinite*.

There is a similar definition of negative definite and negative semi-definite, got by changing the sign in the inequality.

We will only need to consider positive definite forms in what follows. Let's make some observations about positive definite forms. Suppose that $(a, b, c) = ax^2 + bxy + cy^2$ is positive definite. We certainly need $a > 0$, for otherwise substituting $(x, y) = (1, 0)$ gives a negative value. Similarly we must have $c > 0$, for otherwise $(x, y) = (0, 1)$ would give something negative. We can also write (completing the square):

$$ax^2 + bxy + cy^2 = a\left(x + \frac{b}{2a}y\right)^2 + \left(c - \frac{b^2}{4a}\right)y^2,$$

so we can refine the observation that $c > 0$ into the more precise observation that $c - \frac{b^2}{4a} > 0$ (otherwise choose $(x, y) = (-b, 2a)$ to get a negative value of the form). As $a > 0$, this is equivalent to $b^2 - 4ac < 0$.

Corollary 6.10 *The quadratic form $ax^2 + bxy + cy^2$ is positive definite if and only if $a > 0$ and the discriminant $b^2 - 4ac < 0$.*

Let's now hint at the relationship between imaginary quadratic fields and quadratic forms. As an example of a quadratic form, recall that the norm of a complex number $x + iy$ is

$$N(x + iy) = (x + iy)(x - iy) = x^2 + y^2;$$

notice that $x^2 + y^2$ is a positive definite quadratic form. More generally, the norm of the complex number $x + y\sqrt{d}$ (where d is a negative, squarefree integer as usual in this chapter) is $x^2 + (-d)y^2$. Thus, we have started with a general element in an imaginary quadratic field, and in both cases have recovered a positive definite binary quadratic form.

When studying quadratic forms, it is quickly apparent that two apparently different forms may really share many properties. Indeed, two apparently different forms may, in some sense, be the same. For example, the form $x^2 + 2xy + 2y^2$ can be rewritten $(x + y)^2 + y^2$, and a simple change of variable $X = x + y$, $Y = y$ allows us to write this as the form $X^2 + Y^2$. For many applications, we may wish to regard the form $x^2 + 2xy + 2y^2$ as equivalent to the form $x^2 + y^2$. This leads to a more general definition:

Definition 6.11 Two quadratic forms $f(x, y)$ and $g(x, y)$ are *equivalent* if one can be transformed into the other by a substitution of the form

$$(x, y) \mapsto (px + qy, rx + sy),$$

where p, q, r, s are integers with $ps - qr = \pm 1$. That is, $f(x, y)$ and $g(x, y)$ are equivalent if $g(x, y) = f(px + qy, rx + sy)$ for some invertible matrix $\begin{pmatrix} p & q \\ r & s \end{pmatrix} \in$ $\mathrm{GL}_2(\mathbb{Z})$, the *general linear group* of 2×2 matrices with integer entries whose inverse also has integer entries.

If $ps - qr = +1$, we say that $f(x, y)$ and $g(x, y)$ are *properly equivalent* (and in this case the matrix above lies in $\mathrm{SL}_2(\mathbb{Z})$, the *special linear group* of 2×2 matrices with determinant 1).

If we write $f(x, y) = \mathbf{v}^t A \mathbf{v}$, and $M = \begin{pmatrix} p & q \\ r & s \end{pmatrix}$, then $M\mathbf{v} = \begin{pmatrix} p & q \\ r & s \end{pmatrix} \begin{pmatrix} x \\ y \end{pmatrix} =$ $\begin{pmatrix} px + qy \\ rx + sy \end{pmatrix}$, and then

$$f(px + qy, rx + sy) = (M\mathbf{v})^t A (M\mathbf{v}) = \mathbf{v}^t (M^t A M) \mathbf{v}.$$

Thus, in terms of matrices, if f and g correspond to matrices A and B respectively, then f and g are equivalent if there exists a matrix $M \in \mathrm{GL}_2(\mathbb{Z})$ with $B = M^t A M$, and properly equivalent if $M \in \mathrm{SL}_2(\mathbb{Z})$.

Exercise 6.4 Show that two equivalent forms have the same discriminant.

As a special case of the exercise, the same is true for two properly equivalent forms.

Exercise 6.5 Show that (a) equivalence, and (b) proper equivalence, are both equivalence relations on the set of binary quadratic forms, and in particular on the set of binary quadratic forms of any given discriminant.

Before continuing, it is perhaps worth signalling the reasons for specifying these two forms of equivalence.

One motivation for studying quadratic forms comes from considering the numbers they represent—for example, one can ask the question which primes can be written as the sum of two squares. If $p = x^2 + y^2$, then p can also be represented as a value of any quadratic form which is equivalent to $x^2 + y^2$. For this sort of question, it is equivalence, rather than proper equivalence, which is the more useful notion, as it is easy to see that equivalent quadratic forms represent the same integers.

Exercise 6.6 Suppose that f and g are two equivalent quadratic forms. Show that for every natural number n, the number of pairs (x, y) such that $f(x, y) = n$ is equal to the number of pairs (u, v) such that $g(u, v) = n$.

[*Hint. In Definition 6.11, the matrix* $\begin{pmatrix} p & q \\ r & s \end{pmatrix}$ *is invertible, so we can also write* $g(u, v) = f(p'u + q'v, r'u + s'v)$ *for some integers* p', q', r', s' *with* $p's' - q'r' = \pm 1$. *Use these to get inverse bijections between the two sets.*]

However, we will explain in this chapter that there is also a relation between the class group of a given imaginary quadratic field, and the collection of certain quadratic forms of a certain discriminant. Indeed, we will give a bijection between the class group and positive definite binary quadratic forms of this discriminant *up to proper equivalence*. For this bijection, which is the main topic of this chapter, it is therefore proper equivalence which is the better notion. Further, the class group has a group structure, and we will briefly remark at the end of the chapter that this group structure carries across to give a group structure on the collection of proper equivalence classes of quadratic forms. There is no group structure on the collection of equivalence classes of quadratic forms.

6.4 Reduction Theory

Reduction theory is an elegant theory which allows us to determine when two quadratic forms are properly equivalent. If we are given a general (positive definite binary) quadratic form, then we can "reduce" it to a particular "reduced" form, and two forms are properly equivalent precisely when they both reduce to the same form.

Definition 6.12 We say that a form $(a, b, c) = ax^2 + bxy + cy^2$ is *reduced* if either $-a < b \le a < c$, or $0 \le b \le a = c$.

This may seem a slightly odd definition, but we will see that the precise conditions are chosen to ensure that, firstly, every form is properly equivalent to some reduced form, and, secondly, that no two different reduced forms are properly equivalent.

We defined proper equivalence above using matrices of determinant 1. Let's isolate some special cases (we will see later that these generate all of $\mathrm{SL}_2(\mathbb{Z})$).

First, we consider the matrix $\begin{pmatrix} 1 & 1 \\ 0 & 1 \end{pmatrix}$. We therefore consider the transformation

$$\begin{pmatrix} x \\ y \end{pmatrix} \mapsto \begin{pmatrix} 1 & 1 \\ 0 & 1 \end{pmatrix}\begin{pmatrix} x \\ y \end{pmatrix} = \begin{pmatrix} x + y \\ y \end{pmatrix}.$$

The form $ax^2 + bxy + cy^2$ is then properly equivalent to $a(x+y)^2 + b(x+y)y + cy^2$, which expands to $ax^2 + (2a + b)xy + (a + b + c)y^2$. Thus the form (a, b, c) is properly equivalent to the form $(a, b + 2a, c + b + a)$.

The inverse transformation corresponds to the matrix $\begin{pmatrix} 1 & -1 \\ 0 & 1 \end{pmatrix}$, and using this matrix we get that (a, b, c) is properly equivalent to $(a, b - 2a, c - b + a)$.

We will use one final transformation, the one corresponding to $\begin{pmatrix} 0 & 1 \\ -1 & 0 \end{pmatrix}$, which sends (x, y) to $(y, -x)$, and using this matrix we find that (a, b, c) is properly equivalent to $(c, -b, a)$.

Let's explain that, using these three transformations only, any binary positive definite quadratic form can be seen to be equivalent to a reduced form.

Take a positive definite binary quadratic form (a, b, c) (recall that this stands for $ax^2 + bxy + cy^2$ and that a and c are necessarily positive), and apply the following rules repeatedly:

- If $a > c$, or if $a = c$ and $b < 0$, apply the third rule, and perform the transformation $(a, b, c) \mapsto (c, -b, a)$.
- Otherwise, we are in one of two situations:

  - We could have $a < c$.
    If (a, b, c) is not reduced, it must be because $b \leq -a$ or $b > a$. If $b \leq -a$, apply the first rule, $(a, b, c) \mapsto (a, b + 2a, c + b + a)$ and if $b > a$, apply the second rule $(a, b, c) \mapsto (a, b - 2a, c - b + a)$. The result of this should be a form for which the absolute value $|b|$ of the middle coefficient gets smaller (except in the case $b = -a$, when $|b|$ may remain constant for one step only).
  - Alternatively, $a = c$ and $b \geq 0$.
    If (a, b, c) is not reduced, it must be because $b > a$. Then apply the second rule $(a, b, c) \mapsto (a, b - 2a, c - b + a)$. The result of this should again be a form for which the absolute value $|b|$ of the middle coefficient gets smaller.

Let's see an example of this procedure in action. Consider the form $7x^2 - 24xy + 21y^2$, or $(7, -24, 21)$ in our abbreviated notation.

This is not reduced, but we do have $a < c$, as $7 < 21$. But we do not have $-a < b \leq a$. We therefore apply the first rule, to get $(7, -24, 21) \mapsto (7, -10, 4)$ (that is, our original form is properly equivalent to $7x^2 - 10xy + 4y^2$).

Now we have a form which is not reduced because $c < a$. We therefore apply the third rule $(a, b, c) \mapsto (c, -b, a)$, and find $(7, -10, 4) \mapsto (4, 10, 7)$ (so now our original form is properly equivalent to $4x^2 + 10yx + 7y^2$).

Now $(4, 10, 7)$ is not reduced as $b > a$, so we apply the third rule $(a, b, c) \mapsto$ $(a, b - 2a, c - b + a)$ to get $(4, 10, 7) \mapsto (4, 2, 1)$.

You can see that the numbers are getting smaller! But we still don't have a reduced form, as $a > c$. So we apply the third rule to get $(4, 2, 1) \mapsto (1, -2, 4)$, and (finally!) the first rule gives $(1, -2, 4) \mapsto (1, 0, 3)$, which is reduced.

We could have kept track of the changes of variable required as we were going along, but we can recover this by thinking about the matrix transformations involved.

Indeed, as we noted above, each equivalence is represented by a matrix. The first time that we applied a rule $((a, b, c) \mapsto (a, b + 2a, c + b + a))$, we essentially made the change of variable $\begin{pmatrix} x \\ y \end{pmatrix} \mapsto \begin{pmatrix} 1 & 1 \\ 0 & 1 \end{pmatrix} \begin{pmatrix} x \\ y \end{pmatrix}$. This means that we make the transformation $(x, y) \mapsto (x + y, y)$. One can easily check that

$$7(x + y)^2 - 24(x + y)y + 21y^2 = 7x^2 - 10xy + 4y^2.$$

If f_0 denotes the original form $(7, -24, 21)$ and f_1 denotes the result $(7, -10, 4)$ after one step, with $M_1 = \begin{pmatrix} 1 & 1 \\ 0 & 1 \end{pmatrix}$, then this calculation shows that

$$f_0 \left(M_1 \begin{pmatrix} x \\ y \end{pmatrix} \right) = f_1 \left(\begin{pmatrix} x \\ y \end{pmatrix} \right).$$

If we now write

$$\begin{pmatrix} x_1 \\ y_1 \end{pmatrix} = M_1 \begin{pmatrix} x \\ y \end{pmatrix} = \begin{pmatrix} 1 & 1 \\ 0 & 1 \end{pmatrix} \begin{pmatrix} x \\ y \end{pmatrix} = \begin{pmatrix} x + y \\ y \end{pmatrix},$$

then we can apply the inverse matrix to write (x, y) in terms of (x_1, y_1), and

$$\begin{pmatrix} x \\ y \end{pmatrix} = M_1^{-1} \begin{pmatrix} x_1 \\ y_1 \end{pmatrix} = \begin{pmatrix} 1 & -1 \\ 0 & 1 \end{pmatrix} \begin{pmatrix} x_1 \\ y_1 \end{pmatrix} = \begin{pmatrix} x_1 - y_1 \\ y_1 \end{pmatrix}.$$

We therefore see

$$f_0 \left(\begin{pmatrix} x \\ y \end{pmatrix} \right) = f_1 \left(M_1^{-1} \begin{pmatrix} x \\ y \end{pmatrix} \right),$$

i.e.,

$$7x_1^2 - 24x_1y_1 + 21y_1^2 = 7(x_1 - y_1)^2 - 10(x_1 - y_1)y_1 + 4y_1^2.$$

In summary, if $\mathbf{v}$ is the vector $\begin{pmatrix} x \\ y \end{pmatrix}$, we should have

$$f_0(M_1\mathbf{v}) = f_1(\mathbf{v}) \quad \text{and} \quad f_0(\mathbf{v}) = f_1(M_1^{-1}\mathbf{v}).$$

Next we applied the third rule $((a, b, c) \mapsto (c, -b, a))$ to f_1 but we already noted that this corresponds to the matrix $\begin{pmatrix} 0 & 1 \\ -1 & 0 \end{pmatrix}$. As above, we can write f_2 for the form $(4, 10, 7)$, and we should have

$$f_1(M_2\mathbf{v}) = f_2(\mathbf{v}) \quad \text{and} \quad f_1(\mathbf{v}) = f_2(M_2^{-1}\mathbf{v}).$$

But we already knew that

$$f_0(M_1\mathbf{v}) = f_1(\mathbf{v}) \quad \text{and} \quad f_0(\mathbf{v}) = f_1(M_1^{-1}\mathbf{v}),$$

so we can combine these to see that

$$f_2(\mathbf{v}) = f_1(M_2\mathbf{v}) = f_0(M_1(M_2\mathbf{v})).$$

Recall that $v = \begin{pmatrix} x \\ y \end{pmatrix}$, $M_1 = \begin{pmatrix} 1 & 1 \\ 0 & 1 \end{pmatrix}$, and $M_2 = \begin{pmatrix} 0 & 1 \\ -1 & 0 \end{pmatrix}$, so we get

$$f_2(x, y) = f_0\left(\begin{pmatrix} 1 & 1 \\ 0 & 1 \end{pmatrix} \begin{pmatrix} 0 & 1 \\ -1 & 0 \end{pmatrix} \begin{pmatrix} x \\ y \end{pmatrix} \right) = f_0(-x + y, -x),$$

which can easily be verified by a simple calculation.

This pattern continues; $M_3 = \begin{pmatrix} 1 & -1 \\ 0 & 1 \end{pmatrix}$, $M_4 = \begin{pmatrix} 0 & 1 \\ -1 & 0 \end{pmatrix}$ and $M_5 = \begin{pmatrix} 1 & 1 \\ 0 & 1 \end{pmatrix}$ are the matrices for the remaining three steps, and because we saw that $f_5 = (1, 0, 3) = x^2 + 3y^2$ is the final step, it follows that

$$f_5(x, y) = f_0(M_1 M_2 M_3 M_4 M_5(x, y)^t).$$

It is easy to calculate that

$$M = M_1 M_2 M_3 M_4 M_5 = \begin{pmatrix} -2 & -3 \\ -1 & -2 \end{pmatrix},$$

and to verify that

$$f_0\left(M \begin{pmatrix} x \\ y \end{pmatrix}\right) = 7(-2x-3y)^2 - 24(-2x-3y)(-x-2y) + 21(-x-2y)^2 = x^2 + 3y^2,$$

as expected.

Exercise 6.7 Reduce the following quadratic forms $f(x, y)$, and hence find matrices M such that $f(M(x, y)^T) = g(x, y)$, where $g(x, y)$ is a reduced quadratic form.

1. $6x^2 - 7xy + 8y^2$;
2. $13x^2 + 12xy + 11y^2$;
3. $43x^2 + 71xy + 67y^2$.

Invent your own examples—remember that (a, b, c) must be chosen to be positive definite, so $a > 0$, $c > 0$ and $b^2 - 4ac < 0$.

Exercise 6.8 We can also use this to solve one of the questions in Chap. 1. Given a prime number $p \equiv 1 \pmod 4$, how can we write it as the sum of two squares? First, find a value of m with $m^2 \equiv -1 \pmod p$ as in Remark 1.22. So $m^2 = kp - 1$ for some value of k. Then the form $f(x, y) = px^2 + 2mxy + ky^2$ has discriminant -4, and $f(1, 0) = p$. We now reduce this form, keeping track of the changes of variable, finding a reduced form $g(x, y) = f(M(x, y)^t)$ for some matrix M. But since discriminants are unchanged by equivalence, the discriminant of $g(x, y)$ is still -4, and we will see that the only reduced form of discriminant -4 is $g(x, y) = x^2 + y^2$. Now we can find values x_0 and y_0 such that $g(x_0, y_0) = f(1, 0) = p$, so $p = x_0^2 + y_0^2$, as required.

Carry out this procedure for $p = 1009$, using your value of m from Exercise 1.12.

We have now explained why we might believe the following theorem:

Theorem 6.13 *Every positive definite binary quadratic form is properly equivalent to a reduced form.*

We will actually prove something stronger.

We haven't really used the full strength of the definition of what it means for a form to be reduced. The point of the definition is that any positive definite binary quadratic form should be properly equivalent to *precisely one* reduced form, or, equivalently, that every proper equivalence class of positive definite binary quadratic forms contains a unique reduced form. This gives the enhanced version of the above result:

Theorem 6.14 *Every positive definite binary quadratic form is properly equivalent to a unique reduced form.*

Before we prove this, we need a lemma.

Lemma 6.15 *Suppose that $f(x, y) = ax^2 + bxy + cy^2$ is a reduced form. If $a < c$, then the smallest non-zero values taken by $f(x, y)$ for x and y coprime are a and c; furthermore, the only values of (x, y) with $f(x, y) = a$ are $(\pm1, 0)$. If $a = c$, then the smallest non-zero value of $f(x, y)$ is a, and there are either 4 (if $0 \le b < a = c$) or 6 (if $a = b = c$) pairs (x, y) with $f(x, y) = a$.*

Proof As $(x, y) = 1$, if $y = 0$, then $x = \pm1$, and $f(\pm1, 0) = a$. If $|y| = 1$ and $|x| \ge 2$, then

$$|2ax + by| \ge |2ax| - |by| \ge 4a - |b| \ge 3a,$$

so

$$4af(x, y) = (2ax + by)^2 - dy^2 \geq 9a^2 - d$$
$$= 4ac + (9a^2 - b^2) = 4ac + 8a^2 + (a^2 - b^2) \geq 4ac$$

as $|b| \leq a$. So $f(x, \pm 1) > c$ if $|x| \geq 2$. If $|y| \geq 2$, then

$$4af(x, y) = (2ax + by)^2 - dy^2 \geq -dy^2 \geq -4d = 16ac - 4b^2$$
$$\geq 12ac + 4(ac - b^2) \geq 12ac \geq 4ac,$$

as $|b| \leq a \leq c$, and again $f(x, y) > c$ if $|y| \geq 2$. In summary, $f(x, y) > c$ if $|x| \geq 2$ or $|y| \geq 2$. It remains to consider $f(\pm 1, 0) = a$, $f(0, \pm 1) = c$, $f(\pm 1, \pm 1) = a + b + c > c$ and $f(\pm 1, \mp 1) = a - b + c \geq c$. The result follows easily in each case.   □

Now we prove Theorem 6.14.

Proof (of Theorem) Firstly, we should verify that the algorithmic process for reducing quadratic forms really does terminate after a finite number of steps with a reduced form. This is rather easy: at each step, none of the operations increase the coefficient of x^2. As this is always a natural number, eventually it must become constant. At this point, the remaining operations do not increase the natural number $|b|$, and again eventually $|b|$ must become constant. Suppose that one rule maps (a, b, c) to another form with the same values of a and $|b|$. An examination of the rules shows that this is only possible if $a = b$ (and then c is also fixed, as the discriminant $b^2 - 4ac$ is fixed) or if $a = -b$, and the first rule $(a, b, c) \mapsto (a, b + 2a, c + b + a)$ is applied. Then $a = b \leq c$, and the form is now reduced.

Thus every form is equivalent to some reduced form. If a form were equivalent to two different reduced forms, then these reduced forms would be equivalent to each other. The final step will be to show that two distinct reduced forms cannot be equivalent to each other. So suppose that $f(x, y) = ax^2 + bxy + cy^2$ and $g(x, y) = Ax^2 + Bxy + Cy^2$ are two reduced forms which are properly equivalent. We claim that $f = g$, using Lemma 6.15 repeatedly.

The smallest positive number represented by $f(x, y)$ is a, and the smallest positive number represented by $g(x, y)$ is A. On the other hand, equivalent forms represent the same numbers (Exercise 6.6). So $a = A$.

If $c > a$, then there are precisely two pairs $(\pm 1, 0)$ with $f(x, y) = a$. As f and g are equivalent, the same will be true of g, so $C > A$. As c is the second smallest positive number represented by $f(x, y)$ and C is the second smallest positive number represented by $g(x, y)$, then $c = C$. Now f and g have the same discriminant, so $b = \pm B$. We want to see that $b = B$. However, $g(x, y) = f(px + qy, rx + sy)$ for $\begin{pmatrix} p & q \\ r & s \end{pmatrix}$ of determinant 1, and both have the same coefficient of x^2, we see that $p = \pm 1$ and $r = 0$. We can assume $p = 1$ by multiplying all of p, q, r and s by -1 if needed. As $ps - qr = 1$, we see that $s = p = 1$. Then $g(x, y) = f(x + qy, y)$.

If $f(x, y)$ is (a, b, c), then $f(x + qy, y)$ is $(a, b + 2qa, c + qb + q^2a)$. However, both are reduced, so we need $-a < b \leq a$ and $-a < b + 2qa \leq a$, and this is only possible with $q = 0$. So $B = b$ also.

If $c = a$, then there are either 4 or 6 pairs (x, y) with $f(x, y) = a$, and so the same is true of g. Thus $C = A$ also. By the definition of reduced form, $0 \leq b \leq a$ and $0 \leq B \leq A$. As the discriminants of f and g coincide, we deduce that $b = \pm B$, but as both are non-negative, we see that $b = B$ as required. $\qquad \square$

The proof has an interesting consequence, which should remind you of some of the results on units at the start of the chapter.

Definition 6.16 Let $f(x, y)$ be a binary quadratic form. We say that a matrix $M = \begin{pmatrix} p & q \\ r & s \end{pmatrix} \in SL_2(\mathbb{Z})$ is an *automorph of* f if $f(x, y) = f(px + qy, rx + sy)$.

Corollary 6.17 *Suppose that $f(x, y)$ is a reduced binary quadratic form. Then if*
$f(x) = x^2 + y^2$, *there are 4 automorphs of f, given by* $\left\{\pm I, \pm \begin{pmatrix} 0 & 1 \\ -1 & 0 \end{pmatrix}\right\}$; *if $f(x) =$*
$x^2 + xy + y^2$, *there are 6 automorphs of f, given by* $\left\{\pm I, \pm \begin{pmatrix} 1 & 1 \\ -1 & 0 \end{pmatrix}, \pm \begin{pmatrix} 0 & -1 \\ 1 & 1 \end{pmatrix}\right\}$;
in all other cases, there are just 2 automorphs, given by $\{\pm I\}$.

Proof Suppose first that $a < c$. Then Lemma 6.15 shows that the only pairs (x, y) with $f(x, y) = a$ are given by $(\pm 1, 0)$. Thus a matrix M which is an automorph must map $\begin{pmatrix} 1 \\ 0 \end{pmatrix}$ into $\begin{pmatrix} \pm 1 \\ 0 \end{pmatrix}$. So the first column of M must be $\begin{pmatrix} \pm 1 \\ 0 \end{pmatrix}$. As M has determinant 1, this shows that $M = \pm \begin{pmatrix} 1 & m \\ 0 & 1 \end{pmatrix}$ for some $m \in \mathbb{Z}$; however, $f(x + my, y)$ can only be the same as $f(x, y)$ if $m = 0$ (by looking at the coefficient of xy say). Thus $M = \pm I$.

Next, consider the cases where $a = c$. If $0 \leq b < a = c$, then Lemma 6.15 gives 4 pairs (x, y) with $f(x, y) = a$; these are clearly $(\pm 1, 0)$ and $(0, \pm 1)$. Then we argue as above, to conclude that $M = \pm \begin{pmatrix} 1 & m \\ 0 & 1 \end{pmatrix}$ (and $m = 0$ for the same reasons as when $a < c$) or that $M = \pm \begin{pmatrix} 0 & -1 \\ 1 & m \end{pmatrix}$. In this latter case, $M\mathbf{v} = \begin{pmatrix} -y \\ x + my \end{pmatrix}$, and if $f(x, y) = ax^2 + bxy + cy^2$, then

$$f(-y, x + my) = cx^2 + (2mc - b)xy + (a - mb + m^2c)y^2.$$

Comparing coefficients of xy gives $b = 2mc - b$, i.e., $b = mc$. However, $0 \leq b < c$, and the only possibility is $b = 0$, giving $m = 0$, and then there are 4 automorphs, as claimed.

Finally, there is the case $b = a = c$; again the proof of Lemma 6.15 gives 6 pairs (x, y) with $f(x, y) = a$, given by $(\pm 1, 0)$, $(0, \pm 1)$ and $(\pm 1, \mp 1)$. These are again the first columns of possible matrices M giving automorphs, and a similar argument to the above gives the 6 different matrices of the statement. $\qquad \square$

Before discussing the relation to class numbers, it is worth noting the following interesting corollary:

Corollary 6.18 $SL_2(\mathbb{Z})$ *is generated by the two matrices* $\begin{pmatrix} 1 & 1 \\ 0 & 1 \end{pmatrix}$ *and* $\begin{pmatrix} 0 & 1 \\ -1 & 0 \end{pmatrix}$.

Proof Let $M \in SL_2(\mathbb{Z})$. Let $f(x, y)$ be the reduced form $x^2 + 2y^2$ say (any other form with 2 automorphs will do). Then consider $f'(x, y) = f(M(x, y)^t)$, another quadratic form which is properly equivalent to f. Reduce this form by the above method; we must end up with the reduced form in the same class as f'—but this must be f itself, since it is a reduced form, properly equivalent to f', and we have explained that there is a unique such form. We have also explained that the reduction steps correspond to application of the matrices

$$\begin{pmatrix} 1 & 1 \\ 0 & 1 \end{pmatrix}, \quad \begin{pmatrix} 1 & -1 \\ 0 & 1 \end{pmatrix}, \quad \begin{pmatrix} 0 & 1 \\ -1 & 0 \end{pmatrix},$$

but notice that

$$\begin{pmatrix} 1 & -1 \\ 0 & 1 \end{pmatrix} = \begin{pmatrix} 1 & 1 \\ 0 & 1 \end{pmatrix}^{-1},$$

so that all the reduction steps can be expressed in terms of

$$T = \begin{pmatrix} 1 & 1 \\ 0 & 1 \end{pmatrix} \quad \text{and} \quad S = \begin{pmatrix} 0 & 1 \\ -1 & 0 \end{pmatrix},$$

and their inverses (note that $S^{-1} = -S$). The reduction of f' to f involves writing

$$f'(M_1 M_2 \ldots M_t (x, y)^t) = f((x, y)^t)$$

where $M_1, \ldots M_t$ are T, T^{-1} or S. But by definition of f', this means that

$$f(M M_1 M_2 \ldots M_t (x, y)^t) = f((x, y)^t),$$

and this can only happen (see Corollary 6.17) if

$$M M_1 M_2 \ldots M_t = \pm I.$$

But this means that
$$M = M_t^{-1} \ldots M_1^{-1},$$

and so we have written M as a product of matrices which are T, T^{-1} or $S^{-1} = -S = S^3$.   □

6.5 Class Numbers and Quadratic Forms

Quadratic forms are interesting objects in their own right, and the reader could omit this section and still discover an interesting theory of quadratic forms. However, it will be more complete if we relate quadratic forms to class groups of imaginary quadratic fields.

The main aim of this section is to prove the following theorem. Recall that d is a negative squarefree integer:

Theorem 6.19 *The class number of $K = \mathbb{Q}(\sqrt{d})$ is equal to the number of reduced quadratic forms with discriminant D, where $D = D_K$ is given by*

$$D = \begin{cases} 4d, & \text{if } d \equiv 2, 3 \,(\mathrm{mod}\,4), \\ d, & \text{if } d \equiv 1 \,(\mathrm{mod}\,4). \end{cases}$$

To prove this theorem we will show that there exists a bijection between the ideal classes of $\mathbb{Q}(\sqrt{d})$ and reduced quadratic forms with discriminant D. To show this bijection exists we will give a mapping from ideals to quadratic forms and an inverse mapping from quadratic forms to ideals. This means we must show that every ideal generates a quadratic form and that every quadratic form comes from an ideal. We also then need to show that any ideals in the same ideal class generate properly equivalent quadratic forms, and that two properly equivalent quadratic forms are come from ideals in the same ideal class.

Before we prove the theorem, let's give a motivating example.

Recall that $\mathbb{Q}(\sqrt{-5})$ does not have unique factorisation; the number 6 can be factorised as 2×3, and as $(1 + \sqrt{-5})(1 - \sqrt{-5})$. These were genuinely distinct factorisations, but we could resolve the non-uniqueness of factorisation by introducing ideals in the ring of integers $\mathbb{Z}[\sqrt{-5}]$:

$$\mathfrak{a}_1 = \langle 2, 1 + \sqrt{-5} \rangle;$$
$$\mathfrak{a}_2 = \langle 2, 1 - \sqrt{-5} \rangle = \mathfrak{a}_1;$$
$$\mathfrak{a}_3 = \langle 3, 1 + \sqrt{-5} \rangle;$$
$$\mathfrak{a}_4 = \langle 3, 1 - \sqrt{-5} \rangle;$$

Then

$$\langle 2 \rangle = \mathfrak{a}_1 \mathfrak{a}_2;$$
$$\langle 3 \rangle = \mathfrak{a}_3 \mathfrak{a}_4;$$
$$\langle 1 + \sqrt{-5} \rangle = \mathfrak{a}_1 \mathfrak{a}_3;$$
$$\langle 1 - \sqrt{-5} \rangle = \mathfrak{a}_2 \mathfrak{a}_4,$$

and the two distinct factorisations are resolved when we use non-principal ideals.

Notice next that $\mathbb{Z}[\sqrt{-5}] \subset \mathbb{C}$. We can plot elements of ideals in the complex plane. For example, if we consider the ideal $\mathfrak{a}_1 = \langle 2, 1 + \sqrt{-5} \rangle$, its elements are given by

$$\{2(a + b\sqrt{-5}) + (1 + \sqrt{-5})(c + d\sqrt{-5}) \mid a, b, c, d \in \mathbb{Z}\}$$
$$= \{(2a + c - 5d) + (2b + c + d)\sqrt{-5} \mid a, b, c, d \in \mathbb{Z}\}$$
$$= \{A + B\sqrt{-5} \mid A, B \in \mathbb{Z}, 2 \mid A - B\}$$
$$= \{2x + (1 + \sqrt{-5})y \mid x, y \in \mathbb{Z}\}.$$

Not only are 2 and $1 + \sqrt{-5}$ generators for this ideal as a $\mathbb{Z}[\sqrt{-5}]$-module, but also as a $\mathbb{Z}$-module, meaning that any element of the ideal can be written as $2x + (1 + \sqrt{-5})y$ with $x, y \in \mathbb{Z}$. So the general element of this ideal is $(2x + y) + y\sqrt{-5}$ for $x, y \in \mathbb{Z}$. The norm of this element is

$$(2x + y)^2 + 5y^2 = 4x^2 + 4xy + 6y^2 = 2(2x^2 + 2xy + 3y^2).$$

After taking out the common factor of 2, we have the quadratic form $(2, 2, 3)$, which is a positive definite quadratic form of discriminant -20.

Let's try $\mathfrak{a}_3 = \langle 3, 1 + \sqrt{-5} \rangle$. Again, we will try to give a set of 2 generators for $\mathfrak{a}$ over $\mathbb{Z}$, and consider the norm of a general element of the ideal. We note that elements of $\mathfrak{a}_3$ are given by:

$$\{3(a + b\sqrt{-5}) + (1 + \sqrt{-5})(c + d\sqrt{-5}) \mid a, b, c, d \in \mathbb{Z}\}$$
$$= \{(3a + c - 5d) + (3b + c + d)\sqrt{-5} \mid a, b, c, d \in \mathbb{Z}\}$$
$$= \{A + B\sqrt{-5} \mid A, B \in \mathbb{Z}, 3 \mid A - B\}$$
$$= \{3x + (1 + \sqrt{-5})y \mid x, y \in \mathbb{Z}\}.$$

Once again, generators for this ideal as a $\mathbb{Z}$-module are given by 3 and by $1 + \sqrt{-5}$, so a general element of the ideal is $3x + (1 + \sqrt{-5})y$ with $x, y \in \mathbb{Z}$, and this element has norm

$$(3x + y)^2 + 5y^2 = 9x^2 + 6xy + 6y^2 = 3(3x^2 + 2xy + 2y^2),$$

which is 3 times the non-reduced form $(3, 2, 2)$. We can reduce $(3, 2, 2)$ in the usual way:

$$(3, 2, 2) \mapsto (2, -2, 3) \mapsto (2, 2, 3),$$

using $(a, b, c) \mapsto (c, -b, a)$ and then $(a, b, c) \mapsto (a, b + 2a, c + b + a)$.

Here is the surprise: if we start with *any* pairs of generators α and β for a non-principal ideal $\mathfrak{a}$ in $\mathbb{Z}[\sqrt{-5}]$, and do the same thing, writing a general element in the form $\alpha x + \beta y$, and taking the norm, the result is *always* (a multiple of) a form properly equivalent to $(2, 2, 3)$.

Now let's take a principal ideal; the easiest is the whole ring of integers: $\langle 1 \rangle$. So the general element is $1.(x + y\sqrt{-5})$ for $x, y \in \mathbb{Z}$, and the norm of $x + y\sqrt{-5}$ is $x^2 + 5y^2$.

Another example might be the principal ideal $\langle 1 + \sqrt{-5} \rangle$. The general element of this ideal is $(1 + \sqrt{-5})(x + y\sqrt{-5}) = (x - 5y) + (x + y)\sqrt{-5}$, and this has norm

$$(x - 5y)^2 + 5(x + y)^2 = 6x^2 + 30y^2 = 6(x^2 + 5y^2).$$

Again, we see the quadratic form $(1, 0, 5)$ appearing.

You should find that if one starts with a principal ideal $\mathfrak{a}$ of norm N, when one writes a general element in the form $\alpha x + \beta y$ for generators α and β, and then takes the norm of this general element, one obtains something of the form $N.f(x, y)$, where $f(x, y)$ is a quadratic form which reduces to $(1, 0, 5)$.

Exercise 6.9 Consider $K = \mathbb{Q}(\sqrt{-6})$, and the factorisation $6 = 2 \times 3 = (\sqrt{6})^2$. Then $\langle 2 \rangle = \mathfrak{p}_2^2$, $\langle 3 \rangle = \mathfrak{p}_3^2$ and $\langle \sqrt{-6} \rangle = \mathfrak{p}_2\mathfrak{p}_3$ where $\mathfrak{p}_2 = \langle 2, \sqrt{-6} \rangle$ and $\mathfrak{p}_3 = \langle 3, \sqrt{-6} \rangle$.

What are the quadratic forms associated to the non-principal ideals $\mathfrak{p}_2$ and $\mathfrak{p}_3$?

What are the quadratic forms associated to the principal ideals $\langle 2 \rangle$, $\langle 3 \rangle$ and $\langle \sqrt{-6} \rangle$?

Remark 6.20 Actually, these two examples are rather fortunate in one sense: in both examples, the order of the generators didn't matter. In general, if we switch the generators α and β, this is equivalent to interchanging x and y. We already know that the quadratic form (c, b, a) is not generally *properly equivalent* to (a, b, c). In the case of $\mathbb{Q}(\sqrt{-5})$, and quadratic forms of discriminant -20, it is the case that $(5, 0, 1)$ is properly equivalent to $(1, 0, 5)$, and that $(3, 2, 2)$ is properly equivalent to $(2, 2, 3)$.

This means that we shall need to make a choice of the order in which we choose our generators. We will always pick α and β so that, in the Argand diagram, the angle of clockwise rotation from β to α is greater than from α to β. The reader should try to convince themselves that this translates into the condition that β/α should lie in the upper-half complex plane, i.e., should have positive imaginary part.

Definition 6.21 A pair (α, β) of complex numbers is *ordered* if $\text{im}(\beta/\alpha) > 0$.

The reader should invent some more examples of ideals in rings of integers in other imaginary quadratic fields, taking care over the ordering of generating pairs for ideals.

But the basic result should be the same. Given any ideal in a ring of integers of an imaginary quadratic field $\mathbb{Q}(\sqrt{d})$, we can find an ordered pair of generators, (α, β), for the ideal as a $\mathbb{Z}$-module, so that every element of the ideal is written as $\alpha x + \beta y$ for $x, y \in \mathbb{Z}$. Then the norm of this general element turns out to be the norm of the ideal multiplied by a quadratic form in x and y, and this quadratic form has discriminant D. Different choices of ordered generators give properly equivalent quadratic forms, and this gives us a map from ideals to proper equivalence classes of quadratic forms. It will also turn out that two ideals which are in the same ideal class will map to two properly equivalent forms. The reader should be able to try examples of principal ideals, and see that principal ideals in the same field always map to the same quadratic form.

This will give a map from the class group to the set of proper equivalence classes of positive definite binary quadratic forms of discriminant D, and we will see that this is a bijection.

Throughout we work in the imaginary quadratic field $K = \mathbb{Q}(\sqrt{d})$ with d square-free and negative. Write D for d or $4d$, as in Theorem 6.19 above. For simplicity, the complete proof will be given for $d \equiv 2, 3 \pmod 4$, where $\mathbb{Z}_K = \mathbb{Z}[\sqrt{d}]$, in what follows; the arguments for $d \equiv 1 \pmod 4$ are very similar, but almost all the details need minor amendment; these will be left as an exercise. The two cases can be treated together (see [3], for example), but once the easier case is understood, the other case really is similar, but the formulae are just a little more intricate.

6.5.1 $d \equiv 2, 3$ (mod 4)

We fix $d \equiv 2, 3 \pmod 4$, and write $D = 4d$ for the discriminant of $\mathbb{Q}(\sqrt{d})$.

We will show that there is a bijection between classes of ideals in $\mathbb{Z}[\sqrt{d}]$ and proper equivalence classes of (positive definite) quadratic forms of discriminant $D = 4d$.

This will involve several steps.

1. Given an ideal $\mathfrak{a}$ in $\mathbb{Z}_K$, we start by choosing a particular ordered basis, and observe that this basis gives a quadratic form of discriminant D.
2. In fact, changing the ordered basis of the ideal produces a properly equivalent quadratic form. So our map can be viewed as a map from ideals to proper equivalence classes of quadratic forms of discriminant D.
3. Two ideals in the same equivalence class map to the same proper equivalence class of quadratic forms, so we get a map Φ from ideal classes to proper equivalence classes of quadratic forms.
4. Finally, we write down a map Ψ from proper equivalence classes of quadratic forms of discriminant D to ideal classes, and check $\Psi = \Phi^{-1}$.

Stage 1: Ordered bases of ideals

The following lemma is a stronger version of the results of Sect. 3.4. We find a special choice of ordered basis of an ideal with which we can make computations.

Lemma 6.22 *Let $\mathfrak{a}$ be an ideal in the ring of integers $\mathbb{Z}_K$. Then there are positive integers $a, b, c \in \mathbb{Z}$ with $c \mid a$ and $c \mid b$ such that*

$$\mathfrak{a} = a\mathbb{Z} + (b + c\sqrt{d})\mathbb{Z}.$$

Proof Let a be the smallest positive integer in $\mathfrak{a}$ and let $b + c\sqrt{d}$ be any element in $\mathfrak{a}$ with a positive integer c as small as possible.

We first claim that $\mathfrak{a} = a\mathbb{Z} + (b + c\sqrt{d})\mathbb{Z}$.

For this, we want to show that the only elements in $\mathfrak{a}$ are of this form. Suppose $m + n\sqrt{d} \in \mathfrak{a}$.

Choose $s \in \mathbb{Z}$ so that the coefficient of $\sqrt{d}$ of $(m + n\sqrt{d}) - s(b + c\sqrt{d})$ satisfies $0 \leq n - sc < c$; by our choice of $b + c\sqrt{d}$, we must have $n - sc = 0$, so $n = sc$, and so $(m + n\sqrt{d}) - s(b + c\sqrt{d}) = m - sb$, a non-negative integer, still in $\mathfrak{a}$. We can now subtract a multiple of a so that $0 \leq (m - sb) - ta < a$; by minimality of a, we have $(m - sb) - ta = 0$. Combining these gives

$$(m + n\sqrt{d}) - s(b + c\sqrt{d}) - ta = 0,$$

so that $m + n\sqrt{d} \in a\mathbb{Z} + (b + c\sqrt{d})\mathbb{Z}$.

We next want to see that $c \mid a$. If not, we have $a \in \mathfrak{a}$, and so $a\sqrt{d} \in \mathfrak{a}$ by the multiplicative property of ideals. We also have $b + c\sqrt{d} \in \mathfrak{a}$, and if $c \nmid a$, then we can subtract some multiple of $b + c\sqrt{d}$ from $a\sqrt{d}$ to get

$$a\sqrt{d} - t(b + c\sqrt{d}) = -b + (a - tc)\sqrt{d} \in \mathfrak{a},$$

and we can choose t so that $0 < a - tc < c$, contradicting the minimality of c. So $c \mid a$.

Finally, we claim that $c \mid b$. We argue in a similar way to the last claim. So we suppose that $c \nmid b$. As $b + c\sqrt{d} \in \mathfrak{a}$, we also have $(b + c\sqrt{d})\sqrt{d} = b\sqrt{d} + dc \in \mathfrak{a}$. Then, as $c \nmid b$, we can subtract some multiple of $b + c\sqrt{d}$ to get

$$b\sqrt{d} + dc - t(b + c\sqrt{d}) = (dc - tb) + \sqrt{d}(b - tc) \in \mathfrak{a}$$

and if $c \nmid b$, we can choose t so that $0 < b - tc < c$, again contradicting the minimality of c. So $c \mid b$ in this case.

This completes the proof of the lemma. $\qquad\square$

Let's remark that when we write $\sqrt{d}$, we will choose the square root of d lying in the upper-half complex plane; then the pair $(a, b + c\sqrt{d})$ of the lemma is ordered.

Lemma 6.23 *Suppose that the ideal* $\mathfrak{a}$ *of* $\mathbb{Z}_K$ *is written*

$$\mathfrak{a} = a\mathbb{Z} + (b + c\sqrt{d})\mathbb{Z},$$

as in Lemma 6.22. *Then* $N_{K/\mathbb{Q}}(\mathfrak{a}) = ac$.

Proof $N_{K/\mathbb{Q}}(\mathfrak{a})$ is the index of $\mathfrak{a}$ in $\mathbb{Z}[\sqrt{d}]$. But it is clear that coset representatives for the quotient $\mathbb{Z}[\sqrt{d}]/\mathfrak{a}$ are given by $\{x + y\sqrt{d} \mid 0 \leq x < a, 0 \leq y < c\}$ and that this set has cardinality ac. $\qquad\square$

We will also need a partial converse:

Lemma 6.24 *Let a, b and c be in $\mathbb{Z}$. Then the $\mathbb{Z}$-module $\mathfrak{a} = a\mathbb{Z} + (b + c\sqrt{d})\mathbb{Z}$ is an ideal in $\mathbb{Z}_K$ if and only if $c|a$, $c|b$ and $ac|c^2d - b^2$.*

Proof The difference between a $\mathbb{Z}$-module and a $\mathbb{Z}_K$-ideal is the following. To be a $\mathbb{Z}$-module, we need to be able to multiply members of the set by elements of $\mathbb{Z}$ and remain in the set. To be an ideal of $\mathbb{Z}_K$, however, we need to be able to multiply members of the set by elements of $\mathbb{Z}_K$ and remain in the set.

The extra condition we need follows from the requirement that if $\alpha \in \mathfrak{a}$, then $\alpha\sqrt{d} \in \mathfrak{a}$.

Let $\alpha = ax + (b + c\sqrt{d})y = (ax + by) + c\sqrt{d}y$. Then

$$\alpha\sqrt{d} = cdy + (ax + by)\sqrt{d}.$$

For all choices of $x, y \in \mathbb{Z}$, we need that this is in $\mathfrak{a}$. That is, for all $x, y \in \mathbb{Z}$, we need $\alpha\sqrt{d} = as + (b + c\sqrt{d})t$ for some $s, t \in \mathbb{Z}$.

Comparing coefficients of 1 and $\sqrt{d}$, we need that the equations

$$as + bt = cdy$$
$$ct = ax + by$$

have solutions with $s, t \in \mathbb{Z}$, for all $x, y \in \mathbb{Z}$. We can read off the value of t from the second equation:

$$t = \frac{ax + by}{c},$$

and $t \in \mathbb{Z}$ for all $x, y \in \mathbb{Z}$ if $c|a$ and $c|b$. Conversely, if $t \in \mathbb{Z}$, for all $x, y \in \mathbb{Z}$, we see that $c|a$ (choose $x = 1$, $y = 0$) and $c|b$ (choose $x = 0$, $y = 1$). So the condition that $t \in \mathbb{Z}$ is equivalent to $c|a$ and $c|b$.

Having solved for t, we can read off

$$s = \frac{cdy - bt}{a} = \frac{cdy - b(\frac{ax+by}{c})}{a} = \frac{-abx + (c^2d - b^2)}{ac},$$

and this is an integer for all $x, y \in \mathbb{Z}$ if and only if $ac|ab$ (which follows if $c|b$) and $ac|c^2d - b^2$.

The result follows.   □

Now we want to associate a quadratic form to each ideal.

Proposition 6.25 *Let $\mathfrak{a} = \mathbb{Z}a + \mathbb{Z}(b + c\sqrt{d})$ be an ideal of $\mathbb{Z}_K$. Then*

$$\frac{N_{K/\mathbb{Q}}(ax + (b + c\sqrt{d})y)}{N_{K/\mathbb{Q}}(\mathfrak{a})}$$

is a quadratic form with integer coefficients of discriminant D.

Proof Notice that

$$N_{K/\mathbb{Q}}(ax + by + c\sqrt{d}y) = (ax + by)^2 - dc^2y^2$$
$$= a^2x^2 + 2ab.xy + (b^2 - c^2d)y^2$$

giving the quadratic form $(a^2, 2ab, b^2 - c^2d)$ of discriminant $4a^2c^2d$. Each of a^2, $2ab$ and $b^2 - c^2d$ is divisible by ac by Lemma 6.24, and so the quadratic form can be written

$$ac\left(\frac{a}{c}x^2 + 2\frac{b}{c}xy + \left(\frac{b^2 - c^2d}{ac}\right)y^2\right) = N_{K/\mathbb{Q}}(\mathfrak{a})\Phi(\mathfrak{a}).$$

Now

$$\frac{a}{c}x^2 + 2\frac{b}{c}xy + \left(\frac{b^2 - c^2d}{ac}\right)y^2$$

has integer coefficients, and its discriminant is indeed $D = 4d$.   □

We also need to see that this is a positive definite form. We verify this in a slightly more general situation. As above, let $\mathfrak{a}$ be an ideal of $\mathbb{Z}[\sqrt{d}]$, and let (α, β) be an ordered basis for $\mathfrak{a}$. If $(x, y) \neq (0, 0)$, we see $N_{K/\mathbb{Q}}(\alpha x + \beta y) > 0$ is the square of the modulus of a non-zero complex number; thus $N_{K/\mathbb{Q}}(\alpha x + \beta y)$ is positive definite.

Further, in this situation, the same method as above shows that

$$N_{K/\mathbb{Q}}(\alpha x + \beta y) = (\alpha x + \beta y)(\overline{\alpha}x + \overline{\beta}y)$$
$$= (\alpha\overline{\alpha})x^2 + (\alpha\overline{\beta} + \overline{\alpha}\beta)xy + (\beta\overline{\beta})y^2$$
$$= N_{K/\mathbb{Q}}(\alpha)x^2 + T_{K/\mathbb{Q}}(\alpha\overline{\beta})xy + N_{K/\mathbb{Q}}(\beta)y^2,$$

and this is clearly a quadratic form; as $\alpha, \beta \in \mathbb{Z}[\sqrt{d}]$, we see that the coefficients are all in $\mathbb{Z}$ by Corollary 3.17.

Stage 2: The effect of changing ordered generators

The aim here is to see how the quadratic form changes when we change the ordered generating set. So suppose that (α, β) is one ordered generating set for $\mathfrak{a}$; as above, we could choose this to be of the form $(a, b + c\sqrt{d})$.

Suppose that $\gamma, \delta \in \mathbb{Z}[\sqrt{d}]$ is another basis for $\mathfrak{a}$ as a free abelian group, and (γ, δ) is ordered. Then

$$\begin{pmatrix} \alpha \\ \beta \end{pmatrix} = M \begin{pmatrix} \gamma \\ \delta \end{pmatrix}$$

for some matrix $M = \begin{pmatrix} p & r \\ q & s \end{pmatrix}$ with entries in $\mathbb{Z}$ of determinant ± 1. We need a little lemma:

Lemma 6.26 *Suppose that z is in the upper-half complex plane, and $M = \begin{pmatrix} p & r \\ q & s \end{pmatrix} \in$ GL$_2(\mathbb{Z})$. Then $\frac{q+sz}{p+rz}$ is in the upper-half complex plane if and only if $M \in$ SL$_2(\mathbb{Z})$.*

Proof An easy calculation gives the imaginary part of $\frac{q+sz}{p+rz}$:

$$\frac{q+sz}{p+rz} = \frac{(q+sz)(p+r\bar{z})}{(p+rz)(p+r\bar{z})} = \frac{(pq+rsz\bar{z})+(psz+qr\bar{z})}{|p+rz|^2},$$

and so the imaginary part is

$$\frac{\text{im}(z).(ps-qr)}{|p+rz|^2},$$

which is positive if and only if $ps - qr = \det M > 0$. $\qquad\qquad\qquad\square$

Notice that

$$\frac{\beta}{\alpha} = \frac{q\gamma + s\delta}{p\gamma + r\delta} = \frac{q + s(\delta/\gamma)}{p + r(\delta/\gamma)}.$$

As (α, β) and (γ, δ) are both ordered, both β/α and δ/γ lie in the upper-half complex plane, and we conclude from the lemma that $\det M = 1$, so $M \in$ SL$_2(\mathbb{Z})$.

We get a quadratic form from the ordered basis (α, β), given by

$$f_{\alpha,\beta}(x, y) = \frac{N_{K/\mathbb{Q}}(\alpha x + \beta y)}{N_{K/\mathbb{Q}}(\mathfrak{a})};$$

we have already seen that this is integral and positive definite in the particular case $(\alpha, \beta) = (a, b + c\sqrt{d})$ above.

If (γ, δ) is another ordered basis as above, we get a positive definite quadratic form

$$f_{\gamma,\delta}(x, y) = \frac{N_{K/\mathbb{Q}}(\gamma x + \delta y)}{N_{K/\mathbb{Q}}(\mathfrak{a})}$$

in the same way. We have

$$N_{K/\mathbb{Q}}(\mathfrak{a}).f_{\alpha,\beta}(x,y) = (\alpha x + \beta y)(\overline{\alpha}x + \overline{\beta}y)$$

$$= (x\ y)\begin{pmatrix}\alpha\\\beta\end{pmatrix}(\overline{\alpha}\ \overline{\beta})\begin{pmatrix}x\\y\end{pmatrix}$$

$$= (x\ y)\begin{pmatrix}p&r\\q&s\end{pmatrix}\begin{pmatrix}\gamma\\\delta\end{pmatrix}(\overline{\gamma}\ \overline{\delta})\begin{pmatrix}p&r\\q&s\end{pmatrix}^{T}\begin{pmatrix}x\\y\end{pmatrix}$$

$$= (px+qy\ \ rx+sy)\begin{pmatrix}\gamma\\\delta\end{pmatrix}(\overline{\gamma}\ \overline{\delta})\begin{pmatrix}px+qy\\rx+sy\end{pmatrix}$$

$$= N_{K/\mathbb{Q}}(\mathfrak{a}).f_{\gamma,\delta}(px+qy,rx+sy),$$

and since det $M = 1$, the two quadratic forms are properly equivalent, as required. Since $f_{\alpha,\beta}(x,y)$ is integral and positive definite of discriminant D in the particular case $(\alpha,\beta) = (a, b + c\sqrt{d})$ as above, we conclude that $f_{\gamma,\delta}(x,y)$ is integral and positive definite of discriminant D for *any* ordered basis for $\mathfrak{a}$.

If $\mathfrak{a}$ is an ideal of $\mathbb{Z}_K$, we know that we can write it in the form $\mathbb{Z}a + \mathbb{Z}(b + c\sqrt{d})$ and that its norm is ac. Write

$$\Phi(\mathfrak{a}) = \left[\frac{N_{K/\mathbb{Q}}(ax + (b + c\sqrt{d})y)}{N_{K/\mathbb{Q}}(\mathfrak{a})}\right],$$

sending $\mathfrak{a}$ to a proper equivalence class of quadratic forms. (Note that the square brackets indicate that the image is the proper equivalence class of forms). The discussion above shows that we could use any ordered basis for $\mathfrak{a}$, and still get the same proper equivalence class, so Φ really does only depend on $\mathfrak{a}$, not on the basis.

So Φ maps an ideal $\mathfrak{a}$ to the proper equivalence class of quadratic forms $[f_{\alpha,\beta}(x,y)]$ of discriminant D, where $\mathfrak{a}$ has ordered basis (α,β).

Stage 3: From ideal classes to proper equivalence classes of quadratic forms...

Next we need to see that ideals in the same class have the same image under Φ.

Proposition 6.27 *If* $\mathfrak{a}$ *and* $\mathfrak{b}$ *belong to the same ideal class, then* $\Phi(\mathfrak{a})$ *and* $\Phi(\mathfrak{b})$ *are properly equivalent.*

Proof Suppose that $\mathfrak{a}$ and $\mathfrak{b}$ are two ideals in the same ideal class. Then there exists $\theta \in K$ such that $\mathfrak{b} = \theta\mathfrak{a}$; if we write $\theta = \gamma/\delta$ for $\gamma, \delta \in \mathbb{Z}_K = \mathbb{Z}[\sqrt{d}]$, then this is equivalent to $\delta\mathfrak{b} = \gamma\mathfrak{a}$.

Suppose that $\mathfrak{a}$ can be written $\mathbb{Z}\alpha + \mathbb{Z}\beta$. Then $\gamma\mathfrak{a}$ can be written $\mathbb{Z}(\gamma\alpha) + \mathbb{Z}(\gamma\beta)$; by the multiplicativity of the norm,

$$N_{K/\mathbb{Q}}(\alpha\gamma x + \beta\gamma y) = N_{K/\mathbb{Q}}(\gamma)N_{K/\mathbb{Q}}(\alpha x + \beta y).$$

Further, the multiplicativity of the norm of ideals (Theorem 5.37) gives

$$N_{K/\mathbb{Q}}(\langle\gamma\rangle\mathfrak{a}) = N_{K/\mathbb{Q}}(\langle\gamma\rangle)N_{K/\mathbb{Q}}(\mathfrak{a}) = |N_{K/\mathbb{Q}}(\gamma)|N_{K/\mathbb{Q}}(\mathfrak{a}).$$

Since K is imaginary quadratic, $N_{K/\mathbb{Q}}(\gamma) > 0$. Consequently we have an equality

$$f_{\gamma\alpha,\gamma\beta}(x, y) = f_{\alpha,\beta}(x, y),$$

and so $\Phi(\gamma\mathfrak{a}) = \Phi(\mathfrak{a})$.

Similarly $\Phi(\delta\mathfrak{b}) = \Phi(\mathfrak{b})$. But γ and δ were chosen so that $\delta\mathfrak{b} = \gamma\mathfrak{a}$; it follows that $\Phi(\mathfrak{a}) = \Phi(\mathfrak{b})$. $\square$

So Φ can be viewed as a map from the set of equivalence classes of ideals of $\mathbb{Z}_K$ to the set of proper equivalence classes of positive definite quadratic forms of discriminant D_K.

Stage 4: ...and back again—the main result

Conversely, we can associate an ideal to a quadratic form of discriminant $D = 4d$ by the map

$$\Psi((a, b, c)) = \mathbb{Z}a + \mathbb{Z}\left(\frac{b}{2} + \sqrt{d}\right).$$

(Notice that since $D = b^2 - 4ac$ is even, so is b). We can check that this is an ideal using Lemma 6.24; as $c = 1$, only the final condition requires checking, but this is just $a\mid\frac{b^2}{4} - d$. As $D = b^2 - 4ac = 4d$, this is immediate.

Proposition 6.28 *If (a, b, c) and (a', b', c') are properly equivalent, then the ideals $\Psi((a, b, c))$ and $\Psi((a', b', c'))$ lie in the same ideal class.*

Proof Proper equivalences are built up from the basic equivalences

$$(a, b, c) \sim (a, b \pm 2a, c \pm b + a), \qquad (a, b, c) \sim (c, -b, a).$$

We have

- $\Psi((a, b \pm 2a, c \pm b + a)) = \mathbb{Z}a + \mathbb{Z}\left(\frac{b\pm 2a}{2} + \sqrt{d}\right) = \mathbb{Z}a + \mathbb{Z}\left(\frac{b}{2} + \sqrt{d}\right) = \Psi((a, b, c))$.
- $\Psi((c, -b, a)) = \mathbb{Z}c + \mathbb{Z}\left(-\frac{b}{2} + \sqrt{d}\right)$. We want to see that this is an ideal which is in the same ideal class as $\Psi((a, b, c))$. But

$$\left(\frac{b + 2\sqrt{d}}{2c}\right)\Psi((c, -b, a)) = \left(\frac{b + 2\sqrt{d}}{2c}\right)\left[\mathbb{Z}c + \mathbb{Z}\left(-\frac{b}{2} + \sqrt{d}\right)\right]$$

$$= \left[\mathbb{Z}\left(\frac{b + 2\sqrt{d}}{2}\right) + \mathbb{Z}\left(\frac{d - \frac{b^2}{4}}{c}\right)\right]$$

$$= \mathbb{Z}\left(\frac{b}{2} + \sqrt{d}\right) + \mathbb{Z}(-a)$$

$$= \Psi((a, b, c))$$

where the second equality is just scaling the generators by $\frac{b+2\sqrt{d}}{2c} \in \mathbb{Z}[\sqrt{d}]$, and where we use $b^2 - 4ac = 4d$ and $\mathbb{Z}a = \mathbb{Z}(-a)$. Thus the ideal $\Psi((c, -b, a))$ is just a multiple of the ideal $\Psi((a, b, c))$ by some constant, and they therefore lie in the same ideal class. $\qquad\square$

Thus Ψ gives a map from proper equivalence classes of quadratic forms of discriminant $D_K = 4d$ to ideal classes in $\mathbb{Z}_K = \mathbb{Z}[\sqrt{d}]$.

Theorem 6.29 *The maps Φ and Ψ give inverse bijections between the set of proper equivalence classes of quadratic forms of discriminant $4d$ and the set of ideal classes in $\mathbb{Z}[\sqrt{d}]$.*

Proof We need to prove that the maps are two-sided inverses, i.e., that if (a, b, c) is a quadratic form of discriminant $4d$, then $\Phi(\Psi((a, b, c)))$ is a quadratic form properly equivalent to (a, b, c), and that if $\mathfrak{a}$ is an ideal in $\mathbb{Z}[\sqrt{d}]$, then $\Psi(\Phi(\mathfrak{a}))$ is an ideal which is in the same ideal class as $\mathfrak{a}$.

So suppose that (a, b, c) is a quadratic form of discriminant $4d$. Then $\Psi((a, b, c)) = \mathbb{Z}a + \mathbb{Z}\left(\frac{b}{2} + \sqrt{d}\right)$ is an ideal of norm a and

$$\frac{N_{K/\mathbb{Q}}\left(ax + \left(\frac{b}{2} + \sqrt{d}\right)y\right)}{a} = \frac{1}{a}\left(a^2 x^2 + abxy + \left(\frac{b^2}{4} - d\right)y^2\right)$$
$$= ax^2 + bxy + \left(\frac{b^2 - 4d}{4a}\right)y^2$$
$$= ax^2 + bxy + cy^2,$$

the last equality following as $4d = b^2 - 4ac$. We see that $\Phi(\Psi((a, b, c)))$ is not only properly equivalent to (a, b, c), but is actually equal to it.

Finally, let $\mathfrak{a}$ denote an ideal in $\mathbb{Z}[\sqrt{d}]$, and write it $\mathbb{Z}a + \mathbb{Z}(b + c\sqrt{d})$ with $c|a$ and $c|b$, so that $N_{K/\mathbb{Q}}(\mathfrak{a}) = ac$. Then

$$\Phi(\mathfrak{a}) = \frac{N_{K/\mathbb{Q}}(ax + (b + c\sqrt{d})y)}{N_{K/\mathbb{Q}}(\mathfrak{a})} = \frac{a}{c}x^2 + 2\frac{b}{c}xy + \frac{b^2 - c^2 d}{ac}y^2,$$

and

$$\Psi(\Phi(\mathfrak{a})) = \mathbb{Z}\frac{a}{c} + \mathbb{Z}\left(\frac{b}{c} + \sqrt{d}\right) = \frac{1}{c}\left(\mathbb{Z}a + \mathbb{Z}(b + c\sqrt{d})\right),$$

and therefore in the same ideal class as $\mathfrak{a}$, as required. $\qquad\square$

6.5.2 $d \equiv 1$ (mod 4)

The arguments for the case where $d \equiv 1$ (mod 4) are identical to the case where $d \equiv 2, 3$ (mod 4), but one needs to make minor changes to almost all the details, arising from the differences in the rings of integers. We will leave the verification of these details as an exercise for the reader.

Let us fix $d \equiv 1$ (mod 4), and write $D = d$ for the discriminant of $K = \mathbb{Q}(\sqrt{d})$.

At Stage 1, there are several adjustments to make to the structure theory of ideals.

The statement of Lemma 6.22 will be the same, except to replace $\sqrt{d}$ by $\rho_d = \frac{1+\sqrt{d}}{2}$:

Lemma 6.30 *Let $\mathfrak{a}$ be an ideal in the ring of integers $\mathbb{Z}_K$. Then there are integers $a, b, c \in \mathbb{Z}$ with $c \mid a$ and $c \mid b$ such that*

$$\mathfrak{a} = a\mathbb{Z} + (b + c\rho_d)\mathbb{Z}.$$

The proof is very similar to that of Lemma 6.22.

The norm of the ideal $\mathfrak{a} = a\mathbb{Z} + (b + c\rho_d)$ is ac, just as in Lemma 6.23.

The condition (see Lemma 6.24) that such a set is an ideal needs some amendment (although the proof is essentially the same):

Lemma 6.31 *Suppose that $d \equiv 1 (\mathrm{mod}\ 4)$.*

Let a, b and c be in $\mathbb{Z}$. Then the $\mathbb{Z}$-module $\mathfrak{a} = a\mathbb{Z} + (b + c\rho_d)\mathbb{Z}$ is an ideal in $\mathbb{Z}_K$ if and only if $c \mid a$, $c \mid b$ and $ac \mid c^2(\frac{d-1}{4}) - b^2 - bc$.

Given an ideal $\mathfrak{a} = \mathbb{Z}a + \mathbb{Z}(b + c\rho_d)$, one easily sees that

$$\Phi(\mathfrak{a}) = \frac{N_{K/\mathbb{Q}}(ax + (b + c\rho_d)y)}{N_{K/\mathbb{Q}}(\mathfrak{a})}$$

is a quadratic form with integer coefficients of discriminant $D = d$. Indeed,

$$N_{K/\mathbb{Q}}(ax + by + c\rho_d y)$$
$$= (ax + by + \tfrac{c}{2}y)^2 - \tfrac{c^2}{4}dy^2$$
$$= a^2 x^2 + (2ab + ac)xy + \left(b^2 + bc + c^2\left(\frac{1-d}{4}\right)\right)y^2$$

giving the quadratic form $(a^2, 2ab + ac, b^2 + bc + (1 - d)c^2/4)$ of discriminant $a^2 c^2 d$. Again, extracting the common factor of ac from each coefficient gives

$$\Phi(\mathfrak{a}) = \frac{a}{c}x^2 + \left(\frac{2b}{c} + 1\right)xy + \left(\frac{b^2 + bc + c^2\left(\frac{1-d}{4}\right)}{ac}\right)y^2,$$

which has integer coefficients by the conditions just noted, and is easily computed to have discriminant $D = d$.

The rest of this stage is unchanged, as this just depends on the definition of the quadratic form as associated to an ordered pair of generators for an ideal. Stage 2 is unchanged, as this is essentially just a result in linear algebra; Stage 3 is also unchanged. In Stage 4, the inverse map Ψ is defined by

$$\Psi((a,b,c)) = \mathbb{Z}a + \mathbb{Z}\left(\frac{b-1}{2} + \rho_d\right) = \mathbb{Z}a + \mathbb{Z}\left(\frac{b+\sqrt{d}}{2}\right)$$

(recall that b is odd, as $d = b^2 - 4ac \equiv 1 \pmod 4$). The proof of Proposition 6.28 is unchanged, except that with the amended definition of Ψ, we see that

$$\left(\frac{b+\sqrt{d}}{2c}\right)\Psi((c,-b,a)) = \Psi((a,b,c)).$$

6.6 Counting Quadratic Forms

We know:

- There is a bijection from the class group in an imaginary quadratic number field to the collection of positive definite binary quadratic forms with the appropriate discriminant;
- Every positive definite binary quadratic form is equivalent to a unique reduced form with the same discriminant.

This means that the size of the class group of $K = \mathbb{Q}(\sqrt{d})$ is the same as the number of reduced quadratic forms of discriminant D_K, and to calculate the class number of $\mathbb{Q}(\sqrt{d})$, it suffices to count the number of reduced quadratic forms of discriminant $D = D_K$.

This, however, turns out to be rather straightforward.

Let's first note that if (a,b,c) is reduced, then $0 \le |b| \le a \le c$, so certainly $0 \le b^2 \le ac$. Then $-4ac \le b^2 - 4ac \le ac - 4ac = -3ac$, and so

$$-4ac \le D \le -3ac.$$

This gives a finite range for ac:

$$\frac{-D}{4} \le ac \le \frac{-D}{3}.$$

There are finitely many possibilities for a, since $a^2 \le ac$ (as $a \le c$), and, in particular,

$$a^2 \le ac \le \frac{-D}{3}.$$

So a is bounded; as $|b| \leq a$, so is b, and for each choice of a and b, there is at most one value of c with $b^2 - 4ac = D$.

We have therefore proven the following result:

Theorem 6.32 *There are only finitely many reduced quadratic forms of discriminant D.*

Since we have a bijection between this set and the class group, we conclude:

Theorem 6.33 *The class group of an imaginary quadratic field is finite.*

We will give another proof of this theorem for more general number fields later. However, the relation to binary quadratic forms is very useful; it gives us a simple way to compute the class number.

Let's give an example.

Example 6.34 Suppose we want to compute the class number of $\mathbb{Q}(\sqrt{-13})$. We note that as $-13 \equiv 3 \pmod 4$, we have $D = -52$, and the class group of $\mathbb{Q}(\sqrt{-13})$ is in bijection with the collection of reduced quadratic forms of discriminant -52.

We are looking for triples (a, b, c) with $b^2 - 4ac = -52$, and satisfying the reduction conditions. Above, we noted that we must have $\frac{52}{4} \leq ac \leq \frac{52}{3}$, so that $13 \leq ac \leq 17$. We try each possibility.

If $ac = 13$, then $b = 0$ as we need $b^2 - 4ac = -52$. We have $ac = 13$, and $0 \leq a \leq c$, so the only possibility is $a = 1, c = 13$, and we find the triple $(1, 0, 13)$.

If $ac = 14$, then $b^2 = 4$ to get $b^2 - 4ac = -52$. So $b = \pm 2$ and $ac = 14$, with $0 < a \leq c$, so the possibilities for (a, c) are $(a, c) = (1, 14)$ and $(2, 7)$. This gives 4 possible triples:

- $(a, b, c) = (1, 2, 14)$; not reduced, as $|b| > a$ (applying $(a, b, c) \mapsto (a, b - 2a, c - b + a)$ gives $(1, 2, 14) \mapsto (1, 0, 13)$).
- $(a, b, c) = (1, -2, 14)$; not reduced, as $|b| > a$ (applying $(a, b, c) \mapsto (a, b + 2a, c + b + a)$ gives $(1, -2, 14) \mapsto (1, 0, 13)$).
- $(a, b, c) = (2, 2, 7)$, reduced.
- $(a, b, c) = (2, -2, 7)$; not reduced, as $b = -a$ (applying $(a, b, c) \mapsto (a, b + 2a, c + b + a)$ gives $(2, -2, 7) \mapsto (2, 2, 7)$).

If $ac = 15$, then b^2 must be 8, but this is not a square.

If $ac = 16$, then b^2 must be 12, but this is not a square.

If $ac = 17$, then b^2 must be 16. So $b = \pm 4$—but there are no solutions to $ac = 17$ which satisfy $4 = |b| \leq a \leq c$.

We conclude that there are two reduced forms of discriminant -52, namely $(1, 0, 13)$ and $(2, 2, 7)$. The class number of $\mathbb{Q}(\sqrt{-13})$ is therefore 2.

Exercise 6.10 Show that the class number of $\mathbb{Q}(\sqrt{-14})$ is 4. Try some more examples for yourself.

Example 6.35 Now let us compute the class number of $\mathbb{Q}(\sqrt{-19})$.

This time, $-19 \equiv 1 \pmod 4$, so $D = -19$. We need to count the number of reduced forms with discriminant -19.

We can do this as above; we see that $\frac{19}{4} \leq ac \leq \frac{19}{3}$, so $ac = 5$ or $ac = 6$. If $ac = 5$, we need $b^2 = 1$ so that $b^2 - 4ac = -19$; we get $(1, 1, 5)$ and $(1, -1, 5)$ as the only possibilities. The first is reduced, but the second is not—it is properly equivalent to $(1, 1, 5)$ by applying $(a, b, c) \mapsto (a, b + 2a, c + b + a)$. If $ac = 6$, we need $b^2 = 5$, but this is not a square, so there are no reduced forms with $ac = 6$.

We conclude that there is 1 reduced form of discriminant -19, namely $(1, 1, 5)$. Because of the bijection with the class group, we see that the class number of $\mathbb{Q}(\sqrt{-19})$ is 1.

This last example is a very interesting one. We have found a quadratic field with class number 1—so it has unique factorisation! It was not on our list of fields which were Euclidean, so we have found an example of a field with unique factorisation which is not Euclidean.

Exercise 6.11 For each of $\mathbb{Q}(\sqrt{-43})$, $\mathbb{Q}(\sqrt{-67})$ and $\mathbb{Q}(\sqrt{-163})$, use the above method to show that the class number is 1.

It is a famous observation of Euler that the polynomial $X^2 + X + 41$ is prime for all $X = 0, \ldots, 39$. What is the discriminant of this quadratic? Explain why Euler's observation implies that $\mathbb{Q}(\sqrt{-163})$ has class number 1.

Verify similar results for $X^2 + X + 11$ and $X^2 + X + 17$.

We have now seen the following examples where the class group of an imaginary quadratic field is trivial, so that the field has unique factorisation: $\mathbb{Q}(\sqrt{-1})$, $\mathbb{Q}(\sqrt{-2})$, $\mathbb{Q}(\sqrt{-3})$, $\mathbb{Q}(\sqrt{-7})$, $\mathbb{Q}(\sqrt{-11})$, $\mathbb{Q}(\sqrt{-19})$, $\mathbb{Q}(\sqrt{-43})$, $\mathbb{Q}(\sqrt{-67})$ and $\mathbb{Q}(\sqrt{-163})$.

All these examples were known to Gauss. Gauss also predicted that these were the only such fields, and this was for a long time a central problem in algebraic number theory, known as the "Class Number One Problem". It was not until the 1960s that a proof was given, by Alan Baker and Harold Stark independently; subsequently, it was observed that an earlier, rather obscurely written, attempt by Kurt Heegner (dating from the early 1950s) was also valid. Baker won the Fields Medal (the mathematical equivalent of the Nobel Prize) for the techniques he introduced in his solution of the problem.

The proof of the full theorem is therefore rather beyond an undergraduate textbook. Nevertheless, we can prove easily the following partial result:

Theorem 6.36 *Suppose that $d \equiv 2, 3 \pmod 4$ is a negative squarefree integer. Then $\mathbb{Q}(\sqrt{d})$ has unique factorisation if and only if $d = -1$ or $d = -2$.*

In view of the bijection between class numbers and quadratic forms, we can restate our result in terms of quadratic forms:

Theorem 6.37 *Suppose that $d \equiv 2, 3 \pmod 4$ is negative and squarefree, and write $D = 4d$. The only cases where there is only one reduced quadratic form of discriminant D is where $d = -1$ or $d = -2$.*

Actually, we shall prove a result which applies to more general values of d (note that -4 is divisible by 4, and that -3 and -7 are both congruent to 1 (mod 4)):

Theorem 6.38 *Suppose that d is negative, and write $D = 4d$. The only cases where there is only one reduced quadratic form of discriminant D is where $d = -1, -2, -3, -4, -7$.*

Proof In each case, $(1, 0, -d)$ is one reduced quadratic form of discriminant D. For all values of d except those listed in the theorem, we will simply write down another one, thus proving the theorem.

If $-d$ is not a prime power, then we can write $-d = ac$ with $(a, c) = 1$, and $1 < a < c$ (for example, if $-d = 45$, we can choose $a = 5, c = 9$). Then $(a, 0, c)$ is a reduced form of discriminant D and is different from $(1, 0, -d)$.

If $-d = 2^r$, then if $r \geq 4$, we can use $(4, 4, 2^{r-2} + 1)$, and this is easily checked to be reduced, as $4 < 2^{r-2} + 1$. For $r = 3$, when $d = -8$, we also have $(3, 2, 3)$. This just leaves $-d = 1, 2, 4$, which are all in the statement of the theorem.

If $-d = p^r$, where p is an odd prime and $r \geq 1$, then consider $p^r + 1$, which will be even. If $p^r + 1 = ac$ with $2 \leq a < c$ and $(a, c) = 1$, we can use $(a, 2, c)$. This can be done whenever $p^r + 1$ is not a power of 2. (For example, if $-d = 27$, then use $28 = 4 \times 7$, and use $(4, 2, 7)$.) If $p^r + 1 = 2^s$ with $s \geq 6$, use $(8, 6, 2^{s-3} + 1)$, again reduced. If $p^r + 1 = 32$ (so that $p = 31, r = 1$), use $(5, 4, 7)$ or $(5, -4, 7)$. The equation $p^r + 1 = 16$ has no solutions, since 15 is not a prime power. The possibilities $p^r + 1 = 8$ (so $p = 7, r = 1$ giving $d = -7$), $p^r + 1 = 4$ (giving $d = -3$), $p^r + 1 = 2$ (so $d = -1$) are given in the statement of the theorem.   □

Let us end this chapter by remarking that there is even more to the bijection between the class group and the collection of reduced quadratic forms of the appropriate discriminant. After all, the class group has a group structure, as it is the quotient of the abelian group of all fractional ideals (see Theorem 5.30) by the normal subgroup of all principal fractional ideals. The bijection from the class group to the set of all reduced quadratic forms of the appropriate discriminant certainly gives us some group structure on the collection of all reduced quadratic forms of the appropriate discriminant, but we might wonder whether there is a more natural description of this group structure. This is really beyond the scope of this book, but we remark that the group structure can indeed be seen with quadratic forms. Indeed, the identity

$$(a^2 + b^2)(c^2 + d^2) = (ac - bd)^2 + (ad + bc)^2$$

from Lemma 1.18 is an example where we multiply two copies of the unique quadratic form $x^2 + y^2$ of discriminant -4. We expect the answer to be the unique quadratic form of discriminant -4, and indeed this appears on the right-hand side of the identity.

For a slightly more interesting example of this identity, consider the product of forms of discriminant -20. Here there are two classes, the form $x^2 + 5y^2$ corresponding to the principal ideal class, and the form $2x^2 + 2xy + 3y^2$ corresponding to the non-principal ideals. The class group has order 2, and multiplying the identity by the other element will give the other element. This is indeed demonstrated with the quadratic forms:

$$(a^2 + 5b^2)(c^2 + 5d^2) = m^2 + 5n^2,$$

where $m = ac - 5bd, n = ad + bc$;

$$(2a^2 + 2ab + 3b^2)(c^2 + 5d^2) = 2m^2 + 2mn + 3n^2,$$

where $m = ac - ad - 3bd, n = 2ad + bc + bd$, and

$$(2a^2 + 2ab + 3b^2)(2c^2 + 2cd + 3d^2) = m^2 + 5n^2,$$

where $m = 2ac + ad + bc - 2bd, n = ad + bc + bd$.

Exercise 6.12 In Exercise 6.10, you should have found 4 reduced quadratic forms f_1, f_2, f_3 and f_4 of discriminant -56. Thus the class group of $\mathbb{Q}(\sqrt{-14})$ has order 4. It must be either cyclic or isomorphic to the Klein 4-group, where each non-identity element has order 2. For each $i = 1, \ldots, 4$, try to write each $f_i(a, b)f_i(c, d)$ as $f_k(m, n)$ for suitable m and n, and hence decide whether the class group is cyclic or not.

If you wish to make the exercise more lengthy still, you could verify your answer by writing each product $f_i(a, b)f_j(c, d)$ as $f_k(m, n)$ for suitable m and n, so confirming the complete group multiplication table.

Chapter 7
Lattices and Geometrical Methods

In this chapter, we will prove two fundamental results in algebraic number theory: the finiteness of the class group, and Dirichlet's Unit Theorem, which gives the structure of the group of units in the rings of integers of number fields.

Surprisingly, both use geometrical techniques, based around the geometry of lattices. Before we consider the special case of lattices arising from number fields, we will begin the chapter with an overview of the main results of the theory of lattices.

7.1 Lattices

Definition 7.1 Let V be an n-dimensional real vector space. A *lattice* in V is a subgroup of the form

$$\Gamma = \mathbb{Z}v_1 + \cdots + \mathbb{Z}v_m,$$

where $\{v_1, \ldots, v_m\}$ is a linearly independent set of vectors in V. The lattice is called *complete* if $m = n$. To Γ (or rather, to its generating set $\{v_1, \ldots, v_m\}$) is associated its *fundamental mesh* or *fundamental region*, Φ_Γ, defined as

$$\Phi_\Gamma = \{\alpha_1 v_1 + \cdots + \alpha_m v_m \mid 0 \le \alpha_i < 1\}.$$

Note that completeness is equivalent to

$$V = \bigcup_{\gamma \in \Gamma} (\Phi_\Gamma + \gamma);$$

the right-hand side is easily seen to be equal to the real vector space spanned by $v_1, \ldots, v_m$. In order words, Γ is complete if every element of V is a translate of an element in the fundamental region by a lattice point.

F. Jarvis, *Algebraic Number Theory*, Springer Undergraduate
Mathematics Series, DOI: 10.1007/978-3-319-07545-7_7,
© Springer International Publishing Switzerland 2014

Definition 7.2 A subset Γ of $\mathbb{R}^n$ is said to be *discrete* if, for any radius $r \geq 0$, Γ contains only finitely many points at a radius at most r from 0.

The next proposition gives a characterisation of discrete subgroups of $\mathbb{R}^n$. The proof is somewhat analytical, and readers may prefer to take it on trust.

Proposition 7.3 *A subgroup $\Gamma \subset V$ is a lattice if and only if it is discrete.*

Proof If Γ is a lattice, choose a basis $v_1, \ldots, v_m$, and consider the vector space V_0 spanned by these vectors. By linear independence of the set, every vector $v \in V_0$ can be expressed uniquely as a linear combination of the basis, and we can define a continuous map $\phi : \mathbb{R}^m \longrightarrow \mathbb{R}^m$ by $\phi(a_1 v_1 + \cdots + a_m v_m) = (a_1, \ldots, a_m)$. Now pick some radius $r \geq 0$, and consider the closed ball B of radius r around 0. Since B is closed and bounded, it is compact, and thus $\phi(B)$ is also compact, and thus is a subset of the closed ball of some radius M, say. If $v = a_1 v_1 + \cdots + a_m v_m \in B$, then we must have $\|\phi(v)\| \leq M$, and so $\|(a_1, \ldots, a_m)\| \leq M$, which implies that $|a_i| \leq M$ for all i. There are thus only finitely many points in Γ with this property, and so Γ is discrete.

Conversely, suppose Γ is discrete. Again let V_0 be the $\mathbb{R}$-span of Γ, of some dimension m. Let $\{u_1, \ldots, u_m\}$ be a basis of V_0 formed of elements in Γ, and let $\Gamma_0 = \mathbb{Z}u_1 + \cdots + \mathbb{Z}u_m \subseteq \Gamma$. Suppose

$$\Gamma = \bigcup_{i \in I}(\Gamma_0 + \gamma_i),$$

a (disjoint) union of cosets of Γ_0 in Γ. Since Γ_0 is complete in V_0, and $\gamma_i \in V_0$, it is the translate of some element $\mu_i \in \Phi_{\Gamma_0}$ by an element of Γ_0. Then $\Gamma_0 + \gamma_i = \Gamma_0 + \mu_i$, and so

$$\Gamma = \bigcup_{i \in I}(\Gamma_0 + \mu_i).$$

However, $\mu_i \in \Gamma$ as well as $\mu_i \in \Phi_{\Gamma_0}$. As Γ is discrete, $\Gamma \cap \Phi_{\Gamma_0}$ is finite, since Φ_{Γ_0} certainly lies inside some closed ball. It follows that I is finite. Thus, if q denotes the index of Γ_0 in Γ, we have $q\Gamma \subset \Gamma_0$. Then

$$\Gamma \subset \mathbb{Z}\left(\frac{1}{q}u_1\right) + \cdots + \mathbb{Z}\left(\frac{1}{q}u_m\right).$$

But now we can apply Proposition 4.33; Γ is a subset of a free abelian group of rank m, and so Γ admits a $\mathbb{Z}$-basis, i.e.,

$$\Gamma = \mathbb{Z}v_1 + \cdots + \mathbb{Z}v_r$$

for some $r \leq m$. The set $\{v_1, \ldots, v_r\}$ is linearly independent as the vectors span V_0 (so that in fact $r = m$). $\qquad \square$

To illustrate what can go wrong, think about $\mathbb{Z}.1 + \mathbb{Z}.\sqrt{2}$ inside $\mathbb{R}$; this is a perfectly good subgroup of $\mathbb{R}$ (under addition), but is not a lattice, as the basis elements are linearly dependent—each is a multiple of the other. Nor is it discrete, as we can find $a, b \in \mathbb{Z}$ such that $a + b\sqrt{2}$ is arbitrarily close to 0.

Next we give a criterion for a lattice to be complete.

Proposition 7.4 *A lattice $\Gamma \subset V$ is complete if and only if there exists a bounded $B_V \subset V$ such that*

$$V = \bigcup_{\gamma \in \Gamma} (B_V + \gamma).$$

Proof If Γ is complete, taking $B_V = \Phi_\Gamma$ suffices.

Now we prove the converse. Since B_V is bounded, there exists a constant d such that every point of B_V lies at a distance of at most d from 0. Then the collection of translates $\{B_V + \gamma \mid \gamma \in \Gamma\}$ contains no point which lies at a distance greater than d from some element of Γ. However, if V_0 denotes the span of Γ, and V_0 is not all of V, then it is of strictly smaller dimension, and there exist points in V which lie arbitrarily far from V_0. (Think about the situation where V_0 is a 1-dimensional subset of the 2-dimensional vector space $\mathbb{R}^2$.) $\qquad\square$

Having defined lattices in $\mathbb{R}^n$, we need next to compute the volumes of their fundamental domains. Since $\mathbb{R}^n$ has a natural inner product, we can define lengths, areas, volumes and so on for subsets of the vector space $\mathbb{R}^n$ (not every subset has a volume, but we shall only need to work with nice sets where we do have a well-defined volume). In particular, we can compute the *volume* for a lattice in $\mathbb{R}^n$: if Γ is a lattice, write $\mathrm{vol}(\Gamma)$ for $\mathrm{vol}(\Phi_\Gamma)$.

Proposition 7.5 *Suppose that $\Gamma = \mathbb{Z}v_1 + \cdots + \mathbb{Z}v_n$ is a lattice in $\mathbb{R}^n$. If $v_i = (a_{i1} \ldots a_{in})$, then*

$$\mathrm{vol}(\Gamma) = |\det(a_{ij})|.$$

Proof Write $\{e_1, \ldots, e_n\}$ for the standard basis of $\mathbb{R}^n$, so that $v_i = \sum_{j=1}^n a_{ij} e_j$. Writing $x_1, \ldots, x_n$ for the co-ordinates of a general point of $\mathbb{R}^n$ with respect to the standard basis, we have

$$\mathrm{vol}(\Gamma) = \int_{\Phi_\Gamma} 1 \, dx_1 \, dx_2 \ldots dx_n.$$

We want to change basis from the standard basis to $\{v_1, \ldots, v_n\}$, as Φ_Γ is given very simply in co-ordinates with respect to this basis: Φ_Γ is the set of points $\alpha_1 v_1 + \cdots + \alpha_n v_n$ whose co-ordinates satisfy $0 \le \alpha_i < 1$.

As $v_i = \sum_{i=1}^n a_{ij} e_j$, the change of basis matrix from $\{v_1, \ldots, v_n\}$ to $\{e_1, \ldots, e_n\}$ is given by $A = (a_{ij})$; if $x \in \mathbb{R}^n$ is equal to $\sum_{i=1}^n x_i e_i$ and $\sum_{i=1}^n y_i v_i$, then the coefficients are transformed by the matrix A. The co-ordinates of Φ_Γ in the basis $\{v_1, \ldots, v_n\}$ are $0 \le y_i < 1$ by definition. The formula for changing the variable in multiple integrals involves the Jacobian of the transformation, which is just $|\det A|$. More precisely

$$
\begin{aligned}
\mathrm{vol}(\Gamma) &= \int_{\Phi_\Gamma} 1 \, dx_1 \, dx_2 \dots dx_n \\
&= \int_{\Phi_\Gamma} |\det A| \, dy_1 \, dy_2 \dots dy_n \\
&= |\det A| \int_0^1 \cdots \int_0^1 1 \, dy_1 \, dy_2 \dots dy_n \\
&= |\det A|,
\end{aligned}
$$

as claimed.   □

The main result we will need is Minkowski's theorem on lattice points. For this, we take a vector space V, which we identify with $\mathbb{R}^n$ so that we can define a volume for subsets of V.

Definition 7.6 A region $X \subset V$ is *centrally symmetric* if $x \in X$ implies $-x \in X$.

Definition 7.7 A region $X \subset V$ is *convex* if given $x, y \in X$, and $t \in [0, 1]$, then $tx + (1 - t)y \in X$. (That is, if x and y lie in X, so does the line joining them.)

Minkowski's theorem guarantees the existence of non-zero lattice points inside certain subsets of V.

Theorem 7.8 (Minkowski) *Let Γ be a complete lattice in V. Let X be a centrally symmetric convex subset of V. Suppose*

$$
\mathrm{vol}(X) > 2^n \mathrm{vol}(\Gamma).
$$

Then X contains at least one non-zero lattice point of V.

Proof It suffices to prove that there exist distinct $\gamma_1, \gamma_2 \in \Gamma$ such that

$$
\left(\tfrac{1}{2}X + \gamma_1 \right) \cap \left(\tfrac{1}{2}X + \gamma_2 \right) \neq \emptyset.
$$

For, if so, $\gamma_1 - \gamma_2 \in X \cap \Gamma$; if

$$
\gamma_1 + \tfrac{1}{2}x_1 = \gamma_2 + \tfrac{1}{2}x_2,
$$

then $\gamma_1 - \gamma_2 = \tfrac{1}{2}x_2 + \tfrac{1}{2}(-x_1) \in X$, using the convexity and central symmetricity of X.

If $\{\tfrac{1}{2}X + \gamma\}_{\gamma \in \Gamma}$ are pairwise disjoint, the same holds for the intersections $\{\Phi_\Gamma \cap (\tfrac{1}{2}X + \gamma)\}_{\gamma \in \Gamma}$ with Φ_Γ. In particular, since these sets are all contained in Φ_Γ, we have $\mathrm{vol}(\Gamma) \geq \sum_{\gamma \in \Gamma} \mathrm{vol}(\Phi_\Gamma \cap (\tfrac{1}{2}X + \gamma))$.

But $\Phi_\Gamma \cap (\frac{1}{2}X + \gamma)$ and $(\Phi_\Gamma - \gamma) \cap \frac{1}{2}X$ have the same volume (the second is a translation of the first by $-\gamma$).

The set $\{\Phi_\Gamma - \gamma\}_{\gamma \in \Gamma}$ covers V, so $\{(\Phi_\Gamma - \gamma) \cap \frac{1}{2}X\}_{\gamma \in \Gamma}$ covers $\frac{1}{2}X$. Then

$$\mathrm{vol}(\Gamma) \geq \sum_{\gamma \in \Gamma} \mathrm{vol}\left((\Phi_\Gamma - \gamma) \cap \frac{1}{2}X\right) = \mathrm{vol}\left(\frac{1}{2}X\right) = \frac{1}{2^n}\mathrm{vol}(X),$$

contradicting the hypothesis on $\mathrm{vol}(X)$.

7.2 Geometry of Number Fields

We want to apply geometrical techniques, and Minkowski's Theorem particularly, to number fields. For this, we need to work with lattices in vector spaces; we will begin by defining some of the appropriate vector spaces and lattices.

Let K be a number field. Inside $\mathbb{Z}_K$, we have the units $\mathbb{Z}_K^\times$. Recall that $\epsilon \in \mathbb{Z}_K$ is a unit if and only if $N_{K/\mathbb{Q}}(\epsilon) = \pm 1$ and that non-zero elements $x_1, x_2 \in \mathbb{Z}_K$ are *associates* if $\frac{x_1}{x_2} \in \mathbb{Z}_K^\times$.

Suppose $[K : \mathbb{Q}] = n$. From Proposition 3.7, we recall that there are n embeddings of K into $\mathbb{C}$.

Definition 7.9 If $\sigma : K \hookrightarrow \mathbb{C}$ has $\sigma(K) \subset \mathbb{R}$, then σ is said to be *real*. Otherwise σ is said to be *complex*, and, in this case, its *conjugate*, $\overline{\sigma}$, defined by

$$\overline{\sigma}(k) = \overline{\sigma(k)},$$

is also an embedding.

Thus, if there are r_1 real embeddings and r_2 conjugate pairs of complex embeddings, one has $r_1 + 2r_2 = n$.

We will tend to write ρ for a real embedding, σ and $\overline{\sigma}$ for complex pairs, and τ when discussing an arbitrary embedding. So the real embeddings are $\{\rho_1, \ldots, \rho_{r_1}\}$, and the complex embeddings will be $\{\sigma_1, \overline{\sigma}_1, \ldots, \sigma_{r_2}, \overline{\sigma}_{r_2}\}$. That is, we choose one of each complex pair to be denoted σ_i; then the other is $\overline{\sigma}_i$.

Define a map

$$i : K \hookrightarrow \mathbb{R}^{r_1} \times \mathbb{C}^{r_2}$$
$$\alpha \mapsto (\rho_1(\alpha), \ldots, \rho_{r_1}(\alpha), \sigma_1(\alpha), \ldots, \sigma_{r_2}(\alpha))$$

If addition and multiplication on the right-hand side are defined componentwise, then this i preserves the additive and multiplicative structure of K. Write

$$K_\mathbb{R} = \mathbb{R}^{r_1} \times \mathbb{C}^{r_2},$$

and note that $K_\mathbb{R}$ is an n-dimensional real vector space, as $\mathbb{C} \cong \mathbb{R}^2$ and $r_1 + 2r_2 = n$. The map i embeds K into $K_\mathbb{R}$. (Readers aware of the tensor product will realise that $K_\mathbb{R} = K \otimes_\mathbb{Q} \mathbb{R}$.)

One has the norm map $N_{K/\mathbb{Q}}$ on K, where $N_{K/\mathbb{Q}}(\alpha)$ was defined as the determinant of the map $m_\alpha : x \mapsto \alpha x$ on K (using any basis for K as a $\mathbb{Q}$-vector space), and we can similarly define a map $N : K_\mathbb{R} \longrightarrow \mathbb{R}$ so that if $\alpha \in K_\mathbb{R}$, then $N(\alpha)$ is defined as the determinant of the multiplication map $x \mapsto \alpha x$ on the space $K_\mathbb{R}$. It is easy to see that the map is given explicitly by

$$N : K_\mathbb{R} \longrightarrow \mathbb{R}$$

$$(x_1, \ldots, x_{r_1}, z_1, \ldots, z_{r_2}) \mapsto \prod_{i=1}^{r_1} x_i \cdot \prod_{i=1}^{r_2} |z_i|^2$$

and that the following diagram commutes:

$$
\begin{array}{ccc}
K & \xrightarrow{\ i\ } & K_\mathbb{R} \\
{\scriptstyle N_{K/\mathbb{Q}}}\downarrow & & \downarrow{\scriptstyle N} \\
\mathbb{Q} & \longrightarrow & \mathbb{R}
\end{array}
$$

In other words, given $\alpha \in K$, there is an equality

$$N(i(\alpha)) = N_{K/\mathbb{Q}}(\alpha).$$

In order to apply the theory of lattices to $K_\mathbb{R}$, we need some notion of volume, which we can explicitly compute in order to apply Minkowski's Theorem.

In fact, there are two natural notions of volume one can use; the first one we will give is more natural, but the second is easier to use for making explicit computations.

First, think of $K_\mathbb{R}$ as a subset of $\mathbb{C}^n$ by sending

$$(x_1, \ldots, x_{r_1}, z_1, \ldots, z_{r_2}) \mapsto (x_1, \ldots, x_{r_1}, z_1, \overline{z}_1, \ldots, z_{r_2}, \overline{z}_{r_2}).$$

Write $i_\mathbb{C} : K_\mathbb{R} \hookrightarrow \mathbb{C}^n$ for this embedding. Notice that the map $K \hookrightarrow K_\mathbb{R} \hookrightarrow \mathbb{C}^n$ is then given by

$$x \mapsto (\rho_1(x), \ldots, \rho_{r_1}(x), \sigma_1(x), \overline{\sigma}_1(x), \ldots, \sigma_{r_2}(x), \overline{\sigma}_{r_2}(x)).$$

(Note that this is a ring homomorphism, although we won't require this.)

Then $\mathbb{C}^n$ has a natural inner product; given two elements $z = (z_1, \ldots, z_n)$ and $z' = (z'_1, \ldots, z'_n)$, define

$$(z, z') = \sum_{i=1}^{n} z_i \overline{z}'_i.$$

This gives us a length on $\mathbb{C}^n$, where $\|z\| = (z, z)^{1/2}$. As $K_{\mathbb{R}}$ is a subset, we get a length, distance, etc., on $K_{\mathbb{R}}$. In particular, we can define the volume $\mathrm{vol}(X)$ of a subset X of $K_{\mathbb{R}}$ (not every subset has a volume, but the ones we meet will do).

This looks natural because all the embeddings are used (so no arbitrary choice of one embedding in each pair is made) and because the map preserves the multiplicative structure of $K_{\mathbb{R}}$, amongst others. But it is not quite so easy to use in practice as the second we will give.

For the second method, we can identify $K_{\mathbb{R}}$ with $\mathbb{R}^n$ using the isomorphism

$$i_{\mathbb{R}} : (x_1, \ldots, x_{r_1}, z_1, \ldots, z_{r_2}) \mapsto (x_1, \ldots, x_{r_1}, u_1, v_1, \ldots, u_{r_2}, v_{r_2}),$$

where $z_k = u_k + iv_k$ for $k = 1, \ldots, r_2$. (You should notice that $i_{\mathbb{R}}$ is a linear map, but the multiplicative structure of $K_{\mathbb{R}}$ is not preserved.) The space $\mathbb{R}^n$ has a natural inner product, essentially just the usual vector dot product; given two elements $(a_1, \ldots, a_n)$ and $(b_1, \ldots, b_n)$ in $\mathbb{R}^n$, the inner product is given by $\langle a, b \rangle = \sum_{i=1}^{n} a_i b_i$. This gives a notion of length of a vector $x \in \mathbb{R}^n$:

$$\|x\| = \sqrt{\langle x, x \rangle},$$

the usual (Euclidean) length, and then we get the usual notions of distances, areas, volumes etc. We will write $\mathrm{vol}_{\mathbb{R}}(X)$ for the volume of a subset $X \subset K_{\mathbb{R}}$ using this definition.

We are going to be able to make computations of volumes of lattices in $K_{\mathbb{R}}$ by mapping them to $\mathbb{R}^n$ using $i_{\mathbb{R}}$, and then using Proposition 7.5.

Because we can use Proposition 7.5 to compute volumes of lattices in $\mathbb{R}^n$, it is often easier to make explicit computations in $\mathbb{R}^n$. However, the statements of most of the results look nicer if we use $i_{\mathbb{C}}$ to map $K_{\mathbb{R}}$ into $\mathbb{C}^n$. We start by explaining the relation between the two volumes.

Proposition 7.10 *If Γ is a lattice in $K_{\mathbb{R}}$, then $\mathrm{vol}(\Gamma) = 2^{r_2} \mathrm{vol}_{\mathbb{R}}(\Gamma)$.*

Rather than give a general proof, we illustrate it with a short example; the general case is exactly the same, but the determinants involved are larger.

Example 7.11 Suppose $[K : \mathbb{Q}] = 3$, and we have one real embedding ρ, and one pair of complex embeddings σ and $\overline{\sigma}$. Given 3 elements ω_1, ω_2 and ω_3 in K, the volume of the lattice Γ they generate in $K_{\mathbb{R}}$ is got by taking the embedding $i_{\mathbb{C}} : K_{\mathbb{R}} \hookrightarrow \mathbb{C}^n$ and computing (as in Proposition 7.5):

$$\mathrm{vol}(\Gamma) = \begin{vmatrix} \rho(\omega_1) & \sigma(\omega_1) & \overline{\sigma}(\omega_1) \\ \rho(\omega_2) & \sigma(\omega_2) & \overline{\sigma}(\omega_2) \\ \rho(\omega_3) & \sigma(\omega_3) & \overline{\sigma}(\omega_3) \end{vmatrix}.$$

On the other hand, if we use $i_{\mathbb{R}}$ to regard $K_{\mathbb{R}}$ as a subset of $\mathbb{R}^n$, the relevant determinant is computed (by Proposition 7.5) as

$$\text{vol}_{\mathbb{R}}(\Gamma) = \begin{vmatrix} \rho(\omega_1) & \text{Re}(\sigma(\omega_1)) & \text{Im}(\sigma(\omega_1)) \\ \rho(\omega_2) & \text{Re}(\sigma(\omega_2)) & \text{Im}(\sigma(\omega_2)) \\ \rho(\omega_3) & \text{Re}(\sigma(\omega_3)) & \text{Im}(\sigma(\omega_3)) \end{vmatrix}.$$

Write $\sigma(\omega_j) = u_j + iv_j$; simple column operations give

$$\text{vol}(\Gamma) = \begin{vmatrix} \rho(\omega_1) & u_1 + iv_1 & u_1 - iv_1 \\ \rho(\omega_2) & u_2 + iv_2 & u_2 - iv_2 \\ \rho(\omega_3) & u_3 + iv_3 & u_3 - iv_3 \end{vmatrix} = \begin{vmatrix} \rho(\omega_1) & 2u_1 & u_1 - iv_1 \\ \rho(\omega_2) & 2u_2 & u_2 - iv_2 \\ \rho(\omega_3) & 2u_3 & u_3 - iv_3 \end{vmatrix}$$

$$= 2 \begin{vmatrix} \rho(\omega_1) & u_1 & u_1 - iv_1 \\ \rho(\omega_2) & u_2 & u_2 - iv_2 \\ \rho(\omega_3) & u_3 & u_3 - iv_3 \end{vmatrix} = 2 \begin{vmatrix} \rho(\omega_1) & u_1 & -iv_1 \\ \rho(\omega_2) & u_2 & -iv_2 \\ \rho(\omega_3) & u_3 & -iv_3 \end{vmatrix}$$

$$= |-2i| \begin{vmatrix} \rho(\omega_1) & u_1 & v_1 \\ \rho(\omega_2) & u_2 & v_2 \\ \rho(\omega_3) & u_3 & v_3 \end{vmatrix}$$

$$= 2 \, \text{vol}_{\mathbb{R}}(\Gamma).$$

Exactly the same happens in the general case; every pair of complex conjugate embeddings gives an extra factor of 2 in the volume computation.

Now let's compute the volume of the ring of integers and other ideals.

Proposition 7.12 $\Gamma = i(\mathbb{Z}_K)$ *is a complete lattice in* $K_{\mathbb{R}}$ *and* $\text{vol}(\Gamma) = |D_K|^{\frac{1}{2}}$.

Proof Let

$$\mathbb{Z}_K = \mathbb{Z}\omega_1 + \cdots + \mathbb{Z}\omega_n,$$

so that

$$\Gamma = \mathbb{Z}i(\omega_1) + \cdots + \mathbb{Z}i(\omega_n) \subset K_{\mathbb{R}}.$$

Let M be the matrix $(\tau_i \omega_j)$ as τ_i runs over all embeddings of K into $\mathbb{C}$ (note that this is exactly the same matrix as in Sect. 3.3). By Definition 3.18,

$$D_K = \Delta\{\omega_1, \ldots, \omega_n\} = \det(M)^2,$$

and so $|D_K|^{\frac{1}{2}} = |\det(M)|$.

But the same argument as Proposition 7.5 (the same argument works in $\mathbb{C}^n$ rather than $\mathbb{R}^n$) shows that $\text{vol}(\Gamma) = |\det(\tau_i \omega_j)|$, and so we conclude that $\text{vol}(\Gamma) = |D_K|^{\frac{1}{2}}$.   □

Recall (from Sect. 4.8) that if $\mathfrak{a}$ is an integral ideal of K, then $\mathfrak{a}$ admits a $\mathbb{Z}$-basis,

$$\mathfrak{a} = \mathbb{Z}\alpha_1 + \cdots + \mathbb{Z}\alpha_n,$$

where $n = [K : \mathbb{Q}]$ is as above. Define the *discriminant* of the ideal $\mathfrak{a}$ to be

$$D(\mathfrak{a}) = \Delta\{\alpha_1, \ldots, \alpha_n\} = \det(\tau_i \alpha_j)^2,$$

where τ_i runs over all of the embeddings of K into $\mathbb{C}$. The same argument as in Proposition 3.28 shows that it is independent of the choice of $\mathbb{Z}$-basis. By definition, $D_K = D(\mathbb{Z}_K)$.

The previous result easily generalises to ideals. It could be proven in exactly the same way, but we might as well deduce it from the previous proposition.

Proposition 7.13 *If $\mathfrak{a}$ is a non-zero ideal of $\mathbb{Z}_K$, then $\Gamma = i(\mathfrak{a})$ is a complete lattice in $K_{\mathbb{R}}$. Further, $D(\mathfrak{a}) = N_{K/\mathbb{Q}}(\mathfrak{a})^2 D_K$, and Φ_Γ has volume*

$$\mathrm{vol}(\Gamma) = |D(\mathfrak{a})|^{\frac{1}{2}} = N_{K/\mathbb{Q}}(\mathfrak{a}).|D_K|^{\frac{1}{2}}.$$

Proof By definition of the ideal norm, $\mathbb{Z}_K$ is the (disjoint) union of $N_{K/\mathbb{Q}}(\mathfrak{a})$ cosets of $\mathfrak{a}$. Then $i(\mathfrak{a})$ has volume $\mathrm{vol}(i(\mathbb{Z}_K))/N_{K/\mathbb{Q}}(\mathfrak{a})$. The result now follows from the calculation of the volume of $i(\mathbb{Z}_K)$ in Proposition 7.12. $\qquad\square$

We now apply Minkowski's Theorem to give a result which is the key result for the finiteness of the class number for any number field K.

Proposition 7.14 *Let Γ be a lattice in $K_{\mathbb{R}}$, and let $c_1, \ldots, c_{r_1}, C_1, \ldots, C_{r_2} \in \mathbb{R}_{>0}$ satisfy*

$$c_1 \cdots c_{r_1} (C_1 \cdots C_{r_2})^2 > \left(\frac{2}{\pi}\right)^{r_2} \mathrm{vol}(\Gamma).$$

Then there exists a non-zero $v = (x_1, \ldots, x_{r_1}, z_1, \ldots, z_{r_2}) \in \Gamma$ such that $|x_j| < c_j$ for all $j = 1, \ldots, r_1$, and $|z_k| < C_k$ for all $k = 1, \ldots, r_2$.

Proof Let X denote the set of all elements $(x_1, \ldots, x_{r_1}, u_1 + iv_1, \ldots, u_{r_2} + iv_{r_2})$ in $K_{\mathbb{R}} = \mathbb{R}^{r_1} \times \mathbb{C}^{r_2}$ such that $|x_j| < c_j$ for $j = 1, \ldots, r_1$ and $|u_k + iv_k|^2 = u_k^2 + v_k^2 < C_k^2$ for $k = 1, \ldots, r_2$. Then X is centrally symmetric and convex. It is the Cartesian product of r_1 intervals $-c_i < x_i < c_i$ and r_2 circles $u_k^2 + v_k^2 < C_k^2$, and so

$$\mathrm{vol}_{\mathbb{R}}(X) = (2c_1)\ldots(2c_{r_1})(\pi C_1^2)\ldots(\pi C_{r_2}^2) = 2^{r_1}\pi^{r_2} c_1 \ldots c_{r_1}(C_1^2 \ldots C_{r_2}^2).$$

Under the hypothesis of the statement, we see that

$$\mathrm{vol}(X) = 2^{r_2}\mathrm{vol}_{\mathbb{R}}(X) > 2^{r_1+r_2}\pi^{r_2}.\left(\frac{2}{\pi}\right)^{r_2}\mathrm{vol}(\Gamma) = 2^n\mathrm{vol}(\Gamma)$$

as $r_1 + 2r_2 = n$. The result now follows from Minkowski's Theorem (Theorem 7.8). $\qquad\square$

We will be particularly interested in the special case where $\Gamma = i(\mathfrak{a})$ is the lattice associated to an ideal.

Proposition 7.15 *Let $\mathfrak{a}$ be a non-zero integral ideal of $\mathbb{Z}_K$. Then there exists a non-zero $\alpha \in \mathfrak{a}$ such that*

$$|N_{K/\mathbb{Q}}(\alpha)| \leq \left(\frac{2}{\pi}\right)^{r_2} N_{K/\mathbb{Q}}(\mathfrak{a})|D_K|^{1/2}.$$

Proof Choose any

$$M > \left(\frac{2}{\pi}\right)^{r_2} N_{K/\mathbb{Q}}(\mathfrak{a})|D_K|^{1/2} = \left(\frac{2}{\pi}\right)^{r_2} \text{vol}(\mathfrak{a}),$$

the last equality coming from Proposition 7.13. Then choose $c_1, \ldots, c_{r_1}, C_1, \ldots,$ $C_{r_2} \in \mathbb{R}_{>0}$ satisfying $c_1 \ldots c_{r_1}(C_1 \ldots C_{r_2})^2 = M$. By the previous proposition, there is a non-zero element $\alpha \in \mathfrak{a}$ such that $|\rho_1(\alpha)| < c_1, \ldots, |\rho_{r_1}(\alpha)| < c_{r_1}$, $|\sigma_1(\alpha)| < C_1, \ldots, |\sigma_{r_2}(\alpha)| < C_{r_2}$. Note that this also implies that $|\overline{\sigma}_j(\alpha)| < C_j$. As $N_{K/\mathbb{Q}}(\alpha)$ is formed from the product over all embeddings (including the complex conjugates), we conclude that

$$N_{K/\mathbb{Q}}(\alpha) < c_1 \ldots c_{r_1}(C_1 \ldots C_{r_2})^2 = M.$$

We can do this for any M bigger than the given bound, and so we conclude that there exists a non-zero $\alpha \in \mathfrak{a}$ as in the statement. $\square$

7.3 Finiteness of the Class Number

One of the two main applications of this theory comes with the finiteness of the class number. Recall that the *class group* is the group of all fractional ideals of $\mathbb{Z}_K$, modulo the principal fractional ideals. We had already shown this finiteness in Chap. 6 in the case of imaginary quadratic fields; using the geometrical methods above, we can now show this for any number field. We write $C(K)$ for the class group, and its order, h_K, is the *class number*.

Theorem 7.16 *The class group $C(K)$ is finite.*

Proof We will first show that every ideal class $[\mathfrak{a}]$ contains an integral ideal $\mathfrak{c}$ of norm at most $M = \left(\frac{2}{\pi}\right)^{r_2} |D_K|^{\frac{1}{2}}$.

We first take any representative $\mathfrak{b}$ of the class $[\mathfrak{a}^{-1}]$; we assume that $\mathfrak{b}$ is contained in $\mathbb{Z}_K$ (otherwise we can multiply through by a suitable element in $\mathbb{Z}_K$). Proposition 7.15 shows that there exists $\beta \in \mathfrak{b}$, with $\beta \neq 0$, such that

$$|N_{K/\mathbb{Q}}(\beta)| \leq \left(\frac{2}{\pi}\right)^{r_2} |D_K|^{\frac{1}{2}} N_{K/\mathbb{Q}}(\mathfrak{b}).$$

Then let $\mathfrak{c} = \langle \beta \rangle \mathfrak{b}^{-1} \in [\mathfrak{a}]$; also since $\beta \in \mathfrak{b}$, every element of $\mathfrak{c}$ is integral, and so $\mathfrak{c} \subseteq \mathbb{Z}_K$.

Finally,

$$N_{K/\mathbb{Q}}(\mathfrak{c}) = |N_{K/\mathbb{Q}}(\beta)| N_{K/\mathbb{Q}}(\mathfrak{b})^{-1} \leq \left(\frac{2}{\pi} \right)^{r_2} |D_K|^{\frac{1}{2}} = M.$$

But there are only finitely many integral ideals whose norm is at most any given bound M (consider the factorisation of an integral ideal into primes, use the multiplicativity of the norm, and observe that there can only be finitely many primes whose norm is bounded). Thus there can only be finitely many ideal classes. □

For the finiteness of the class number, it sufficed to show that every ideal class contains an ideal of norm at most some fixed bound M. But the method of proof also suggests a way to construct the class group, at least for number fields of small discriminant. Given a number field K, compute D_K and the constant M. Then we know that every ideal of $\mathbb{Z}_K$ is equivalent to an ideal with norm at most M. We also know that ideals factor uniquely into prime ideals, and that the norm is multiplicative. We therefore list all prime ideals whose norm is bounded by M, and then all products of those whose norm is at most M. Every integral ideal will be equivalent to at least one ideal on this list, and so the class number is bounded by the number of these ideals.

In fact, the claim that every ideal class contains an ideal of norm at most $M = \left(\frac{2}{\pi} \right)^{r_2} |D_K|^{\frac{1}{2}}$ is far from being best possible. When it comes to finding the class group explicitly, it is helpful to have a much better bound.

The main problem is that the convex shape X is defined as a "hypercube", and something more spherical gives better bounds. For example, in $\mathbb{R}^2$, there is an open square of area 4 given by $\{(x, y) \mid |x| < 1, |y| < 1\}$ with no lattice point other than $(0, 0)$. But the bound for circles is much better; there is no circle of area more than π containing no lattice point other than $(0, 0)$. Better still is to use a square with sides which are diagonal, parallel to $y = \pm x$; every such square of area more than 2 contains a lattice point other than $(0, 0)$.

We will now spend some time finding a better bound, and using it to compute some class groups of fields with small discriminants.

If instead, for $t > 0$, we consider the subset of $K_{\mathbb{R}}$ defined by

$$X_t = \left\{ (x_1, \ldots, x_{r_1}, z_1, \ldots, z_{r_2}) \mid |x_1| + \cdots + |x_{r_1}| + 2|z_1| + \cdots + 2|z_{r_2}| < t \right\},$$

we get a region of a different shape. It is clearly bounded and centrally symmetric.

Exercise 7.1 Show that X_t is also convex.

Lemma 7.17 *The volume of X_t is* $\mathrm{vol}(X_t) = 2^{r_1} \pi^{r_2} \dfrac{t^n}{n!}$.

Proof Use $i_{\mathbb{R}}$ to write X_t as a subset of $\mathbb{R}^n$.

Then $i_{\mathbb{R}}(X_t)$ is the set of points $\{(x_1, \ldots, x_{r_1}, u_1, v_1, \ldots, u_{r_2}, v_{r_2})\}$ satisfying

$$|x_1| + \cdots + |x_{r_1}| + 2\sqrt{u_1^2 + v_1^2} + \cdots + 2\sqrt{u_{r_2}^2 + v_{r_2}^2} < t,$$

where $z_j = u_j + iv_j$ for $j = 1, \ldots, r_2$.

We compute $\mathrm{vol}(X) = 2^{r_2}\mathrm{vol}_{\mathbb{R}}(X)$. Make a change of variable to put $(u_j, v_j) = \left(\frac{R_j}{2}\cos\theta_j, \frac{R_j}{2}\sin\theta_j\right)$; the usual formula for change of variables to polar co-ordinates gives $4du_j dv_j = R_j dR_j d\theta_j$. Then

$$\mathrm{vol}_{\mathbb{R}}(X_t) = \int\limits_{X_t} 1\, dx_1 \ldots dx_{r_1} du_1 dv_1 \ldots du_{r_2} dv_{r_2}$$

$$= 2^{r_1} \int\limits_{X_t,\, x_i \geq 0} 1\, dx_1 \ldots dx_{r_1} du_1 dv_1 \ldots du_{r_2} dv_{r_2}$$

$$= 2^{r_1} 4^{-r_2} \int\limits_{X_t,\, x_i \geq 0} R_1 \cdots R_{r_2}\, dx_1 \ldots dx_{r_1} dR_1 d\theta_1 \ldots dR_{r_2} d\theta_{r_2}$$

$$= 2^{r_1} 4^{-r_2} (2\pi)^{r_2} \int\limits_{Y_t} R_1 \cdots R_{r_2}\, dx_1 \ldots dx_{r_1} dR_1 \ldots dR_{r_2}$$

where

$$Y_t = \{(x_1, \ldots, x_{r_1}, R_1, \ldots, R_{r_2}) \mid x_j, R_k \geq 0,\ x_1 + \cdots + x_{r_1} + R_1 + \cdots + R_{r_2} < t\}.$$

Write

$$I_{r_1, r_2}(t) = \int\limits_{Y_t} R_1 \cdots R_{r_2}\, dx_1 \ldots dx_{r_1} dR_1 \ldots dR_{r_2}.$$

Then simple changes of variables show that

$$I_{r,s}(t) = t^{r+2s} I_{r,s}(1), \quad I_{r,s}(1) = \frac{I_{r-1,s}(1)}{r+2s}, \quad I_{0,s}(1) = \frac{I_{0,s-1}(1)}{2s(2s-1)}. \qquad (7.1)$$

Using the second repeatedly gives

$$I_{r,s}(1) = \frac{(2s)!}{(r+2s)!} I_{0,s}(1),$$

and then the third gives

$$I_{0,s}(1) = \frac{1}{(2s)!}$$

using $I_{0,0}(1) = 1$. Then

$$I_{r_1,r_2}(t) = t^n I_{r_1,r_2}(1) = t^n \frac{(2r_2)!}{n!} I_{0,r_2}(1) = \frac{1}{n!} t^n.$$

Combining these shows that

$$\begin{aligned} \mathrm{vol}(X_t) &= 2^{r_2} \mathrm{vol}_{\mathbb{R}}(X_t) \\ &= 2^{r_2}.2^{r_1} 4^{-r_2} (2\pi)^{r_2} I_{r_1,r_2}(t) \\ &= 2^{r_1} \pi^{r_2} \frac{t^n}{n!} \end{aligned}$$

as required. □

Exercise 7.2 Verify the relations given in (7.1).

With this new region, we can use the method of Proposition 7.15 and get a better result:

Proposition 7.18 *Every ideal class of K contains an integral ideal $\mathfrak{c}$ of norm at most*

$$\frac{n!}{n^n} \left(\frac{4}{\pi} \right)^{r_2} |D_K|^{1/2}.$$

Proof We first remark that the set X_t, whose definition above might have looked a bit unmotivated, has a more natural interpretation when we consider those elements $\alpha \in K$ such that $i(\alpha) \in X_t$. If $\alpha \in K$, and $i(\alpha) = (x_1, \ldots, x_{r_1}, z_1, \ldots, z_{r_2})$, then $x_j = \rho_j(\alpha)$, and $z_k = \sigma_k(\alpha)$. As $2|z_k| = |z_k| + |\bar{z}_k|$, the expression $|x_1| + \cdots + |x_{r_1}| + 2|z_1| + \cdots + 2|z_{r_2}|$ can be viewed as $\sum_\tau |\tau(\alpha)|$, where τ runs over all embeddings of K into $\mathbb{C}$.

We argue as above. Let $[\mathfrak{a}]$ be any ideal class, and take any integral representative $\mathfrak{b}$ of $[\mathfrak{a}^{-1}]$. Then in order to apply Minkowski's Theorem, we choose a value of t for which the volume of X_t is at least $2^n N_{K/\mathbb{Q}}(\mathfrak{b})|D_K|^{1/2}$. This simply requires choosing t so that

$$2^{r_1} \pi^{r_2} \frac{t^n}{n!} > 2^n \mathrm{vol}(\mathfrak{b}) = 2^n N_{K/\mathbb{Q}}(\mathfrak{b})|D_K|^{1/2},$$

or, as $n = r_1 + 2r_2$,

$$t^n > n! \left(\frac{4}{\pi} \right)^{r_2} N_{K/\mathbb{Q}}(\mathfrak{b})|D_K|^{1/2}.$$

Then there exists a non-zero element $\beta \in \mathfrak{b}$ with $i(\beta)$ in X_t, by Minkowski's Theorem. Since this is valid for any t satisfying this inequality, we deduce that there is a non-zero element $\beta \in \mathfrak{b}$ in X_t, where

$$t^n = n! \left(\frac{4}{\pi} \right)^{r_2} N_{K/\mathbb{Q}}(\mathfrak{b})|D_K|^{1/2}.$$

The arithmetic mean–geometric mean inequality implies that $\left(\prod_\tau |\tau(\beta)| \right)^{1/n} \le \frac{1}{n} \sum_\tau |\tau(\beta)|$. The left-hand side is just $N_{K/\mathbb{Q}}(\beta)$, and the right-hand side is at most $\frac{1}{n} t$, by definition of X_t. Thus

$$|N_{K/\mathbb{Q}}(\beta)|^{1/n} \le \frac{t}{n},$$

i.e., that

$$|N_{K/\mathbb{Q}}(\beta)| < \left(\frac{t}{n} \right)^n.$$

By picking t^n as above, we conclude that there exists $\beta \in i^{-1}(X_t) \cap \mathfrak{b}$ such that

$$|N_{K/\mathbb{Q}}(\beta)| < \frac{n!}{n^n} \left(\frac{4}{\pi} \right)^{r_2} N_{K/\mathbb{Q}}(\mathfrak{b})|D_K|^{1/2}.$$

So if $\mathfrak{c} = \langle \beta \rangle \mathfrak{b}^{-1} \in [\mathfrak{a}]$, then

$$N_{K/\mathbb{Q}}(\mathfrak{c}) = |N_{K/\mathbb{Q}}(\beta)| N_{K/\mathbb{Q}}(\mathfrak{b})^{-1},$$

and the result follows.   □

This bound is known as the *Minkowski bound*.

We can use this to compute class groups for fields with reasonably small discriminant.

Example 7.19 For $K = \mathbb{Q}(\sqrt{5})$, the discriminant is 5, and $r_2 = 0$. So the Minkowski bound is $\sqrt{5}/2 = 1.118\dots$. So every ideal is equivalent to one with norm 1, but the only ideal of norm 1 is the full ring of integers, which is principal. Thus the class number is 1.

Example 7.20 For $K = \mathbb{Q}(\sqrt[3]{2})$, the discriminant is 108, and $r_2 = 1$. So the Minkowski bound is

$$\frac{6}{27} \left(\frac{4}{\pi} \right) \sqrt{108} = 2.940\dots.$$

So every ideal is equivalent to one whose norm is at most 2. The only ideal of norm 1 is the full ring of integers, which is principal; the ideal $\langle 2 \rangle = \mathfrak{p}_2^3$, where $\mathfrak{p}_2 = \langle \sqrt[3]{2} \rangle$ is also principal. Thus every ideal is equivalent to a principal ideal, so the class group is trivial.

Example 7.21 For $K = \mathbb{Q}(\sqrt{-5})$, the discriminant is -20, and $r_2 = 1$, so the Minkowski bound is $\frac{2}{\pi}\sqrt{20} = 2.84\dots$. So every ideal is equivalent to an integral ideal of norm at most 2. The full ring of integers is the only ideal of norm 1. An ideal of norm 2 must divide the prime 2, and

$$\langle 2 \rangle = \mathfrak{p}_2^2,$$

where $\mathfrak{p}_2 = \langle 2, 1 + \sqrt{-5} \rangle$, so there is a unique prime ideal of norm 2, and it is not principal. Thus the class number is 2.

We can use this result to prove a similar result to the main result of Sect. 1.5.

Corollary 7.22 *There are no integer solutions to $x^3 = y^2 + 5$.*

Proof If x is even, then y is odd, but then $y^2 + 5 \equiv 2 \pmod{4}$, which is impossible, as $8 | x^3$. So x is odd.

If $p | (x, y)$, then $p | x^3 - y^2$, so $p | 5$, and the only possible common factor is 5. But if $5 | x$ and $5 | y$, then $5^3 | x^3$ whereas $5^2 \nmid y^2 + 5$. So x and y are coprime.

Suppose that $x^3 = y^2 + 5$. Then

$$x^3 = (y + \sqrt{-5})(y - \sqrt{-5}).$$

Suppose $y + \sqrt{-5}$ and $y - \sqrt{-5}$ both lie in some prime ideal $\mathfrak{p}$ (i.e., they are not coprime). Notice that this implies that $x^3 \in \mathfrak{p}$, and, as $\mathfrak{p}$ is prime, that $x \in \mathfrak{p}$. Then $2y$ is in $\mathfrak{p}$. As x is odd, 2 is not in $\mathfrak{p}$. But $\mathfrak{p}$ is prime, so this implies that $y \in \mathfrak{p}$. This contradicts the coprimality of x and y.

Then

$$\langle y + \sqrt{-5} \rangle = \mathfrak{a}^3$$
$$\langle y - \sqrt{-5} \rangle = \mathfrak{b}^3,$$

and as the class number of $\mathbb{Q}(\sqrt{-5})$ is 2, the fact that $\mathfrak{a}$ is an ideal whose cube is principal is enough to see that $\mathfrak{a}$ is principal. (Similarly, $\mathfrak{b}$ is also principal.)

So $y + \sqrt{-5} = u\alpha^3$ for some unit u. But the units in $\mathbb{Q}(\sqrt{-5})$ are just ± 1, which are both cubes, so $y + \sqrt{-5} = \alpha^3$ for some $\alpha = a + b\sqrt{-5}$. Then

$$y + \sqrt{-5} = (a + b\sqrt{-5})^3,$$

and so $1 = b(3a^2 - 5b^2)$, by considering the coefficients of $\sqrt{-5}$. Then $b = \pm 1$, so $3a^2 - 5 = \pm 1$—but this has no integral solutions for a. $\qquad \square$

Exercise 7.3 More generally, suppose that $k \equiv 1, 2 \pmod{4}$, that k is squarefree, and k is not of the form $3t^2 \pm 1$ for some integer t. If 3 does not divide the class number of $\mathbb{Q}(\sqrt{-k})$, show that $x^3 = y^2 + k$ has no solution in integers.

Exercise 7.4 Show that the Minkowski bound for $\mathbb{Q}(\sqrt{-1}), \mathbb{Q}(\sqrt{-2}), \mathbb{Q}(\sqrt{-3})$ and $\mathbb{Q}(\sqrt{-7})$ is less than 2, and deduce that these fields all have unique factorisation.

Exercise 7.5 Show that the Minkowski bound for $\mathbb{Q}(\sqrt{2}), \mathbb{Q}(\sqrt{3})$ and $\mathbb{Q}(\sqrt{13})$ is less than 2, and deduce that these fields all have unique factorisation.

Exercise 7.6 Find the class numbers of $\mathbb{Q}(\sqrt{6})$ and $\mathbb{Q}(\sqrt{-6})$.

Exercise 7.7 Find the class number of $\mathbb{Q}(\sqrt[3]{3})$.

Exercise 7.8 Find the class numbers of some more fields of small discriminant.

Corollary 7.23 *If K is a number field with $[K : \mathbb{Q}] > 1$, then $|D_K| > 1$.*

Proof The Minkowski bound shows that

$$1 \leq \frac{n!}{n^n} \left(\frac{4}{\pi}\right)^{r_2} |D_K|^{1/2},$$

by choosing any ideal $\mathfrak{a}$ in $\mathbb{Z}_K$ (as its norm is at least 1). So

$$|D_K|^{1/2} \geq \frac{n^n}{n!} \left(\frac{\pi}{4}\right)^{r_2} \geq \frac{n^n}{n!} \left(\frac{\pi}{4}\right)^{n/2}.$$

Let γ_n denote the constant on the right-hand side of this inequality. Then $\gamma_2 = \pi/2 > 1$, and for $n \geq 2$,

$$\frac{\gamma_{n+1}}{\gamma_n} = \left(\frac{\pi}{4}\right)^{1/2} \left(1 + \frac{1}{n}\right)^n > 1$$

(and it tends to $\frac{e\sqrt{\pi}}{2}$), so the γ_n are increasing. Thus $|D_K| > 1$. $\square$

Corollary 7.24 *If K is a number field with $[K : \mathbb{Q}] > 1$, then some prime p ramifies in K.*

Proof The primes that ramify in K include all those dividing the discriminant, and so the result follows from the last corollary. $\square$

7.4 Dirichlet's Unit Theorem

In this section, we prove Dirichlet's unit theorem, which describes the structure of the group of units $\mathbb{Z}_K^\times$ for any number field K. The standard proof nowadays involves Minkowski's theorem, but Dirichlet's proof actually dates from somewhat earlier.

As part of a result which applies for general number fields, we will recover the fact that imaginary quadratic fields have finite groups of units, and we will also see that these are the only fields other than $\mathbb{Q}$ with this property. In Chap. 8, we will say a little more about the case of real quadratic fields, and will briefly mention some other fields of small degree.

Because the unit group is multiplicative, and Minkowski's theorem refers to vector spaces, which are additive, we need some sort of logarithm so that we can work in an additive setting. Define such a logarithm map by

$$\ell : K_{\mathbb{R}}^\times \longrightarrow \mathbb{R}^{r_1 + r_2}$$

$$(x_1, \ldots, x_{r_1}, z_1, \ldots, z_{r_2}) \mapsto (\log |x_1|, \ldots, \log |x_{r_1}|, \log |z_1|^2, \ldots, \log |z_{r_2}|^2)$$

and another logarithm map (we will use the same letter, but this should cause no confusion):

$$\ell : \mathbb{R}^{\times} \longrightarrow \mathbb{R}$$
$$x \mapsto \log |x|$$

Then one has the commutative diagram:

$$
\begin{array}{ccccc}
K^{\times} & \xrightarrow{\ i\ } & K_{\mathbb{R}}^{\times} & \xrightarrow{\ \ell\ } & \mathbb{R}^{r_1+r_2} \\
{\scriptstyle N_{K/\mathbb{Q}}}\downarrow & & {\scriptstyle N}\downarrow & & {\scriptstyle \mathrm{tr}}\downarrow \\
\mathbb{Q}^{\times} & \longrightarrow & \mathbb{R}^{\times} & \xrightarrow{\ \ell\ } & \mathbb{R}
\end{array}
$$

Here, tr denotes the map

$$\mathbb{R}^{r_1+r_2} \longrightarrow \mathbb{R}$$
$$(x_1, \ldots, x_{r_1+r_2}) \mapsto x_1 + \cdots + x_{r_1+r_2}$$

Recall that $\mathbb{Z}_K^{\times} = \{\epsilon \in \mathbb{Z}_K \mid N_{K/\mathbb{Q}}(\epsilon) = \pm 1\}$, and put

$$S = \{y \in K_{\mathbb{R}}^{\times} \mid N(y) = \pm 1\},$$
$$H = \{x \in \mathbb{R}^{r_1+r_2} \mid \mathrm{tr}\,(x) = 0\}.$$

Notice that i maps $\mathbb{Z}_K^{\times}$ into S, and that ℓ maps $S \subset K_{\mathbb{R}}^{\times}$ into $H \subset \mathbb{R}^{r_1+r_2}$. Thus the composite map takes the units $\mathbb{Z}_K^{\times}$ into the vector space H.

Remark 7.25 Since $\mathbb{R}^{r_1+r_2}$ has dimension $r_1 + r_2$, and H is defined by the vanishing of a single linear function, H is a subspace of dimension $r = r_1 + r_2 - 1$. It is in this vector space that we will work.

Let λ denote the composite map, taking $\mathbb{Z}_K^{\times}$ into H:

$$\lambda : \mathbb{Z}_K^{\times} \xrightarrow{\ i\ } S \xrightarrow{\ \ell\ } H,$$

and let $\Gamma = \lambda(\mathbb{Z}_K^{\times}) \subset H$. We need to understand the kernel of λ and also the structure of Γ.

Proposition 7.26 *The kernel of λ is $\mu(K)$, the group of roots of unity in K.*

Proof Clearly $\mu(K) \subset \ker(\lambda)$, as for all $x \in \mu(K)$, and all embeddings τ of K into $\mathbb{C}$, one has $|\tau(x)| = 1$, so the image of x in $K_{\mathbb{R}}^{\times}$ is killed by ℓ.

Conversely, if $\epsilon \in \ker \lambda$, then $|\tau(\epsilon)| = 1$ for all embeddings τ. Thus $i(\epsilon)$ lies in a bounded region of $K_{\mathbb{R}}$. Also, $i(\epsilon) \in i(\mathbb{Z}_K)$, a lattice in $K_{\mathbb{R}}$. Thus there are finitely many possibilities for $i(\epsilon)$ as lattices are discrete, so that $\ker(\lambda)$ is finite. It is also closed under multiplication, so every element in $\ker(\lambda)$ is of finite order, and thus is a root of unity.   □

Exercise 7.9 If $[K : \mathbb{Q}]$ is odd, show that K has a real embedding, and deduce that $\mu(K) = \{\pm 1\}$.

Having understood the kernel of λ, we now need to understand its image. It will turn out that Dirichlet's Theorem is equivalent to understanding the structure of Γ.

Lemma 7.27 *Γ is a subgroup of H.*

Proof $\mathbb{Z}_K^\times$ is a group, and λ is a homomorphism, so $\Gamma = \lambda(\mathbb{Z}_K^\times)$ is a group, contained in H.   □

Proposition 7.28 *Γ is a lattice in H.*

Proof Since $\Gamma = \lambda(\mathbb{Z}_K^\times)$ is a subgroup of H, we simply need to check that Γ is discrete.

Let B denote a ball of radius $r \geq 0$ in H. We need to see that $\Gamma \cap B$ is finite (Definition 7.2). But $\ell^{-1}(\Gamma \cap B) = \ell^{-1}(\Gamma) \cap \ell^{-1}(B) = i(\mathbb{Z}_K^\times) \cap \ell^{-1}(B)$. But by definition of ℓ, we see that $\ell^{-1}(B)$ is contained in a bounded region in $K_\mathbb{R}$, and thus in a ball of some radius. Also, $i(\mathbb{Z}_K)$ is a lattice in $K_\mathbb{R}$, so is discrete, and so $i(\mathbb{Z}_K^\times) \cap \ell^{-1}(B) \subseteq i(\mathbb{Z}_K) \cap \ell^{-1}(B)$ is finite. Applying ℓ again, we see that $\Gamma \cap B$ is finite as required.   □

The hard part is to prove that Γ is a *complete* lattice, so that its span is all of H. We want to apply the criterion of Proposition 7.4 and find some bounded region $B_H \subset H$ such that every element of H can be expressed as the sum of something in B_H and something in Γ.

However, although the result is stated in H, we will construct our region by working in $S \subset K_\mathbb{R}^\times$, and applying ℓ.

Proposition 7.29 *There is a bounded region $B_S \subset S$ such that*

$$S = \bigcup_{\epsilon \in \mathbb{Z}_K^\times} i(\epsilon) B_S.$$

Proof Let y denote an element of S. We want to write this as $i(\epsilon)x$ for some unit ϵ and some element x in a bounded region B_S of S.

Consider the lattice $i(\mathbb{Z}_K) \subset K_\mathbb{R}$, of volume $|D_K|^{1/2}$. The lattice $yi(\mathbb{Z}_K)$ also has volume $|D_K|^{1/2}$, since multiplication by y has determinant $N(y) = \pm 1$ (recall that $y \in S$).

Choose $c_1, \ldots, c_{r_1}, C_1, \ldots, C_{r_2} \in \mathbb{R}_{>0}$ with $M = c_1 \ldots c_{r_1}(C_1 \ldots C_{r_2})^2 > (\frac{2}{\pi})^{r_2}|D_K|^{\frac{1}{2}}$. Put

$$X = \{(x_1, \ldots, x_{r_1}, z_1, \ldots, z_{r_2}) \in K_\mathbb{R} \mid |x_j| < c_j, |z_k| < C_k\}.$$

Then by Proposition 7.14, X contains a non-zero point $x \in yi(\mathbb{Z}_K)$.

As $x = yi(\alpha)$, for some $\alpha \in \mathbb{Z}_K$, we have $N(x) = N(y)N(i(\alpha)) = \pm N_{K/\mathbb{Q}}(\alpha)$, and so $N_{K/\mathbb{Q}}(\alpha) < M$. Only finitely many ideals of $\mathbb{Z}_K$ have norm at most M; since any element of norm at most M would generate a principal ideal of norm at most M, it follows that there are only finitely many non-associate numbers of norm at most M. Choose a set $\{\alpha_1, \ldots, \alpha_N\}$ consisting of a complete set of non-associate numbers of norm at most M.

So $\alpha = \epsilon^{-1}\alpha_k$ for some k and some unit ϵ. But then $y = xi(\alpha)^{-1} = xi(\alpha_k)^{-1}i(\epsilon)$. Consider the set $B_S = \{s \in S \mid s \in i(\alpha_k)^{-1}X$ for some $k\}$. As X is bounded, and B_S is the union of finitely many translates of X, we conclude that B_S is bounded. Further, every element $y \in S$ is of the form $xi(\epsilon)$ for some $x \in B_S$ and unit ϵ, and so $S = \bigcup_{\epsilon \in \mathbb{Z}_K^\times} i(\epsilon)B_S$, which is what we wanted. $\square$

Corollary 7.30 Γ *is a complete lattice in* H.

Proof In $S \subset K_\mathbb{R}^\times$, there is a bounded region B_S with

$$S = \bigcup_{\epsilon \in \mathbb{Z}_K^\times} i(\epsilon)B_S. \tag{7.2}$$

Then we will apply our logarithm maps, and take $B_H = \ell(B_S)$. Since ℓ is a logarithm map, one does need to verify that B_H is bounded (after all, the usual logarithm sends the bounded interval $(0, 1]$ to the unbounded interval $(-\infty, 0]$).

We defined B_S explicitly in the proof of the proposition; it is a finite set of translates of X. But $\ell(X)$ is bounded, since $X \subset S$, so every element $x = (x_1, \ldots, x_{r_1}, z_1, \ldots, z_{r_2}) \in X$ has $N(x) = \pm 1$; as $|x_j|$ and $|z_k|$ are bounded for each component, so that $\prod_{j=1}^{r_1} |x_j| \cdot \prod_{k=1}^{r_2} |z_k|^2 = 1$, we see that each $|x_j|$ and $|z_k|$ is bounded away from 0 (so there is a constant $c > 0$ such that each $|x_j| > c$ and each $|z_k| > c$). Then it follows easily that $\ell(X)$ is bounded in H. A very similar argument applies to each translate $\ell(i(\alpha_k)^{-1}X)$, and it follows that $B_H = \ell(B_S)$ is bounded.

Applying ℓ to (7.2), the equality becomes

$$H = \bigcup_{\epsilon \in \mathbb{Z}_K^\times} (\lambda(\epsilon) + B_H).$$

But $\Gamma = \lambda(\mathbb{Z}_K^\times)$, so this becomes

$$H = \bigcup_{\gamma \in \Gamma} (\gamma + B_H),$$

and the result follows from Proposition 7.4. $\square$

We can now deduce Dirichlet's Unit Theorem in its usual form. Write $r = r_1 + r_2 - 1$.

Theorem 7.31 (Dirichlet) $\mathbb{Z}_K^\times \cong \mu(K) \times \mathbb{Z}^r$, where $\mu(K)$ denotes the group of roots of unity in K and $r = r_1 + r_2 - 1$. Equivalently, there exist $\epsilon_1, \ldots, \epsilon_r$ such that all $\epsilon \in \mathbb{Z}_K^\times$ can be written uniquely in the form

$$\epsilon = \zeta \epsilon_1^{v_1} \ldots \epsilon_r^{v_r}$$

with $\zeta \in \mu(K)$ and $v_i \in \mathbb{Z}$.

Proof The map $\lambda : K^\times \longrightarrow \mathbb{R}^{r_1+r_2}$ restricts to a map $\lambda : \mathbb{Z}_K^\times \longrightarrow H$. Its kernel is $\mu(K)$, and its image is Γ; by Corollary 7.30, we have $\Gamma \cong \mathbb{Z}^r$ as it is a complete lattice in an r-dimensional vector space. $\square$

Definition 7.32 The ϵ_i are called *fundamental units*.

We have already seen (in Chap. 6) that imaginary quadratic fields have finitely many units; we can also deduce this from the theorem, since $r_1 = 0$ and $r_2 = 1$, and so $r = r_1 + r_2 - 1 = 0$.

Exercise 7.10 Show that imaginary quadratic fields are the only number fields apart from $\mathbb{Q}$ with finitely many units.

Chapter 8
Other Fields of Small Degree

The results of Chapter 6 give a fairly complete description of imaginary quadratic fields. But other fields have some different properties, and we will meet some of these for the first time in this chapter.

The main class of fields we will consider are the real quadratic fields. While we will not treat them in quite the same detail as the imaginary quadratic case, we will nevertheless give fairly complete proofs for the results we prove. After that, we will also consider some aspects of biquadratic fields, obtained by adjoining two square roots to $\mathbb{Q}$, before listing some results for cubic fields; full proofs of results for these cases would take too long, and we merely state the main results.

The main new phenomenon in the case of real quadratic fields is the structure of the units; for $\mathbb{Q}$ and for imaginary quadratic fields, we always have an easily identifiable finite set of units, but, as we shall see, real quadratic fields will have infinitely many units.

In order to treat real quadratic fields, we need a digression into continued fractions and Pell's equation.

Write K for the number field $\mathbb{Q}(\sqrt{d})$, where $d > 0$. As in Chap. 6, we can assume d is a squarefree integer. We will assume that $\sqrt{d}$ is chosen to be the positive square root of d; this is equivalent to choosing an embedding from K into $\mathbb{R}$, and will allow us to regard one element of K as larger or smaller than another.

Let's work out some units for $\mathbb{Q}(\sqrt{2})$. Our previous calculations show that the ring of integers is $\mathbb{Z}[\sqrt{2}]$, and so a general integer is one of the form $a + b\sqrt{2}$ for $a, b \in \mathbb{Z}$.

The norm of $a + b\sqrt{2}$ is given by

$$N_{\mathbb{Q}(\sqrt{2})/\mathbb{Q}}(a + b\sqrt{2}) = (a + b\sqrt{2})(a - b\sqrt{2}) = a^2 - 2b^2,$$

and units have the property that their norm is ± 1. We therefore need to solve the equation

$$a^2 - 2b^2 = \pm 1;$$

this is essentially an example of *Pell's equation*, $x^2 - ny^2 = 1$, and we shall see in Sect. 8.2 how to solve it using continued fractions.

F. Jarvis, *Algebraic Number Theory*, Springer Undergraduate
Mathematics Series, DOI: 10.1007/978-3-319-07545-7_8,
© Springer International Publishing Switzerland 2014

But we can spot some solutions easily; we observe that, other than trivial solutions $a = \pm 1$, $b = 0$, which correspond to the elements ± 1 in $\mathbb{Q}(\sqrt{2})$, we can see that $a = \pm 1$, $b = \pm 1$, also give solutions, corresponding to units $\pm 1 \pm \sqrt{2}$. Notice that

$$-(1+\sqrt{2}) = -1-\sqrt{2}; \qquad (1+\sqrt{2})^{-1} = -1+\sqrt{2}; \qquad -(1+\sqrt{2})^{-1} = 1-\sqrt{2},$$

so all these units are easily generated from $1 + \sqrt{2}$.

In the imaginary quadratic case, it was always true that the units were roots of unity. But we can easily see that $1 + \sqrt{2}$ is not a root of unity: it is a real number greater than 1. As the product of units is again a unit, any power of $1 + \sqrt{2}$ is also a unit; thus, for example, $(1 + \sqrt{2})^2 = 3 + 2\sqrt{2}$ is a unit (it is easy to see that its inverse is $3 - 2\sqrt{2}$).

More generally, this argument shows that $(1 + \sqrt{2})^n$ is a unit for all $n \geq 1$. Since $(1 + \sqrt{2})^{-1} = (-1 + \sqrt{2})$, we can conclude that $(1 + \sqrt{2})^n$ is a unit for every integer n. So, from the single unit $1 + \sqrt{2}$, we can generate infinitely many units $\{\pm(1 + \sqrt{2})^n\}$. We shall see later that these are the only units in $\mathbb{Z}[\sqrt{2}]$.

Let's think about the case where $\mathbb{Z}_K = \mathbb{Z}[\sqrt{d}]$. Then an element $\lambda = a + b\sqrt{d}$ is a unit if and only if its norm is ± 1. The norm is given by

$$N_{K/\mathbb{Q}}(\lambda) = (a + b\sqrt{d})(a - b\sqrt{d}) = a^2 - db^2,$$

and so we need (a, b) to be a solution of the equation

$$x^2 - dy^2 = \pm 1.$$

The equation $x^2 - dy^2 = 1$, where d is a positive integer and not a square, is known as Pell's equation, and it always has infinitely many integral solutions. To find them, one way is to observe that the equation $x^2 - dy^2 = 1$ implies that x^2 and dy^2 are very close, so that x^2/y^2 is approximately d. In particular, x/y is very close to $\sqrt{d}$. Finding rational numbers close to a given real number can be done using the theory of continued fractions; since this is not treated in [7], we will prove the main relevant results here, although it may be part of many undergraduate courses in elementary number theory.

8.1 Continued Fractions

In the first chapter, we considered Euclid's algorithm for the pair 630 and 132:

$$
\begin{aligned}
630 &= 4 \times 132 + 102 \\
132 &= 1 \times 102 + 30 \\
102 &= 3 \times 30 + 12 \\
30 &= 2 \times 12 + 6 \\
12 &= 2 \times 6 + 0
\end{aligned}
$$

We can interpret this as telling us more about the quotient $\frac{630}{132}$. As well as telling us that we can cancel the highest common factor of 6 from the numerator and denominator to get the fraction $\frac{105}{22}$ in lowest terms, we also see that:

$$\frac{630}{132} = 4 + \frac{102}{132}$$
$$\frac{132}{102} = 1 + \frac{30}{102}$$
$$\frac{102}{30} = 3 + \frac{12}{30}$$
$$\frac{30}{12} = 2 + \frac{6}{12}$$
$$\frac{12}{6} = 2 + \frac{0}{6},$$

and we see that the left-hand side of each equation is the reciprocal of the final term on the right-hand side of the previous one. We can combine all these equations into a single expression:

$$\frac{630}{132} = 4 + \frac{102}{132} = 4 + \frac{1}{\frac{132}{102}}$$
$$= 4 + \frac{1}{1 + \frac{30}{102}} = 4 + \frac{1}{1 + \frac{1}{\frac{102}{30}}}$$
$$= \cdots$$
$$= 4 + \cfrac{1}{1 + \cfrac{1}{3 + \cfrac{1}{2 + \cfrac{1}{2}}}}$$

This last expression is the *continued fraction* for $\frac{630}{132}$. The reader should convince themselves that the right-hand side really is equal to the left-hand side, by evaluating the fraction from the bottom, first replacing $2 + \frac{1}{2}$ by $\frac{5}{2}$, then $3 + \frac{2}{5}$ by $\frac{17}{12}$, etc. The algorithm takes the left-hand side as an input, and writes its integer part and remainder on the right-hand side. Then the input to the next line is the reciprocal of the remainder.

The notation above for the continued fraction is rather cumbersome, and we shall use the abbreviated form [4; 1, 3, 2, 2].

We can obtain fractions which approximate the original expression by taking only the initial parts of the expression:

$$[4] = 4$$

$$[4; 1] = 4 + \frac{1}{1} = 5$$

$$[4; 1, 3] = 4 + \cfrac{1}{1 + \frac{1}{3}} = \frac{19}{4}$$

$$[4; 1, 3, 2] = 4 + \cfrac{1}{1 + \cfrac{1}{3 + \frac{1}{2}}} = \frac{43}{9}$$

These fractions approach the original expression very quickly; they are known as the *convergents*. Another way to recover the convergents is to list the numbers appearing in the continued fraction expansion, together with two further rows, in a table as follows:

$$\begin{array}{c|ccccc} & 4 & 1 & 3 & 2 & 2 \\ \hline 0 \; 1 & \\ 1 \; 0 & \end{array}$$

Then we complete the table; each successive column is completed by taking the previous column and multiplying by the integer at the top, and adding the column before that:

$$\begin{array}{c|cccc} & \mathbf{4} & 1 & 3 & 2 & 2 \\ \hline 0 \; 1 & \mathbf{4} \times 1 + 0 \\ 1 \; 0 & \mathbf{4} \times 0 + 1 \end{array}$$

and then

$$\begin{array}{c|cccc} & 4 & \mathbf{1} & 3 & 2 & 2 \\ \hline 0 \; 1 & 4 & \mathbf{1} \times 4 + 1 \\ 1 \; 0 & 1 & \mathbf{1} \times 1 + 0 \end{array}$$

Repeating this gives:

$$\begin{array}{c|cccc} & 4 & 1 & \mathbf{3} & 2 & 2 \\ \hline 0 \; 1 & 4 & 5 & \mathbf{3} \times 5 + 4 \\ 1 \; 0 & 1 & 1 & \mathbf{3} \times 1 + 1 \end{array}$$

and two more iterations of the process leads to

$$\begin{array}{c|ccccc} & 4 & 1 & 3 & 2 & 2 \\ \hline 0 \; 1 & 4 & 5 & 19 & 43 & 105 \\ 1 \; 0 & 1 & 1 & 4 & 9 & 22 \end{array}$$

and the numerator and denominator of the convergents appear as the columns.

Exercise 8.1 Find the continued fraction of $\frac{999}{700}$, and compute all the convergents.

Let us write $\rho_n = p_n/q_n$ for the convergents to a number $\xi \in \mathbb{R}$, with p_0/q_0 corresponding to the entry below the first number, so that $p_0 = \lfloor \xi \rfloor$, $q_0 = 1$. We extend this to the left, so that $p_{-2} = 0$, $q_{-2} = 1$ and $p_{-1} = 1$, $q_{-1} = 0$ represent the first two columns of the table.

If the continued fraction of ξ is $[a_0; a_1, a_2, \ldots]$, then $p_k = a_k p_{k-1} + p_{k-2}$ and $q_k = a_k q_{k-1} + q_{k-2}$.

Lemma 8.1 *If $\frac{p_n}{q_n}$ and $\frac{p_{n+1}}{q_{n+1}}$ are successive convergents, then $p_{n+1} q_n - p_n q_{n+1} = (-1)^n$.*

Proof We prove this by induction on n. For $n = -2$, we have $p_{-1} q_{-2} - p_{-2} q_{-1} = 1$. Now suppose that $p_{k+1} q_k - p_k q_{k+1} = (-1)^k$. As $p_{k+2} = a_{k+2} p_{k+1} + p_k$ and $q_{k+2} = a_{k+2} q_{k+1} + q_k$, it follows that

$$p_{k+2} q_{k+1} - p_{k+1} q_{k+2} = (a_{k+2} p_{k+1} + p_k) q_{k+1} - p_{k+1}(a_{k+2} q_{k+1} + q_k)$$
$$= -(p_{k+1} q_k - p_k q_{k+1})$$

and the result follows. $\qquad\qquad\qquad\qquad\qquad\qquad\qquad\qquad\qquad\qquad\qquad\square$

Exercise 8.2 Verify this for the convergents of $\frac{999}{700}$, computed in Exercise 8.1.

Let's fix some more notation. Given $\xi = [a_0; a_1, a_2, \ldots] = a_0 + \cfrac{1}{a_1 + \cfrac{1}{a_2 + \cdots}}$,

put $\xi_n = [a_n; a_{n+1}, a_{n+2}, \ldots]$, so that, for example,

$$\xi = a_0 + \cfrac{1}{\xi_1} = a_0 + \cfrac{1}{a_1 + \cfrac{1}{\xi_2}} = \cdots .$$

Lemma 8.2 *With the notation above,*

$$\xi = [a_0; a_1, \ldots, a_{n-1}, \xi_n] = \frac{\xi_n p_{n-1} + p_{n-2}}{\xi_n q_{n-1} + q_{n-2}}.$$

Proof The first equality follows by definition. The second is a special case of the general claim that for all x,

$$[a_0; a_1, \ldots, a_n, x] = \frac{x p_n + p_{n-1}}{x q_n + q_{n-1}},$$

which we will prove by induction.

This general statement is clearly true for $n = 0$. Suppose that it is also true when $n = k - 1$. Then

$$[a_0; a_1, \ldots, a_k, x] = [a_0; a_1, \ldots, a_k + \tfrac{1}{x}]$$
$$= \frac{(a_k + \tfrac{1}{x}) p_{k-1} + p_{k-2}}{(a_k + \tfrac{1}{x}) q_{k-1} + q_{k-2}}$$

$$= \frac{x(a_k p_{k-1} + p_{k-2}) + p_{k-1}}{x(a_k q_{k-1} + q_{k-2}) + q_{k-1}}$$
$$= \frac{x p_k + p_{k-1}}{x q_k + q_{k-1}},$$

where the second equality uses the induction hypothesis for $n = k - 1$, and the result follows by induction. □

We show now that if ξ is irrational, the continued fraction convergents are very close rational approximations.

Proposition 8.3 *Suppose that ξ is irrational. For any $n \geq 0$,*

$$\left| \xi - \frac{p_n}{q_n} \right| < \frac{1}{q_n q_{n+1}}.$$

Proof This follows as

$$\xi - \frac{p_n}{q_n} = \frac{\xi_{n+1} p_n + p_{n-1}}{\xi_{n+1} q_n + q_{n-1}} - \frac{p_n}{q_n}$$
$$= \frac{p_{n-1} q_n - p_n q_{n-1}}{q_n (\xi_{n+1} q_n + q_{n-1})}$$
$$= \frac{(-1)^n}{q_n (\xi_{n+1} q_n + q_{n-1})}$$

using Lemma 8.1, and so

$$\left| \xi - \frac{p_n}{q_n} \right| = \frac{1}{q_n (\xi_{n+1} q_n + q_{n-1})} < \frac{1}{q_n (a_{n+1} q_n + q_{n-1})} = \frac{1}{q_n q_{n+1}}.$$

The inequality follows as $a_{n+1} = \lfloor \xi_{n+1} \rfloor$, so $a_{n+1} < \xi_{n+1}$. □

Let's extract one useful result from the proof:

Corollary 8.4 *If $\rho_n = \frac{p_n}{q_n}$ are the convergents to ξ, then if $\xi < \rho_n$, it follows that $\xi > \rho_{n+1}$ and vice versa.*

Proof This follows from the expression

$$\xi - \frac{p_n}{q_n} = \frac{(-1)^n}{q_n (\xi_{n+1} q_n + q_{n-1})}$$

which is clearly alternating in sign. □

We are aiming towards a result in the other direction, so that we have a criterion for when a given rational is definitely a convergent.

Proposition 8.5 *If a/b is a rational number such that $|b\xi - a| < |q_n\xi - p_n|$, then $b \geq q_{n+1}$.*

Proof Suppose that we have $|b\xi - a| < |q_n\xi - p_n|$ for some $b < q_{n+1}$.

Because $p_n q_{n+1} - p_{n+1} q_n = \pm 1$, there are integers x and y such that

$$xp_n + yp_{n+1} = a,$$
$$xq_n + yq_{n+1} = b.$$

Clearly $x \neq 0$, for otherwise $yq_{n+1} = b$, and so $b \geq q_{n+1}$. If $y = 0$, then $a = xp_n$ and $b = xq_n$, so that

$$|b\xi - a| = |x|.|q_n\xi - p_n| \geq |q_n\xi - p_n|,$$

a contradiction.

If $y < 0$, then $xq_n = b - yq_{n+1}$, so $x > 0$. And if $y > 0$, then, as $b < q_{n+1}$, we see that $xq_n = b - yq_{n+1} < 0$, so $x < 0$. So x and y have opposite signs, and then $x(q_n\xi - p_n)$ and $y(q_{n+1}\xi - p_{n+1})$ have the same signs (using Corollary 8.4).

Then

$$|b\xi - a| = |x(q_n\xi - p_n) + y(q_{n+1}\xi - p_{n+1})| > |x(q_n\xi - p_n)| \geq |q_n\xi - p_n|,$$

a contradiction. $\qquad\square$

We can now show that $\frac{p_n}{q_n}$ is the best convergent amongst rationals with denominators of the same size or smaller.

Corollary 8.6 *If a/b is a rational number such that $|\xi - a/b| < |\xi - p_n/q_n|$ for some n, then $b > q_n$.*

Proof Suppose that there is some a/b with $|\xi - a/b| < |\xi - p_n/q_n|$ and $b \leq q_n$, then

$$b\left|\xi - \frac{a}{b}\right| < q_n\left|\xi - \frac{p_n}{q_n}\right|,$$

and so $|b\xi - a| < |q_n\xi - p_n|$, contradicting the proposition. $\qquad\square$

Next, we can give the desired criterion for a good rational approximation to be a convergent.

Proposition 8.7 *Suppose that ξ is irrational and that a/b is a rational with*

$$\left|\xi - \frac{a}{b}\right| < \frac{1}{2b^2}.$$

Then a/b is a continued fraction convergent to ξ.

Proof As before, write p_n/q_n for the convergents to ξ, and suppose that a/b is not a convergent. Then $q_n \leq b < q_{n+1}$ for some n. By Proposition 8.5, we also have $|b\xi - a| \geq |q_n\xi - p_n|$. Then

$$|q_n\xi - p_n| \leq |b\xi - a| < \frac{1}{2b},$$

and so $\left|\xi - \dfrac{p_n}{q_n}\right| < \dfrac{1}{2bq_n}$. But this means that

$$\frac{1}{bq_n} \leq \frac{|bp_n - aq_n|}{bq_n} = \left|\frac{p_n}{q_n} - \frac{a}{b}\right| \leq \left|\xi - \frac{p_n}{q_n}\right| + \left|\xi - \frac{a}{b}\right| < \frac{1}{2bq_n} + \frac{1}{2b^2},$$

which implies that $b < q_n$, a contradiction.   □

8.2 Continued Fractions of Square Roots

Square roots have particularly interesting continued fractions.

Example 8.8 Let's compute the continued fraction of $\sqrt{19}$. It turns out that we just need to know that $4 < \sqrt{19} < 5$.

As usual, start by taking the integer part and remainder:

$$\sqrt{19} = 4 + (\sqrt{19} - 4).$$

Then do the same for the reciprocal of the remainder, rationalising the denominator:

$$\frac{1}{\sqrt{19} - 4} = \frac{\sqrt{19} + 4}{3} = 2 + \frac{\sqrt{19} - 2}{3}.$$

This repeats:

$$\frac{3}{\sqrt{19} - 2} = \frac{\sqrt{19} + 2}{5} = 1 + \frac{\sqrt{19} - 3}{5}$$

$$\frac{5}{\sqrt{19} - 3} = \frac{\sqrt{19} + 3}{2} = 3 + \frac{\sqrt{19} - 3}{2}$$

$$\frac{2}{\sqrt{19} - 3} = \frac{\sqrt{19} + 3}{5} = 1 + \frac{\sqrt{19} - 2}{5}$$

$$\frac{5}{\sqrt{19} - 2} = \frac{\sqrt{19} + 2}{3} = 2 + \frac{\sqrt{19} - 4}{3}$$

$$\frac{3}{\sqrt{19} - 4} = \sqrt{19} + 4 = 8 + (\sqrt{19} - 4)$$

and the process repeats. We get the infinite continued fraction

$$[4; 2, 1, 3, 1, 2, 8, 2, 1, 3, 1, 2, 8, 2, 1, 3, 1, 2, 8, \ldots],$$

which we abbreviate $[4; \overline{2, 1, 3, 1, 2, 8}]$.

Exercise 8.3 Compute the continued fraction expansions of $\sqrt{21}$ and $\sqrt{71}$. (As usual, you should make up some more examples for yourself!)

Hopefully you should have observed that each term in the calculation is of the form $\xi_n = \frac{\sqrt{d}+M_n}{N_n}$ for some integers M_n and N_n. In Example 8.8, we have $\xi_1 = \frac{\sqrt{19}+4}{3}$, so $M_1 = 4$ and $N_1 = 3$. In the same way, $\xi_2 = \frac{\sqrt{19}+2}{5}$, so $M_2 = 2$ and $N_2 = 5$, and so on. Let's prove first that this is always the case.

Proposition 8.9 *Let d be a positive integer, not a square. Put $M_0 = 0$, $N_0 = 1$, $\xi_0 = \sqrt{d}$ and $a_0 = \lfloor \xi_0 \rfloor$. Then define recursively sequences by $M_{n+1} = a_n N_n - M_n$, $N_{n+1} = \frac{d-M_{n+1}^2}{N_n}$, $\xi_{n+1} = \frac{\sqrt{d}+M_{n+1}}{N_{n+1}}$ and $a_{n+1} = \lfloor \xi_{n+1} \rfloor$. Then*

1. *M_n and N_n are integers for all n;*
2. *$\xi_n = a_n + 1/\xi_{n+1}$, and so $\xi = [a_0; a_1, a_2, \ldots]$.*

Proof 1. We'll prove this by induction. Clearly M_0 and N_0 are integers, and our inductive hypothesis will be that M_k and N_k are integers for $k \le n$. By definition, a_n is always an integer, so clearly $M_{n+1} = a_n N_n - M_n$ is an integer. The real content of this proposition is that N_{n+1} should be an integer. But

$$N_{n+1} = \frac{d - M_{n+1}^2}{N_n} = \frac{d - (a_n N_n - M_n)^2}{N_n} = \frac{d - M_n^2}{N_n} + 2a_n M_n - a_n^2 N_n,$$

so we just need to check that $\frac{d-M_n^2}{N_n}$ is an integer. If $n \ge 1$, $N_n = \frac{d-M_n^2}{N_{n-1}}$ and so $N_n | d - M_n^2$. If $i = 0$, $N_0 = 1$, and so $N_0 | d - M_0^2$.
2. This now follows easily, by substituting the expressions for ξ_n and ξ_{n+1} into the expression. $\qquad\square$

Lemma 8.10 *With the notation of the previous result, $N_n > 0$ for all sufficiently large n.*

Proof Write $\xi_n' = \frac{-\sqrt{d}+M_n}{N_n}$ for the conjugate of ξ_n. We know from Lemma 8.2 that

$$\xi = \xi_0 = \frac{\xi_n p_{n-1} + p_{n-2}}{\xi_n q_{n-1} + q_{n-2}},$$

and so

$$\xi_0' = \frac{\xi_n' p_{n-1} + p_{n-2}}{\xi_n' q_{n-1} + q_{n-2}},$$

which rearranges to give

$$\xi'_n = -\frac{q_{n-2}}{q_{n-1}} \left(\frac{\xi'_0 - p_{n-2}/q_{n-2}}{\xi'_0 - p_{n-1}/q_{n-1}} \right).$$

As $k \to \infty$, $p_k/q_k \to \xi_0$, and so the bracket tends to 1. Thus for large enough n, $\xi'_n < 0$. Thus $\xi_n > 0$ and $\xi'_n < 0$ for such n, and

$$\frac{2\sqrt{d}}{N_n} = \xi_n - \xi'_n > 0,$$

showing that $N_n > 0$.   □

Lemma 8.11 *With the notation of the previous results, there exists an integer $k > 0$ with $\xi_j = \xi_{j+k}$ for some j.*

Proof We know that $\xi_n = \frac{\sqrt{d}+M_n}{N_n}$. As $N_n N_{n+1} = d - M_{n+1}^2$, and $N_n > 0$ for sufficiently large n, we see that for all such n, that $M_{n+1}^2 < d$. This means that there are only finitely many possibilities for each M_n. Also, $N_n N_{n+1} < d$, and if $N_n > 0$, this means that $N_{n+1} < d$, so that there are only finitely many possibilities for N_n also. This shows that eventually, $\xi_j = \xi_{j+k}$ for some j and $k > 0$.   □

Theorem 8.12 *The continued fraction of $\sqrt{d}$ has the form $[b_0; \overline{b_1, \ldots, b_k}]$ where $b_k = 2b_0$.*

Proof Take $\xi_0 = \sqrt{d} + \lfloor \sqrt{d} \rfloor$, and we work out the continued fraction $[a_0, a_1, \ldots]$ of ξ_0. Then certainly $\xi_0 > 1$ and $a_0 \geq 1$, and $\xi'_0 = \lfloor \sqrt{d} \rfloor - \sqrt{d}$ satisfies $-1 < \xi'_0 < 0$. In fact, we claim that $-1 < \xi'_n < 0$ for all non-negative integers n, and we prove this by induction.

As $1/\xi_{n+1} = \xi_n - a_n$, we have $1/\xi'_{n+1} = \xi'_n - a_n$. If $\xi'_n < 0$, clearly $1/\xi'_{n+1} < 0$, so $\xi'_{n+1} < 0$, and further, $1/\xi'_{n+1} < -1$ (recall that $a_n \geq 1$ as d is not a square), so that $-1 < \xi'_{n+1}$. Thus $-1 < \xi'_{n+1} < 0$, and the claim follows by induction.

In particular, $-1 < \xi'_n < 0$, so that $-1 < \frac{1}{\xi'_{n+1}} - a_n < 0$, and then

$$a_n = \left\lfloor -\frac{1}{\xi'_{n+1}} \right\rfloor.$$

By the previous lemma, there are integers j and $k > 0$ with $\xi_j = \xi_{j+k}$. But this implies that $\xi'_j = \xi'_{j+k}$, and then

$$a_{j-1} = \left\lfloor -\frac{1}{\xi'_j} \right\rfloor = \left\lfloor -\frac{1}{\xi'_{j+k}} \right\rfloor = a_{j+k-1}.$$

Finally,

$$\xi_{j-1} = a_{j-1} + \frac{1}{\xi_j} = a_{j+k-1} + \frac{1}{\xi_{j+k}} = \xi_{j+k-1}.$$

Applying this repeatedly, we see that $\xi_0 = \xi_k$, and so the continued fraction repeats. We get that

$$\xi_0 = [\overline{a_0, a_1, \ldots, a_{k-1}}].$$

As $\xi_0 = \sqrt{d} + \lfloor\sqrt{d}\rfloor$, the continued fraction $[a_0, a_1, \ldots]$ of ξ_0 is identical to the continued fraction $[b_0, b_1, \ldots]$ of $\sqrt{d}$ except that $b_0 = a_0 - \lfloor\sqrt{d}\rfloor$. But $a_0 = \lfloor\xi_0\rfloor = 2\lfloor\sqrt{d}\rfloor$, so $b_0 = \lfloor\sqrt{d}\rfloor$, and $a_0 = 2b_0$. By the periodicity of the continued fraction for ξ_0, we have $a_0 = a_k = b_k$, and so $b_k = 2b_0$ as claimed. □

Definition 8.13 We say that $k > 0$ is the *period* of $\sqrt{d}$ if it is the smallest index with $\xi_k = \xi_0$.

Recall that if $\xi = \sqrt{d}$, we put $\xi_n = \frac{\sqrt{d}+M_n}{N_n}$. We are going to relate the convergent p_n/q_n of $\sqrt{d}$ with the denominator N_{n+1} of ξ_{n+1}:

Proposition 8.14 *If $\frac{p_n}{q_n}$ denotes the nth convergent to $\sqrt{d}$, then $p_n^2 - dq_n^2 = (-1)^{n+1}N_{n+1}$.*

Proof Put $\xi_0 = \sqrt{d}$, and define ξ_n as above. By Lemma 8.2,

$$\sqrt{d} = \frac{\xi_{n+1}p_n + p_{n-1}}{\xi_{n+1}q_n + q_{n-1}}.$$

We also have $\xi_{n+1} = \frac{\sqrt{d}+M_{n+1}}{N_{n+1}}$, and we substitute this in. After simplifying, we get

$$\sqrt{d} = \frac{M_{n+1}p_n + N_{n+1}p_{n-1} + p_n\sqrt{d}}{M_{n+1}q_n + N_{n+1}q_{n-1} + q_n\sqrt{d}}.$$

Rearranging this gives

$$dq_n + \sqrt{d}(M_{n+1}q_n + N_{n+1}q_{n-1}) = M_{n+1}p_n + N_{n+1}p_{n-1} + \sqrt{d}p_n.$$

Equating coefficients of $\sqrt{d}$ and the remaining terms gives the two equations

$$M_{n+1}p_n + N_{n+1}p_{n-1} = dq_n$$
$$M_{n+1}q_n + N_{n+1}q_{n-1} = p_n.$$

Multiply the first equation by q_n, and the second by p_n, and subtract to get

$$p_n^2 - dq_n^2 = N_{n+1}(p_nq_{n-1} - q_np_{n-1}),$$

and the result follows from Lemma 8.1. □

We can put everything together, and deduce the main result on Pell's equation. If $x^2 - dy^2 = 1$, then x^2/y^2 will be close to d, and so x/y will be close to $\sqrt{d}$. This suggests that we look for solutions amongst the convergents to $\sqrt{d}$.

Theorem 8.15 *Let $d > 0$ be an integer, not a square. Then the equation $x^2 - dy^2 = 1$ has infinitely many solutions. The equation $x^2 - dy^2 = -1$ has infinitely many solutions if the continued fraction for $\sqrt{d}$ has odd period.*

Proof From the previous result, we know that $p_n^2 - dq_n^2 = (-1)^{n+1} N_{n+1}$, where p_n/q_n is a convergent to $\sqrt{d}$, and N_{n+1} is the denominator of ξ_{n+1} as above. We also know that the sequence (ξ_n) repeats with some period k; this means that (N_n) repeats with period k. As $N_0 = 1$, we deduce that $N_{sk} = 1$ for all integers $s \geq 0$. For any n of the form $sk - 1$ with s or k even, then (p_n, q_n) solves $x^2 - dy^2 = 1$, and there are therefore always infinitely many solutions. If k is odd, and $n = sk - 1$ with s odd, then (p_n, q_n) solves $x^2 - dy^2 = -1$. □

To illustrate these, here is the table of convergents for $\sqrt{19}$ (using the calculations of Example 8.8), with an extra row corresponding to the $p_n^2 - dq_n^2$:

		4	2	1	3	1	2	8
0	1	4	9	13	48	61	170	1421
1	0	1	2	3	11	14	39	326
	$p_n^2 - dq_n^2$	−3	5	−2	5	−3	1	−3

Exercise 8.4 Compute the values of $p_n^2 - dq_n^2$ for the convergents of $\sqrt{31}$.

Exercise 8.5 Solve $x^2 - 11y^2 = 1$ and $x^2 - 31y^2 = 1$.

Further details on the theory of continued fractions may be found in various books on elementary number theory, [12] for example.

8.3 Real Quadratic Fields

Now we turn our attention back to the study of number fields, and real quadratic fields in particular.

Write K for the number field $\mathbb{Q}(\sqrt{d})$, where $d > 0$ is squarefree. At the start of the chapter, we observed that numbers of the form $\pm(1 + \sqrt{2})^n$ were all units in $\mathbb{Z}[\sqrt{2}]$, and so $\mathbb{Q}(\sqrt{2})$ has infinitely many units.

Exercise 8.6 Find infinitely many units in $\mathbb{Z}[\sqrt{7}]$.

We suggested that, in general, units $s + t\sqrt{d}$ (for the moment, we will suppose $\mathbb{Z}_K = \mathbb{Z}[\sqrt{d}]$) should have the property that s/t is a continued fraction convergent to $\sqrt{d}$, and in the previous sections we have shown how to compute these. We will now explain that this is indeed the case.

Theorem 8.16 *Let d be a squarefree positive integer, and write $\frac{p_n}{q_n}$ for the convergents to $\sqrt{d}$. Suppose that m is an integer with $|m| < \sqrt{d}$. Then any solution (s, t) to $x^2 - dy^2 = m$ with $(s, t) = 1$ satisfies $s = p_n$ and $t = q_n$ for some n.*

Proof First consider the case $m > 0$.

Suppose that $s^2 - dt^2 = m$. Then $\frac{s}{t} = \sqrt{d + \frac{m}{t^2}} > \sqrt{d}$, and

$$0 < \frac{s}{t} - \sqrt{d} = \frac{m}{t(s + t\sqrt{d})} < \frac{\sqrt{d}}{t(s + t\sqrt{d})} = \frac{1}{t^2(s/(t\sqrt{d}) + 1)}.$$

As $\frac{s}{t} > \sqrt{d}$, the quotient $s/(t\sqrt{d})$ is greater than 1, and so

$$\left| \frac{s}{t} - \sqrt{d} \right| < \frac{1}{2t^2}.$$

The result follows from Proposition 8.7.

A similar argument applies when $m < 0$, but with some complications. It is easy to check that s/t is a convergent to $\sqrt{d}$ precisely if t/s is a convergent to $1/\sqrt{d}$. We rewrite the expression $s^2 - dt^2 = m$ as

$$t^2 - \left(\frac{1}{d} \right) s^2 = \left(-\frac{m}{d} \right),$$

and apply the argument above; after all, $-\frac{m}{d} > 0$ and $\left| \frac{m}{d} \right| < \sqrt{\frac{1}{d}}$. The argument works in the same way to conclude that t/s is a convergent to $\sqrt{\frac{1}{d}}$, and so s/t is a convergent to $\sqrt{d}$. $\qquad\square$

It follows that units $s + t\sqrt{d}$ can be computed by looking through the continued fraction convergents to $\sqrt{d}$, and finding those convergents p_n/q_n with $p_n^2 - dq_n^2 = \pm 1$.

It remains to check that there are solutions to this equation. But we have already remarked that $N_{sk} = 1$ for all values of s, where k denotes the period of $\sqrt{d}$. This means that there are convergents p_n/q_n with $p_n^2 - dq_n^2 = \pm 1$.

The same method works also for the case $d \equiv 1 \pmod 4$. Here, though, integers are of two forms. Some integers are of the form $a + b\sqrt{d}$ with $a, b \in \mathbb{Z}$; we can find units of this form by solving $a^2 - db^2 = \pm 1$ as above. But other integers are of the form $a + b\sqrt{d}$ where a and b are halves of odd integers. In this case, we need also to find solutions to $a^2 - db^2 = \pm 1$ with $a, b \in \frac{1}{2}\mathbb{Z}$. Multiplying through by 4, we need to solve $A^2 - dB^2 = \pm 4$ with $A, B \in \mathbb{Z}$. Again, Theorem 8.16 guarantees that all solutions may be found in the continued fraction convergents to $\sqrt{d}$ (at least for $d \geq 17$; smaller cases can be treated by hand).

Example 8.17 Let's find some units for $\mathbb{Q}(\sqrt{61})$. The continued fraction expansion of $\sqrt{61}$ is given by $[7; \overline{1, 4, 3, 1, 2, 2, 1, 3, 4, 1, 14}]$. We can compute the convergents and the corresponding values of $p_n^2 - 61q_n^2$; we find that the 7th convergent is $39/5$ and $39^2 - 61 \times 5^2 = -4$. It follows that $\frac{39+5\sqrt{61}}{2}$ is a unit, and then so are $\pm(39+5\sqrt{61})^n$ for any integer n.

Let's now return to $\mathbb{Z}[\sqrt{2}]$ and explain that the units $\pm(1 + \sqrt{2})^n$ are the only units.

We first compute the continued fraction:

$$\sqrt{2} = 1 + (\sqrt{2} - 1)$$

$$\frac{1}{\sqrt{2} - 1} = \sqrt{2} + 1 = 2 + (\sqrt{2} - 1),$$

and the process repeats with period 1. Thus the continued fraction is $[1; \overline{2}]$, and the table of convergents begins:

	a_n	1	2	2	2	2	2
0	1	1	3	7	17	41	99
1	0	1	2	5	12	29	70
	$p_n^2 - 2q_n^2$	-1	1	-1	1	-1	1

If η denotes the smallest unit of $\mathbb{Z}[\sqrt{2}]$ satisfying $\eta > 1$, and $\eta = a + b\sqrt{2}$, then a/b must be a continued fraction convergent of $\sqrt{2}$. The calculation above shows that $\eta = 1 + \sqrt{2}$.

Now we can explain that the units in $\mathbb{Z}[\sqrt{2}]$ are all necessarily of the form $\pm\eta^n$.

Suppose that λ is a unit of $\mathbb{Z}[\sqrt{2}]$, and that $\lambda \neq \pm 1$. Then one of λ, $-\lambda$, $1/\lambda$ and $-1/\lambda$ is greater than 1. Suppose it is λ (if not, redefine λ so that it is this unit).

For some n, we have $\eta^n \leq \lambda < \eta^{n+1}$. By multiplying throughout by η^{-n}, we get $1 \leq \lambda\eta^{-n} < \eta$, and so we find a unit $\lambda\eta^{-n}$ strictly less than η, and at least 1. But η was chosen to be the smallest unit which was greater than 1, and so we must have $\lambda\eta^{-n} = 1$. Then $\lambda = \eta^n$, as required.

It is easy to see that the argument above generalises to any real quadratic field.

Theorem 8.18 *Suppose that K is a real quadratic field. Then there exists some unit $\eta > 1$ such that every unit of $\mathbb{Z}_K$ is of the form $\pm\eta^n$ for some integer n.*

Proof Let η denote the smallest unit of $\mathbb{Z}_K$ greater than 1; this can always be found from the continued fraction convergents of $\sqrt{d}$. Let λ be a unit of $\mathbb{Z}_K$, and that $\lambda \neq \pm 1$ (which clearly correspond to the case $n = 0$). Suppose first that $\lambda > 1$.

For some $n \geq 1$, we have $\eta^n \leq \lambda < \eta^{n+1}$. By multiplying throughout by η^{-n}, we get $1 \leq \lambda\eta^{-n} < \eta$, and so we find a unit $\lambda\eta^{-n}$ strictly less than η, and at least 1. But η was chosen to be the smallest unit which was greater than 1, and so we must have $\lambda\eta^{-n} = 1$. Then $\lambda = \eta^n$ for some integer $n \geq 1$.

If $\lambda \neq \pm 1$ is any unit, then one of λ, $-\lambda$, $1/\lambda$ and $-1/\lambda$ is greater than 1, and therefore of the form η^n for some $n \geq 1$, and so $\lambda = \pm\eta^n$ for some $n \in \mathbb{Z}$. $\qquad\square$

Definition 8.19 A unit $\eta > 1$ such that every unit in $\mathbb{Z}_K$ is of the form $\pm\eta^n$ is called a *fundamental unit*.

From the proof of Theorem 8.18, it is clear that η may be chosen to be the smallest unit of $\mathbb{Z}_K$ greater than 1, and we know that these may be found by examining the continued fraction expansion of $\sqrt{d}$.

The units in $\mathbb{Z}_K$, written $U(\mathbb{Z}_K)$ or $\mathbb{Z}_K^\times$, are therefore given by $\{\pm 1\} \times \eta^\mathbb{Z}$, and are isomorphic as an abstract group to $C_2 \times \mathbb{Z}$, the first component corresponding to the choice of sign, and the second to the power of the fundamental unit.

Exercise 8.7 Find the fundamental units for $\mathbb{Q}(\sqrt{11})$, $\mathbb{Q}(\sqrt{51})$ and $\mathbb{Q}(\sqrt{58})$

Exercise 8.8 Find fundamental units for $\mathbb{Q}(\sqrt{29})$ and $\mathbb{Q}(\sqrt{33})$. (Note that that $d \equiv 1 \pmod 4$ in these cases.)

Sometimes it turns out that the continued fraction expansions of $\sqrt{d}$ only repeat with quite a long period. In this case, one has to look a long way to find the fundamental unit.

Exercise 8.9 Find the fundamental unit for $\mathbb{Q}(\sqrt{94})$. You may assume that the continued fraction expansion for $\sqrt{94}$ is $[9; \overline{1, 2, 3, 1, 1, 5, 1, 8, 1, 5, 1, 1, 3, 2, 1, 18}]$.

As in the imaginary quadratic case, there is a link between class numbers of real quadratic fields and quadratic forms. This time, however, the quadratic forms are not positive definite, and counting the forms is not so straightforward in practice; although there is a notion of reduced form, it works much less well.

The proofs are rather similar to those in the imaginary quadratic case (for a sketch of the argument, see [3]). We will simply state the main result.

Suppose that $K = \mathbb{Q}(\sqrt{d})$ is a real quadratic field, with d a squarefree integer. Let $\mathcal{F}_K$ denote the set of proper equivalence classes of quadratic forms (a, b, c) of discriminant D_K where a, b and c do not share a common factor, and let C_K^+ denote the *narrow class group* of K, defined as the quotient of the group of fractional ideals of K by the subgroup of principal fractional ideals which admit a generator with positive norm. Note that the class group C_K is a quotient of the group of fractional ideals by a possibly larger subgroup, so C_K is a quotient of C_K^+.

As in Sect. 6.5, each ideal $\mathfrak{a}$ of $\mathbb{Z}_K$ is of the form $\mathbb{Z}a + \mathbb{Z}(b + c\rho_d)$, where $\rho_d = \sqrt{d}$ or $\frac{1+\sqrt{d}}{2}$ depending on the value of $d \pmod 4$, and we can define

$$\Phi(\mathfrak{a}) = \frac{N_{K/\mathbb{Q}}(ax + (b + c\rho_d)y)}{N_{K/\mathbb{Q}}(\mathfrak{a})},$$

and this is a quadratic form of discriminant D_K. Unlike the imaginary quadratic case, it will not be positive (or negative) definite; some values of (x, y) will produce positive values, and others negative.

Theorem 8.20 Φ *induces a bijection from the narrow class group* C_K^+ *to* $\mathcal{F}_K$. *The inverse is the map* Ψ *which associates to* $(a, b, c) \in \mathcal{F}_K$ *the ideal*

$$\Psi((a,b,c)) = \left(a\mathbb{Z} + \frac{-b+\sqrt{D_K}}{2}\mathbb{Z}\right)\alpha,$$

where $\alpha \in K^\times$ is any element such that $aN_{K/\mathbb{Q}}(\alpha) > 0$.

Write $\overline{\mathcal{F}}_K$ for the quotient of $\mathcal{F}_K$ got from the equivalence relation $(a,b,c) \sim (-a,b,-c)$. Then Φ induces a bijection from the class group C_K to $\overline{\mathcal{F}}_K$.

Rather less is known about the class numbers of real quadratic fields. For example, it does seem that there are a lot of real quadratic fields with class number 1—it is conjectured that there are infinitely many real quadratic fields with class number 1, but this is not known. More precisely, the *Cohen-Lenstra heuristics* suggest that about 75 % of real quadratic fields ought to have class number 1.

There are explicit formulae for class numbers of any quadratic field (see Sect. 10.5); the formula for the class number of a real quadratic field involves the logarithm of the fundamental unit, and, as already noted, the behaviour of the fundamental unit is erratic, and the relevant term in the explicit formula is hard to control.

8.4 Biquadratic Fields

One might imagine that fields of degree 3 might be the next easiest case to consider, but in fact there is an interesting class of degree 4 number fields to consider. Degree 4 extensions fall into a number of classes, and we will consider those of the form $\mathbb{Q}(\sqrt{m}, \sqrt{n})$, where m and n are two squarefree integers with $m \neq n$. Such fields are known as *biquadratic*.

We use the techniques already developed to compute the rings of integers for the biquadratic field $K = \mathbb{Q}(\sqrt{m}, \sqrt{n})$. Note that $\sqrt{k} \in K$, where $k = mn/(m,n)^2$; the three fields $\mathbb{Q}(\sqrt{m})$, $\mathbb{Q}(\sqrt{n})$ and $\mathbb{Q}(\sqrt{k})$ form the three quadratic subfields of K.

There are four embeddings from K into $\mathbb{C}$:

$$a + b\sqrt{m} + c\sqrt{n} + d\sqrt{k} \mapsto \begin{cases} a + b\sqrt{m} + c\sqrt{n} + d\sqrt{k}, \\ a + b\sqrt{m} - c\sqrt{n} - d\sqrt{k}, \\ a - b\sqrt{m} + c\sqrt{n} - d\sqrt{k}, \\ a - b\sqrt{m} - c\sqrt{n} + d\sqrt{k}. \end{cases}$$

Apart from the identity, notice that these can be viewed as "conjugations" fixing each of the three quadratic subfields in turn—for example, the final embedding in the list above conjugates $\sqrt{m}$ and $\sqrt{n}$, and therefore fixes $\sqrt{m}\sqrt{n} = (m,n)\sqrt{k}$, and is a conjugation fixing $\mathbb{Q}(\sqrt{k})$. Notice that each embedding actually has image equal to K, so these embeddings are automorphisms of K. (This would not necessarily happen for more general degree 4 number fields.)

Lemma 8.21 *If $K = \mathbb{Q}(\sqrt{m}, \sqrt{n})$, then we can assume, without loss of generality, that we are in one of the following cases:*

1. $m \equiv 3 \,(mod\ 4),\ k \equiv n \equiv 2 \,(mod\ 4)$;
2. $m \equiv 1 \,(mod\ 4),\ k \equiv n \equiv 2 \,(mod\ 4)$;
3. $m \equiv 1 \,(mod\ 4),\ k \equiv n \equiv 3 \,(mod\ 4)$;
4. $m \equiv 1 \,(mod\ 4),\ k \equiv n \equiv 1 \,(mod\ 4)$.

Proof First, if $2|m$ and $2|n$, clearly k is odd (recall m and n are squarefree, so neither is divisible by 4). Then as $\mathbb{Q}(\sqrt{m}, \sqrt{n}) = \mathbb{Q}(\sqrt{m}, \sqrt{k})$, we see that we can always assume that (at least) one of the two generators is the square root of an odd integer.

If $m \equiv 3 (mod\ 4)$ and $n \equiv 1 \,(mod\ 4)$, we can simply interchange m and n as $\mathbb{Q}(\sqrt{m}, \sqrt{n}) = \mathbb{Q}(\sqrt{n}, \sqrt{m})$.

If $m \equiv 3 \,(mod\ 4)$ and $n \equiv 3 \,(mod\ 4)$, we can replace n by k. Note that $k \equiv mn \,(mod\ 4)$, as $mn = k(m, n)^2$, and (m, n) is odd (as m and n are), so has square congruent to 1 $(mod\ 4)$. This implies that $k \equiv mn \equiv 1 \,(mod\ 4)$, and $\mathbb{Q}(\sqrt{m}, \sqrt{n}) = \mathbb{Q}(\sqrt{m}, \sqrt{k})$.

Thus after permuting m, n and k, we can assume that m and n satisfy the given congruences; then $k \equiv mn \,(mod\ 4)$ by the argument just given (m is always odd, so (m, n) is), and also satisfies the given congruence. $\qquad \square$

Let us begin by finding the rings of integers of K.

Proposition 8.22 *With the numbering of Lemma 8.21, an integral basis for (the rings of integers of)* $\mathbb{Q}(\sqrt{m}, \sqrt{n})$ *are given by*

1. $\{1, \sqrt{m}, \sqrt{n}, \frac{\sqrt{n}+\sqrt{k}}{2}\}$;
2. $\{1, \frac{1+\sqrt{m}}{2}, \sqrt{n}, \frac{\sqrt{n}+\sqrt{k}}{2}\}$;
3. $\{1, \frac{1+\sqrt{m}}{2}, \sqrt{n}, \frac{\sqrt{n}+\sqrt{k}}{2}\}$;
4. $\{1, \frac{1+\sqrt{m}}{2}, \frac{1+\sqrt{n}}{2}, \frac{(1+\sqrt{m})(1+\sqrt{n})}{4}\}$.

Proof We will follow the method of Sect. 3.6.

Let $\alpha \in \mathbb{Z}_K$. Then we can write $\alpha = a + b\sqrt{m} + c\sqrt{n} + d\sqrt{k}$ for some $a, b, c, d \in \mathbb{Q}$.

As $\alpha \in \mathbb{Z}_K$, all of its conjugates

$$\alpha_2 = a - b\sqrt{m} + c\sqrt{n} - d\sqrt{k}$$
$$\alpha_3 = a + b\sqrt{m} - c\sqrt{n} - d\sqrt{k}$$
$$\alpha_4 = a - b\sqrt{m} - c\sqrt{n} + d\sqrt{k}$$

are also algebraic integers. As the set of algebraic integers is closed under addition, the following are also algebraic integers:

$$\alpha + \alpha_2 = 2a + 2c\sqrt{n}$$
$$\alpha + \alpha_3 = 2a + 2b\sqrt{m}$$
$$\alpha + \alpha_4 = 2a + 2d\sqrt{k}.$$

In the first case, where $m \equiv 3 \pmod 4$, $k \equiv n \equiv 2 \pmod 4$, Proposition 2.34 shows that these are integral if $2a, 2b, 2c, 2d \in \mathbb{Z}$. Thus

$$\alpha = \frac{A + B\sqrt{m} + C\sqrt{n} + D\sqrt{k}}{2},$$

for $A, B, C, D \in \mathbb{Z}$, where $A = 2a$, $B = 2b$, $C = 2c$ and $D = 2d$.

Also,

$$\alpha\alpha_3 = (a + b\sqrt{m})^2 - (c\sqrt{n} + d\sqrt{k})^2$$
$$= a^2 + 2\sqrt{m}ab + mb^2 - nc^2 - 2ncd\sqrt{m}/(m, n) - kd^2$$
$$= \frac{A^2 + mB^2 - nC^2 - kD^2}{4} + \frac{AB - nCD/(m, n)}{2}\sqrt{m}$$

is also integral. Thus $4 | A^2 + mB^2 - nC^2 - kD^2$ and $2 | AB - nCD/(m, n)$. The second implies that $2 | AC$ (n is even and m is odd, so (m, n) is odd and $n/(m, n)$ is even), so that at least one of A and B is even. If only one were even, then $A^2 + mB^2 - nC^2 - kD^2$ would be odd, and the first requirement would fail. So both A and B are even.

The second divisibility is automatic, and the first reduces to $4 | nC^2 + kD^2$, or $2 | C^2 + D^2$, so that C and D are both even or both odd.

So integers are all of the form

$$\alpha = a + b\sqrt{m} + c\sqrt{n} + d\sqrt{k}$$

with $a, b \in \mathbb{Z}$ and c and d both integral or both halves of odd integers.

Such elements are integer linear combinations of 1, $\sqrt{m}$, $\sqrt{n}$ and $\frac{\sqrt{n}+\sqrt{k}}{2}$. The first three are obviously integral, and it is a simple check that if $\gamma = \frac{\sqrt{n}+\sqrt{k}}{2}$, then

$$(4\gamma^2 - (n + k))^2 = 4mn^2/(m, n)^2,$$

and that the congruence conditions on m, n and k imply that this simplifies to a monic polynomial with integer coefficients,

$$\gamma^4 - (n + k)\gamma^2/2 + \left(\frac{n + k}{4}\right) = mn^2/4(m, n)^2,$$

so γ is also integral.

Thus an integral basis is $\{1, \sqrt{m}, \sqrt{n}, \frac{\sqrt{n}+\sqrt{k}}{2}\}$.

The remaining cases are similar, and some details will be left to the reader.

In the second and third cases, which can be treated together, $m \equiv 1 \pmod 4$ and $k \equiv n \equiv 2$ or $3 \pmod 4$. Again, suppose $\alpha = a + b\sqrt{m} + c\sqrt{n} + d\sqrt{k} \in \mathbb{Z}_K$; one shows as in the first case that

$$\alpha = a + b\sqrt{m} + c\sqrt{n} + d\sqrt{k}$$

with $2a, 2b, 2c, 2d \in \mathbb{Z}$. Again let α_2, α_3 and α_4 denote the conjugates of α. Considering $\alpha + \alpha_i$ again, one sees that a and b are both integers or both halves of odd integers, and c and d are both integers or both halves of odd integers. Then every integer must be an integer linear combination of $1, \frac{1+\sqrt{m}}{2}, \sqrt{n}, \frac{\sqrt{n}+\sqrt{k}}{2}$, and one shows easily that these are all integral.

The final case, $m \equiv k \equiv n \equiv 1 \pmod 4$ is a little different. Again we consider $\alpha + \alpha_2$, $\alpha + \alpha_3$ and $\alpha + \alpha_4$. By Proposition 2.34, these are integral if $4a, 4b, 4c, 4d \in \mathbb{Z}$, with $2a, 2b, 2c$ and $2d$ all integral or all halves of odd integers. Thus

$$\alpha = \frac{A + B\sqrt{m} + C\sqrt{n} + D\sqrt{k}}{4},$$

for $A, B, C, D \in \mathbb{Z}$, where $A = 4a$, $B = 4b$, $C = 4c$ and $D = 4d$ are all even or all odd. So we can write

$$\alpha = \frac{A' + B'\sqrt{m} + C'\sqrt{n}}{4} + D'\left(\frac{1+\sqrt{m}}{2}\right)\left(\frac{1+\sqrt{n}}{2}\right),$$

with $D' \in \mathbb{Z}$. As

$$\frac{A' + B'\sqrt{m} + C'\sqrt{n}}{4} = \alpha - D'\left(\frac{1+\sqrt{m}}{2}\right)\left(\frac{1+\sqrt{n}}{2}\right),$$

it must be integral, so A', B', C' are all even (the coefficient of $\sqrt{k}$ is 0, which is even). Thus $A' = 2a'$, $B' = 2b'$ and $C' = 2c'$, and we consider $\frac{a'+b'\sqrt{m}+c'\sqrt{n}}{2}$. This is the sum of $b'\left(\frac{1+\sqrt{m}}{2}\right) + c'\left(\frac{1+\sqrt{n}}{2}\right)$ and $\frac{a'-b'-c'}{2}$; it is an integer, so $2|a' - b' - c'$. It follows that the integral basis is as given in the statement.   □

Exercise 8.10 With the numbering of Lemma 8.21, show that the discriminant of K is $64mnk$ in (1), $16mnk$ in (2) and (3), and mnk in (4).

Let us next work out the possible roots of unity in K.

Lemma 8.23 *If $K = \mathbb{Q}(\sqrt{m}, \sqrt{n})$, then the roots of unity in K have order 2, 4, 6, 8 or 12.*

Proof Suppose K contains the rth roots of unity. Then $\mu_r \subset K$, and so $\mathbb{Q}(\mu_r) \subseteq K$. Then we must have $[\mathbb{Q}(\mu_r) : \mathbb{Q}] \le 4$. However, we will see in Corollary 9.9 that $[\mathbb{Q}(\mu_r) : \mathbb{Q}] = \phi(r)$, and so r must satisfy $\phi(r) \le 4$.

If $r = \prod_{p|r} p^{r_p}$, then $\phi(r) = \prod_{p|r} p^{r_p-1}(p-1)$. This shows that no prime $p \ge 7$ can divide n (otherwise $p - 1 \ge 6$ would divide $\phi(r)$); that $5^2 \nmid r$, $3^2 \nmid r$ and $2^4 \nmid r$. This leads to a small list of possibilities for r, and one quickly finds that $r = 1, 2, 3, 4, 5, 6, 8, 10$ or 12. Of course, $-1 \in K$, so we always have square roots, so r will be even. $K = \mathbb{Q}(\mu_{10})$ is ruled out as it is not biquadratic; although it contains $\mathbb{Q}(\sqrt{5})$, there is no other integer d with $\mathbb{Q}(\sqrt{d}) \subseteq \mathbb{Q}(\mu_{10})$, and so $\mathbb{Q}(\mu_{10})$ cannot be written $\mathbb{Q}(\sqrt{5}, \sqrt{n})$ for any n. This just leaves the list in the statement. $\square$

The only possibility with $r = 12$ is $\mathbb{Q}(\mu_{12}) = \mathbb{Q}(\mu_4, \mu_6) = \mathbb{Q}(i, \sqrt{-3})$, and the only possibility with $r = 8$ is $\mathbb{Q}(\mu_8) = \mathbb{Q}(i, \sqrt{2})$.

If both $m > 0$ and $n > 0$, then also $k > 0$, and every embedding of $\sqrt{m}$, $\sqrt{n}$ and $\sqrt{k}$ is real. We shall refer to this case as a *real biquadratic field*. Since the only real roots of unity are ± 1, Dirichlet's Unit Theorem (Theorem 7.31) implies that there are three units, ϵ_1, ϵ_2 and ϵ_3 such that every unit can be written in the form $\pm \epsilon_1^{a_1} \epsilon_2^{a_2} \epsilon_3^{a_3}$, where a_1, a_2 and a_3 are in $\mathbb{Z}$. The fundamental units ϵ_i are, in general, difficult to compute; we have already seen this for real quadratic fields, and the real biquadratic case is considerably harder.

If $m < 0$, say, then each embedding maps $\sqrt{m}$ to $\pm \sqrt{m}$, and this is not real. So all the embeddings are complex, and occur in two complex conjugate pairs. We shall refer to this case as an *imaginary biquadratic field*. Dirichlet's Unit Theorem shows that there is a single unit ϵ such that every unit can be written as $\zeta \epsilon^a$, where ζ is a root of unity in K, and $a \in \mathbb{Z}$. Here, however, the computation of the fundamental unit ϵ is more tractable, and we state the result without proof (see [5], Theorem 42):

Theorem 8.24 *Suppose that $K = \mathbb{Q}(\sqrt{m}, \sqrt{n})$ is an imaginary biquadratic field. Then there is a unique real quadratic subfield $k \subset K$. Let η denote a fundamental unit of k. Then either $\epsilon = \eta$ is also a fundamental unit of K or $\epsilon^2 = \zeta \eta$ for some $\zeta \in \mu_K$. In the first case, the units of U_K are $U_k \mu_K$, and in the second case $U_k \mu_K$ has index 2 in U_K.*

8.5   Cubic Fields

A cubic field is a degree 3 extension of $\mathbb{Q}$, and can therefore be defined as $K = \mathbb{Q}(\gamma)$, where γ is a root of an irreducible cubic equation $f(X) \in \mathbb{Q}[X]$. If $\gamma_1 = \gamma$, γ_2 and γ_3 denote the three complex roots of $f(X)$, the three embeddings from K into $\mathbb{C}$ are given by sending γ to each of the three roots—thus we have

$$\tau_i : \mathbb{Q}(\gamma) \longrightarrow \mathbb{C}.$$
$$\sum_k a_k \gamma^k \mapsto \sum_k a_k \gamma_i^k$$

There is a new phenomenon in the cubic case, not present in the cases considered so far: the image of the embeddings may differ from K. Indeed, the image of the embedding τ_i is $\mathbb{Q}(\gamma_i)$, and we may have $\gamma_i \notin \mathbb{Q}(\gamma)$.

For example, suppose that $f(X) = X^3 - 2$, so that we could have $K = \mathbb{Q}(\sqrt[3]{2})$. Then the other roots of $f(X)$, which are $\omega \sqrt[3]{2}$ and $\omega^2 \sqrt[3]{2}$, where $\omega = e^{2\pi i/3} = \frac{-1+i\sqrt{3}}{2}$, do not belong to K.

We can consider the *splitting field* $L = \mathbb{Q}(\gamma_1, \gamma_2, \gamma_3)$, the field generated over $\mathbb{Q}$ by all of the roots of $f(X)$. (Notice that $\gamma_3 \in \mathbb{Q}(\gamma_1, \gamma_2)$, so that $L = \mathbb{Q}(\gamma_1, \gamma_2)$.)

Lemma 8.25 *If L is the splitting field of an irreducible cubic equation $f(X)$, then $[L : \mathbb{Q}] = 3$ or 6.*

Proof Certainly L contains γ_1, and so $L \supseteq \mathbb{Q}(\gamma_1)$. As the minimal polynomial of γ_1 is a cubic, $[\mathbb{Q}(\gamma_1) : \mathbb{Q}] = 3$. Over $\mathbb{Q}(\gamma_1)$, the cubic $f(X)$ must factor as

$$f(X) = (X - \gamma_1)f_1(X),$$

where $f_1(X) \in \mathbb{Q}(\gamma_1)[X]$ is a quadratic with roots γ_2 and γ_3. If $\gamma_2 \in \mathbb{Q}(\gamma_1)$, then so is γ_3, and $L = \mathbb{Q}(\gamma_1)$, of degree 3.

Otherwise, γ_2 and γ_3 are roots of an irreducible quadratic $f_1(X)$ over $\mathbb{Q}(\gamma_1)$, and so $[\mathbb{Q}(\gamma_1, \gamma_2) : \mathbb{Q}(\gamma_1)] = 2$. The tower law for degrees of field extensions now gives $[L : \mathbb{Q}] = 6$. □

The reason that general cubic number fields are more complicated than biquadratic number fields is that we often have to deal with the splitting field, which is generally of degree 6. The cubic fields whose splitting field are of degree 3 are rather unusual, and none of the natural family of cubic fields $\mathbb{Q}(\sqrt[3]{a})$ have this property, whereas the biquadratic fields are more natural examples of quartic number fields with a degree 4 splitting field.

The cubic equation might have three real roots; in this case, each of the embeddings τ_i are real. Dirichlet's Unit Theorem (see Theorem 7.31) then implies that every unit is $\pm \eta_1^{a_1} \eta_2^{a_2}$, for certain fundamental units η_1 and η_2, where a_1 and a_2 run through integers.

Alternatively, the cubic might have one real root, and one complex conjugate pair of roots. Then there is one real embedding, and one conjugate pair of complex embeddings. Dirichlet's Unit Theorem implies that the units are then of the form $\zeta \eta^a$, where ζ is a root of unity in K, η is a fundamental unit, and $a \in \mathbb{Z}$.

The most natural family of cubics to consider are those of the form $\mathbb{Q}(\sqrt[3]{a})$, where a is an integer not divisible by a cube. Then the minimal polynomial of $\sqrt[3]{a}$ is $X^3 - a$, and the three roots of this are $\sqrt[3]{a}$, $\omega\sqrt[3]{a}$ and $\omega^2\sqrt[3]{a}$, where $\omega = e^{2\pi i/3}$ (note that $-\omega - 1 = \omega^2$). We therefore have one real root, and one complex conjugate pair, so these cubics belong to the second group above.

We have already seen some examples of the computation of the rings of integers of some cubic fields, in Sect. 3.6. The general case can be done in a similar way, and we leave the details as an exercise (see also [9], pp. 49–51). Here is the main result:

Theorem 8.26 *Suppose that $K = \mathbb{Q}(\sqrt[3]{m})$, where $m = m_1 m_2^2$, with m_1 and m_2 coprime and squarefree. Write $m' = m_1^2 m_2$. Then if $m^2 \not\equiv 1 \pmod 9$, the ring of integers has integral basis $\{1, \sqrt[3]{m}, \sqrt[3]{m'}\}$, and K has discriminant $-27m_1^2 m_2^2$. If $m^2 \equiv 1 \pmod 9$, the ring of integers has integral basis $\{1, \sqrt[3]{m}, \frac{m_2 \pm m_2\sqrt[3]{m} + \sqrt[3]{m'}}{3}\}$, and K has discriminant $-3m_1^2 m_2^2$.*

Chapter 9
Cyclotomic Fields and the Fermat Equation

Cyclotomic fields are the number fields generated over $\mathbb{Q}$ by roots of unity. They played (and still play) an important role in developing modern algebraic number theory, most notably because of their connection with Fermat's Last Theorem (see Sect. 9.4). Whole books have been written about cyclotomic fields, but we will just begin to develop a few of their properties.

9.1 Definitions

We have already used the notion of roots of unity at a number of points in the book, but here is a formal definition:

Definition 9.1 An nth *root of unity* is a number $\zeta \in \mathbb{C}$ such that $\zeta^n = 1$, so that $\zeta = e^{2\pi i k/n}$ for some k. We say that ζ is *primitive* if $\zeta^a \neq 1$ for any $0 < a < n$, so that $\zeta = e^{2\pi i k/n}$ for k coprime to n.

It follows that the number of primitive nth roots of unity is

$$\phi(n) = |\{0 \leq k < n \mid k \text{ and } n \text{ are coprime}\}|.$$

Definition 9.2 The nth *cyclotomic field* is the number field $\mathbb{Q}(\zeta)$, where ζ is any primitive nth root of unity.

Exercise 9.1 Check that this definition does not depend on ζ; two different choices of primitive nth roots of unity produce the same number field.

Example 9.3 Let $\zeta \in \mathbb{C}$ be a primitive 5th root of unity. The minimal polynomial of ζ over $\mathbb{Q}$ is $X^4 + X^3 + X^2 + X + 1$. The remaining roots of this polynomial are the other three primitive 5th roots of unity. If ξ is one of them, then $\xi = \zeta^j$ for some j. It follows that $\mathbb{Q}(\xi) = \mathbb{Q}(\zeta)$.

F. Jarvis, *Algebraic Number Theory*, Springer Undergraduate
Mathematics Series, DOI: 10.1007/978-3-319-07545-7_9,
© Springer International Publishing Switzerland 2014

In order to state the most general result, we need to know more about cyclotomic polynomials.

Definition 9.4 Let $n \geq 1$. Define the nth *cyclotomic polynomial* by

$$\lambda_n(X) = \prod_{\text{primitive } n\text{th roots of unity}} (X - \zeta).$$

Let's write down the first few:

$$\lambda_1(X) = X - 1$$
$$\lambda_2(X) = X + 1$$
$$\lambda_3(X) = (X - \omega)(X - \omega^2) = X^2 + X + 1$$
$$\lambda_4(X) = (X + i)(X - i) = X^2 + 1$$
$$\lambda_5(X) = \frac{X^5 - 1}{X - 1} = X^4 + X^3 + X^2 + X + 1$$
$$\lambda_6(X) = (X + \omega)(X + \omega^2) = X^2 - X + 1$$

where ω denotes a primitive cube root of unity. In general, one can see that $\lambda_p(X) = \frac{X^p - 1}{X - 1} = X^{p-1} + \cdots + 1$ when p is a prime.

We have the following lemma:

Lemma 9.5
$$X^n - 1 = \prod_{d|n} \lambda_d(X).$$

Proof Any nth root of unity will be a primitive dth root for some $d|n$. Conversely, if $d|n$, a primitive dth root of unity is an nth root of unity.   □

For example, if $n = 6$, the 6th roots of unity are 1, -1, $\pm\omega$ and $\pm\omega^2$, where $\omega = e^{2\pi i/3} = \frac{-1+\sqrt{-3}}{2}$ is a primitive cube root of unity. We split these into the primitive 1st roots, i.e., 1, the primitive square roots, i.e., -1, the primitive cube roots, i.e., ω and ω^2, and the primitive 6th roots, $-\omega$ and $-\omega^2$. It is clear then that the product of the cyclotomic polynomials λ_d for $d|6$ is $X^6 - 1$. As there is a factor of λ_n for every primitive nth root of unity, it follows that $\deg \lambda_n = \phi(n)$.

The lemma allows us to compute cyclotomic polynomials recursively.

Proposition 9.6 λ_n *is a monic polynomial with integer coefficients.*

Proof We prove this by induction on n. Note $\lambda_1 = X - 1$ satisfies the statement. Let $f(X) = \prod_{d|n, d<n} \lambda_d(X)$. Then by induction, f is monic with integer coefficients. By Lemma 9.5, $X^n - 1 = f\lambda_n$. Now we claim that if $p = qr$ is a product of polynomials, where p and q are monic with integer coefficients, then so is r.

For this, we suppose that

$$p(X) = X^{s+t} + p_1 X^{s+t-1} + \cdots + p_{s+t}$$
$$q(X) = X^s + q_1 X^{s-1} + \cdots + q_s$$
$$r(X) = r_0 X^t + r_1 X^{t-1} + \cdots + r_t$$

By comparing coefficients of X^{s+t}, we see $r_0 = 1$, so r is monic. Also, suppose we have shown that $r_0, \ldots r_{k-1} \in \mathbb{Z}$. Then, comparing coefficients of X^{s+t-k}, we see that

$$p_k = q_k + q_{k-1} r_1 + \cdots + q_1 r_{k-1} + r_k,$$

so we see $r_k \in \mathbb{Z}$. Inductively, each $r_i \in \mathbb{Z}$, so $r \in \mathbb{Z}[X]$. This proves the claim.

Now we apply this with $p = X^n - 1$, $q = f$ and $r = \lambda_n$, to see that $\lambda_n \in \mathbb{Z}[X]$ has integer coefficients. $\qquad \square$

In fact, $\lambda_n(X)$ is irreducible in $\mathbb{Q}[X]$ and hence is the minimal polynomial of any primitive nth root of unity. There's a simple proof if n is a prime number p, which we will give first, using Eisenstein's criterion.

Lemma 9.7 *If p is prime, the polynomial $\lambda_p(X)$ is irreducible.*

Proof In this case, $\lambda_p(X) = \frac{X^p - 1}{X - 1}$, so that

$$\begin{aligned}
\lambda_p(X + 1) &= \frac{(X + 1)^p - 1}{(X + 1) - 1} \\
&= \frac{(X + 1)^p - 1}{X} \\
&= X^{p-1} + \binom{p}{1} X^{p-2} + \binom{p}{2} X^{p-3} + \cdots + \binom{p}{p-2} X + \binom{p}{p-1},
\end{aligned}$$

and all the coefficients except the leading term are divisible by p, with the constant term equal to p. Then Eisenstein's criterion shows that $\lambda_p(X + 1)$ is irreducible, and therefore $\lambda_p(X)$ is also irreducible. $\qquad \square$

The general case is harder; it seems first to have been proven by Kronecker, but the proof we will give is due to Schur (1929).

Proposition 9.8 *The polynomial $\lambda_n(X)$ is irreducible.*

Proof Let $f_n(X) = X^n - 1$, and we work out the discriminant of $f_n(X)$, defined as the product the squares of the differences of roots. The same argument as in the proof of Proposition 3.31 shows that

$$\prod_{i<j} (\zeta^i - \zeta^j)^2 = \prod_{j=1}^n f_n'(\zeta^i).$$

But $f_n'(X) = nX^{n-1}$, and so

$$\prod_{i<j}(\zeta^i - \zeta^j)^2 = n^n \left(\prod_{j=1}^{n}\zeta^i\right)^{n-1} = \pm n^n.$$

Suppose that $g(X)|f_n(X)$, and that ζ is a root of $g(X)$. Then we claim that ζ^p is a root of $g(X)$ for any prime number $p \nmid n$.

Suppose not, so that $g(\zeta^p) \neq 0$. As $g(X)|f_n(X)$, we can factor $g(X) = (X - \zeta_1)\cdots(X - \zeta_d)$ for some d. Then $g(\zeta^p)$ is a product of differences of nth roots of unity, so divides the discriminant $\pm n^n$ already calculated. Modulo p, we have $g(X^p) \equiv g(X)^p \pmod{p}$, and so $p|g(\zeta^p) - g(\zeta)^p$. Thus $p|g(\zeta^p)$ as $g(\zeta) = 0$. But $g(\zeta^p)$ is an algebraic number dividing n^n, and so $p|n$, a contradiction.

So if $g(X)$ is a nontrivial factor of $\lambda_n(X)$, and therefore of $f_n(X)$, and ζ is a primitive nth root of unity which is a root of $g(X)$, then all powers ζ^k must be roots of $g(X)$ for all k coprime to n; simply factor k into primes, and apply the result above successively. In particular, every primitive nth root of unity is a root of $g(X)$, showing that $g(X) = \lambda_n(X)$, i.e., that $\lambda_n(X)$ is irreducible.   □

Corollary 9.9 *If ζ is a primitive nth root of unity, then $[\mathbb{Q}(\zeta) : \mathbb{Q}] = \phi(n)$.*

Exercise 9.2 x$n > 1$ is odd, $\lambda_{2n}(X) = \lambda_n(-X)$.

[*Hint: show that ζ is a primitive $2n$th root of unity if and only if $-\zeta$ is a primitive nth root of unity.*]

Exercise 9.3

1. Show that if $m|n$, then the degree of $\lambda_{mn}(X)$ is the same as the degree of $\lambda_n(X^m)$.
2. Prove also that these two polynomials have the same roots, and deduce that $\lambda_{mn}(X) = \lambda_n(X^m)$.
3. As a special case, deduce that $\lambda_{p^r}(X) = \lambda_p(X^{p^{r-1}})$, and is therefore equal to $\dfrac{X^{p^r} - 1}{X^{p^{r-1}} - 1}$.

We will use the final part in the next section.

Exercise 9.4 Now let p and q be distinct primes. By considering $X^{pq} - 1$, show that $\lambda_q(X^p) = \lambda_q(X)\lambda_{pq}(X)$.

Exercises 1.2 and 1.3 shows that there are infinitely many primes congruent to 1 (mod 4) and to 1 (mod 6) respectively. The reader should observe that the 4th and 6th cyclotomic polynomials play a crucial role in these exercises. The next exercise shows that the argument generalises (with some mild complication) to any n.

Exercise 9.5 Suppose that there are finitely many primes $p_1, \ldots, p_r$ which are congruent to 1 (mod n). Write $x = np_1 \ldots p_r$, and suppose that $p|\lambda_n(x)$.

1. Deduce that $p \mid x^n - 1$, and conclude that $x^n \equiv 1 \pmod{p}$.
2. By considering which roots of unity are roots of the two sides, show that if $n = kl$ with $l > 1$, then $\lambda_n(X) \mid \frac{X^n - 1}{X^k - 1}$.
3. Suppose that $x^k \equiv 1 \pmod{p}$ for some $k < n$ and $k \mid n$. By considering the expansion

$$\frac{x^n - 1}{x^k - 1} = x^{k(l-1)} + x^{k(l-2)} + \cdots + x^k + 1,$$

   modulo p, derive a contradiction.
4. Deduce that there are infinitely many primes congruent to 1 (mod n).

9.2 Discriminants and Integral Bases

In this section, we will compute integral bases for the cyclotomic fields $\mathbb{Q}(\zeta)$, and their discriminants.

More precisely, put $K = \mathbb{Q}(\zeta)$ where ζ is a primitive nth root of unity, and then we will show that $\mathbb{Z}_K = \mathbb{Z}[\zeta]$; this implies that $\{1, \zeta, \ldots, \zeta^{\phi(n)-1}\}$ forms an integral basis. We will use this later (Sect. 9.4) only in the case where n is prime.

We first treat the case when $n = p^r$ is a power of a single prime p. We need a lemma, which describes the ramification behaviour of p in K:

Lemma 9.10 *Let $n = p^r$, let ζ denote a primitive nth root of unity, and put $\pi = 1 - \zeta$. Then*

$$p\mathbb{Z}_K = \langle \pi \rangle^k$$

where $k = [\mathbb{Q}(\zeta) : \mathbb{Q}] = \phi(p^r) = p^{r-1}(p-1)$. Furthermore, $N_{K/\mathbb{Q}}(\pi) = p$.

Proof As above, the minimal polynomial of ζ is the nth cyclotomic polynomial. In the case of a prime power $n = p^r$,

$$\lambda_{p^r}(X) = \frac{X^{p^r} - 1}{X^{p^{r-1}} - 1} = X^{p^{r-1}(p-1)} + X^{p^{r-1}(p-2)} + \cdots + X^{p^{r-1}} + 1.$$

The roots of $\lambda_{p^r}(X)$ are all the primitive nth roots of unity, which are given by ζ^g, with $g \in G = \{1 \le k \le n \mid p \nmid g\}$. So

$$\lambda_{p^r}(X) = \prod_{g \in G} (X - \zeta^g).$$

We now put $X = 1$ in these two expressions for $\lambda_{p^r}(X)$. In the first, the explicit expression on the right-hand side shows that $\lambda_{p^r}(1) = p$, and substituting this into the second gives

$$p = \prod_{g \in G}(1 - \zeta^g),$$

so that

$$p\mathbb{Z}_K = \langle p \rangle = \prod_{g \in G}\langle 1 - \zeta^g \rangle.$$

We claim that the ideals in the factorisation in this product are all the same. This follows as the generators are associates:

$$\frac{1 - \zeta^g}{1 - \zeta} = 1 + \zeta + \cdots + \zeta^{g-1} \in \mathbb{Z}[\zeta],$$

and conversely, we can find $h \in G$ with $gh \equiv 1 \pmod{p^r}$, and then

$$\frac{1 - \zeta}{1 - \zeta^g} = \frac{1 - (\zeta^g)^h}{1 - \zeta^g} = 1 + \zeta^g + \cdots + \zeta^{g(h-1)} \in \mathbb{Z}[\zeta].$$

Thus $\langle 1 - \zeta^g \rangle = \langle 1 - \zeta \rangle$ for all $g \in G$. Then

$$p\mathbb{Z}_K = \prod_{g \in G}\langle 1 - \zeta^g \rangle = \langle 1 - \zeta \rangle^{|G|} = \langle \pi \rangle^k,$$

with k as in the statement of the lemma.

To get the claim about the norm, we simply apply $N_{K/\mathbb{Q}}$ to this equality. We know that $N_{K/\mathbb{Q}}(p) = p^{[K:\mathbb{Q}]} = p^k$. On the other hand, the norm of the right-hand side is $N_{K/\mathbb{Q}}(\pi)^k$, so that $N_{K/\mathbb{Q}}(\pi) = p$. □

Next, we can compute the discriminant of the basis $\{1, \zeta, \ldots, \zeta^{k-1}\}$:

Lemma 9.11 *With notation as in the previous lemma, the discriminant* $\Delta\{1, \zeta, \ldots, \zeta^{k-1}\} = \pm p^s$ *for some exponent s.*

Proof Write $\lambda(X) = \lambda_{p^r}(X) = \frac{X^{p^r}-1}{X^{p^{r-1}}-1}$, and rearrange this as

$$(X^{p^{r-1}} - 1)\lambda(X) = X^{p^r} - 1. \tag{9.1}$$

We will use Proposition 3.31 to compute the discriminant, so we need to compute the norm of $\lambda'(\zeta)$. Differentiate (9.1), to get

$$(p^{r-1}X^{p^{r-1}-1})\lambda(X) + (X^{p^{r-1}} - 1)\lambda'(X) = p^r X^{p^r-1}.$$

Substitute $X = \zeta$:

$$(\zeta^{p^{r-1}} - 1)\lambda'(\zeta) = p^r \zeta^{p^r-1} = p^r \zeta^{-1}.$$

Put $\xi = \zeta^{p^{r-1}}$; this is a pth root of unity, and so $N_{\mathbb{Q}(\xi)/\mathbb{Q}}(\xi - 1) = \pm p$ by Lemma 9.10. Then $N_{\mathbb{Q}(\zeta)/\mathbb{Q}}(\xi - 1) = (\pm p)^{[\mathbb{Q}(\zeta):\mathbb{Q}(\xi)]} = \pm p^{p^{r-1}}$. Take norms of the equality $(\xi - 1)\lambda'(\zeta) = p^r \zeta^{-1}$, and find

$$N_{\mathbb{Q}(\zeta)/\mathbb{Q}}(\xi - 1)N_{\mathbb{Q}(\zeta)/\mathbb{Q}}(\lambda'(\zeta)) = N_{\mathbb{Q}(\zeta)/\mathbb{Q}}(p^r)N_{\mathbb{Q}(\zeta)/\mathbb{Q}}(\zeta)^{-1}.$$

Substituting in the earlier calculations, noting that ζ is a root of unity (and thus has norm ± 1), this becomes:

$$\pm p^{p^{r-1}} N_{\mathbb{Q}(\zeta)/\mathbb{Q}}(\lambda'(\zeta)) = (p^r)^{p^{r-1}(p-1)}.$$

By Proposition 3.31, $\Delta\{1, \zeta, \ldots, \zeta^{k-1}\}$ can be computed as $N_{\mathbb{Q}(\zeta)/\mathbb{Q}}(\lambda'(\zeta))$, which can now be read off as $\pm p^s$, where $s = rp^{r-1}(p - 1) - p^{r-1}$ by Proposition 3.31.
$\square$

In passing, we note that this implies that p is the only prime ramifying in $\mathbb{Q}(\zeta_{p^r})$ (see Proposition 5.44). We can also use this to see that $\mathbb{Q}(\zeta)$ is monogenic, so that it has an integral basis generated by a single element.

Proposition 9.12 *Let $n = p^r$, and let ζ denote a primitive nth root of unity. Then the ring of integers of $K = \mathbb{Q}(\zeta)$ is given by $\mathbb{Z}[\zeta]$.*

Proof Write $\mathbb{Z}_K$ for the ring of integers. As $\Delta\{1, \zeta, \ldots, \zeta^{k-1}\} = \pm p^s$ for some integer s, where $k = [\mathbb{Q}(\zeta) : \mathbb{Q}]$, we know from Lemma 3.32 that

$$p^s \mathbb{Z}_K \subseteq \mathbb{Z}[\zeta] \subseteq \mathbb{Z}_K.$$

As in Lemma 9.10, if $\pi = 1 - \zeta$, then $N_{\mathbb{Q}(\zeta)/\mathbb{Q}}(\pi) = p$. Thus $\mathbb{Z}_K/\pi\mathbb{Z}_K \cong \mathbb{Z}/p\mathbb{Z}$, so that $\mathbb{Z}_K = \mathbb{Z} + \pi\mathbb{Z}_K$, and therefore

$$\mathbb{Z}_K = \mathbb{Z}[\zeta] + \pi\mathbb{Z}_K. \tag{9.2}$$

Multiplying through by π gives

$$\pi\mathbb{Z}_K = \pi\mathbb{Z}[\zeta] + \pi^2\mathbb{Z}_K,$$

and substituting this into (9.2) gives

$$\mathbb{Z}_K = \mathbb{Z}[\zeta] + (\pi\mathbb{Z}[\zeta] + \pi^2\mathbb{Z}_K) = \mathbb{Z}[\zeta] + \pi^2\mathbb{Z}_K. \tag{9.3}$$

We can repeat this procedure to see that

$$\mathbb{Z}_K = \mathbb{Z}[\zeta] + \pi^m\mathbb{Z}_K$$

for all $m \geq 1$. However, if we put $m = s$, we have already observed that $\pi^s\mathbb{Z}_K \subseteq \mathbb{Z}[\zeta]$, and so we conclude that $\mathbb{Z}_K = \mathbb{Z}[\zeta]$, as required.
$\square$

This completes the argument for prime power exponents; given a general exponent, we just need to combine the information from each of its prime powers.

Theorem 9.13 *Let $n \in \mathbb{Z}_{\geq 1}$, and let ζ denote a primitive nth root of unity. Then the ring of integers of $K = \mathbb{Q}(\zeta)$ is given by $\mathbb{Z}[\zeta]$.*

Proof Write $n = p_1^{r_1} \ldots p_s^{r_s}$. For $i = 1, \ldots, s$, write $\zeta_i = \zeta^{n/p_i^{r_i}}$, which are $p_i^{r_i}$th roots of unity. Write $K_i = \mathbb{Q}(\zeta_i) \subseteq \mathbb{Q}(\zeta)$. The K_i are cyclotomic fields, and $\mathbb{Z}_{K_i} = \mathbb{Z}[\zeta_i]$ by Proposition 9.12. So each $\mathbb{Z}_{K_i}$ is generated by powers of ζ_i, which, in turn, are therefore powers of ζ.

As the discriminants of K_1 and K_2 are coprime, we conclude from Proposition 3.34 that the ring of integers of $K_1 K_2 = \mathbb{Q}(\zeta_1, \zeta_2)$ has a basis consisting of powers $\zeta_1^{a_1} \zeta_2^{a_2}$, and that these are again all powers of ζ.

Similarly we conclude that the ring of integers of $K_1 K_2 K_3 = \mathbb{Q}(\zeta_1, \zeta_2, \zeta_3)$ also has a basis consisting of powers of ζ; continuing in this way, we see that the ring of integers of $K_1 \ldots K_s = \mathbb{Q}(\zeta_1, \ldots, \zeta_s) = \mathbb{Q}(\zeta)$ has a basis consisting of powers of ζ, and so $\mathbb{Z}_K = \mathbb{Z}[\zeta]$ as required. $\square$

9.3 Gauss Sums and Quadratic Reciprocity

In this section, we give a proof of quadratic reciprocity using "Gauss sums", which arise as weighted sums of roots of unity.

We will also sketch a proof that cyclotomic fields need not always have unique factorisation. Even in the case $\mathbb{Q}(\zeta)$, with ζ a pth root of unity for some prime p, of most interest for the Fermat equation (see Sect. 9.4), it turns out that $\mathbb{Q}(\zeta)$ does not always have unique factorisation. In fact, $\mathbb{Q}(\zeta)$ has unique factorisation for all $p \leq 19$, but for all $p > 19$, $\mathbb{Q}(\zeta)$ fails to have unique factorisation (and the class group grows quickly; already for $p = 97$, the class number of $\mathbb{Q}(\zeta)$ is greater than 10^{11}). We will sketch the argument that $\mathbb{Q}(\zeta_{23})$ fails to have unique factorisation.

Partly for this reason, we will initially develop some theory of Gauss sums just for this example. We first show that the $\mathbb{Q}(\zeta_{23})$ contains $\mathbb{Q}(\sqrt{-23})$ as a subfield, and the proof will introduce what will be a Gauss sum.

Lemma 9.14 $\mathbb{Q}(\sqrt{-23}) \subset \mathbb{Q}(\zeta_{23})$ *and* $[\mathbb{Q}(\zeta_{23}) : \mathbb{Q}(\sqrt{-23})] = 11$.

Proof Put

$$\tau = \sum_{a=1}^{22} \left(\frac{a}{23}\right) \zeta^a,$$

where $\zeta = \zeta_{23}$, and $\left(\frac{a}{23}\right)$ is the Legendre symbol. We claim that $\tau^2 = -23$, so that $\sqrt{-23} = \pm \tau \in \mathbb{Q}(\zeta_{23})$.

Indeed,

$$\tau^2 = \sum_{a=1}^{22}\sum_{b=1}^{22} \left(\frac{a}{23}\right)\zeta^a \left(\frac{b}{23}\right)\zeta^b.$$

Given a pair (a, b), define c by $b \equiv ac \pmod{23}$, and note that if a is fixed, then when b runs through all values $1, \ldots, 22$, so does c. Thus:

$$\tau^2 = \sum_{a=1}^{22}\sum_{b=1}^{22} \left(\frac{a}{23}\right)\zeta^a \left(\frac{b}{23}\right)\zeta^b$$

$$= \sum_{a=1}^{22}\sum_{c=1}^{22} \left(\frac{a^2 c}{23}\right)\zeta^{a+ac}$$

$$= \sum_{a=1}^{22}\sum_{c=1}^{21} \left(\frac{c}{23}\right)\zeta^{a(1+c)} + \sum_{a=1}^{22}\left(\frac{-1}{23}\right)$$

$$= \sum_{c=1}^{21}\left[\left(\frac{c}{23}\right)\sum_{a=1}^{22}\zeta^{a(1+c)}\right] + 22\cdot(-1)$$

as $\left(\dfrac{-1}{23}\right) = -1$

$$= \left[\sum_{c=1}^{21}\left(\frac{c}{23}\right)\cdot(-1)\right] - 22$$

as $\sum_{a=0}^{22}\zeta^{ka} = 0$ for $k \not\equiv 0 \pmod{23}$, so that $\sum_{a=1}^{22}\zeta^{ak} = -1$

$$= 1\cdot(-1) - 22 = -23$$

using $\sum_{c=1}^{22}\left(\frac{c}{23}\right) = 0$, so that $\sum_{c=1}^{21}\left(\frac{c}{23}\right) = -\left(\frac{-1}{23}\right)$.

The second claim follows from the tower law for field extensions; we know $[\mathbb{Q}(\zeta_{23}) : \mathbb{Q}] = \phi(23) = 22$, and $[\mathbb{Q}(\sqrt{-23}) : \mathbb{Q}] = 2$. ☐

Exercise 9.6 Generalise Lemma 9.14 to show that in $\mathbb{Q}(\zeta_p)$, with p prime, the *Gauss sum* $\tau = \sum_{a=1}^{p-1}\left(\frac{a}{p}\right)\zeta_p^a$ satisfies

$$\tau^2 = \begin{cases} p, & \text{if } p \equiv 1 \pmod 4, \\ -p, & \text{if } p \equiv 3 \pmod 4. \end{cases}$$

and so $\tau^2 = \left(\frac{-1}{p}\right)p.$

Gauss sums allow us to give a fairly simple proof of the famous Quadratic Reciprocity Theorem, using the generalisation Exercise 9.6 of Lemma 9.14:

Theorem 9.15 *Suppose that p and q are distinct odd primes. Then*

$$\left(\frac{p}{q}\right)\left(\frac{q}{p}\right) = (-1)^{(p-1)(q-1)/4}.$$

Proof As above, write $\zeta = \zeta_p$, and consider again the Gauss sum $\tau(\zeta) = \sum_{a=1}^{p-1}\left(\frac{a}{p}\right)\zeta^a$, and consider also $\tau(\zeta^q) = \sum_{a=1}^{p-1}\left(\frac{a}{p}\right)(\zeta^q)^a$. Observe that

$$\left(\frac{q}{p}\right)\tau(\zeta^q) = \sum_{a=1}^{p-1}\left(\frac{aq}{p}\right)\zeta^{aq} = \tau(\zeta), \qquad (9.4)$$

using the multiplicativity of the Legendre symbol.

We can also evaluate $\tau(\zeta^q)$ by working modulo $q\mathbb{Z}[\zeta_p]$:

$$\begin{aligned}
\tau(\zeta^q) &\equiv \tau(\zeta)^q \\
&= \tau(\zeta)\left(\tau(\zeta)^2\right)^{(q-1)/2} \\
&= \tau(\zeta)p^{*(q-1)/2} \\
&= \tau(\zeta)\left(\frac{p^*}{q}\right)
\end{aligned}$$

where we write $p^* = \left(\frac{-1}{p}\right)p$ as in Exercise 9.6, using Euler's criterion $a^{(q-1)/2} \equiv \left(\frac{a}{q}\right)$ (mod q). Comparing this with (9.4), we see that $\left(\frac{q}{p}\right)\left(\frac{p^*}{q}\right) = 1$. But

$$\left(\frac{p^*}{q}\right) = \left(\frac{(-1)^{(p-1)/2}p}{q}\right) = \left(\frac{-1}{q}\right)^{(p-1)/2}\left(\frac{p}{q}\right) = (-1)^{(p-1)(q-1)/4}\left(\frac{p}{q}\right),$$

and the result follows.   □

Now let's sketch the argument that $\mathbb{Q}(\zeta_{23})$ fails to have unique factorisation. This uses a new concept which will not be used elsewhere in the book; we will merely give the definition, and state its main properties. The interested reader will find proofs in [10] and many other graduate level textbooks in the subject.

Definition 9.16 Suppose that $L \supseteq K$ is an extension of number fields, and that $\mathfrak{A}$ is an ideal in $\mathbb{Z}_L$. Then the *relative ideal norm* $N_{L/K}(\mathfrak{A})$ is the ideal in $\mathbb{Z}_K$ generated by all of the elements $N_{L/K}(A)$, where $A \in \mathfrak{A}$.

The relative ideal norm has the following properties.

1. $N_{L/K}(\mathfrak{A}\mathfrak{B}) = N_{L/K}(\mathfrak{A})N_{L/K}(\mathfrak{B})$ for ideals $\mathfrak{A}$ and $\mathfrak{B}$ in $\mathbb{Z}_L$;
2. if $\mathfrak{P}$ is a prime ideal in $\mathbb{Z}_L$, then $N_{L/K}(\mathfrak{P}) = \mathfrak{p}^f$, where $\mathfrak{p} = \mathfrak{P} \cap \mathbb{Z}_K$, and where f is the degree of the residue field extension $\mathbb{Z}_L/\mathfrak{P} \supseteq \mathbb{Z}_K/\mathfrak{p}$;
3. if $\mathfrak{a}$ is an ideal of $\mathbb{Z}_K$, then $N_{L/K}(\mathfrak{a}\mathbb{Z}_L) = \mathfrak{a}^{[L:K]}$;
4. if $\mathfrak{A} = \langle \alpha \rangle$ is a principal ideal of $\mathbb{Z}_L$, then $N_{L/K}(\mathfrak{A}) = \langle N_{L/K}(\alpha) \rangle$ is a principal ideal of $\mathbb{Z}_K$;
5. if $M \supseteq L \supseteq K$ are extensions of number fields, then $N_{M/K}(\mathfrak{A}) = N_{L/K}(N_{M/L}(\mathfrak{A}))$, if $\mathfrak{A}$ is an ideal of $\mathbb{Z}_M$.

Recall that the ring of integers of $\mathbb{Q}(\sqrt{-23})$ is $\mathbb{Z}[\rho]$, where $\rho = \frac{1+\sqrt{-23}}{2}$. We can compute the class number of $\mathbb{Q}(\sqrt{-23})$ using the quadratic forms method of Chap. 6; we find that there are three distinct reduced forms of discriminant -23, namely

$$x^2 + xy + 6y^2, \quad 2x^2 + xy + 3y^2, \quad 2x^2 - xy + 3y^2.$$

Thus the class number of $\mathbb{Q}(\sqrt{-23})$ is 3, and the class group is therefore isomorphic to C_3, the cyclic group of order 3.

We will also consider the factorisation of the prime 2 in $\mathbb{Q}(\sqrt{-23})$ by using Proposition 5.42. The minimal polynomial of ρ is $X^2 - X + 6$, and

$$X^2 - X + 6 \equiv X^2 + X = X(X + 1) \pmod{2},$$

so that in $\mathbb{Z}[\rho]$, we have $2\mathbb{Z}[\rho] = \mathfrak{p}\mathfrak{p}'$. Again, the general techniques earlier (see Remark 5.43) show that we can take

$$\mathfrak{p} = \langle 2, \rho \rangle, \qquad \mathfrak{p}' = \langle 2, \rho - 1 \rangle.$$

Note that the methods of Chap. 4 show that $\mathfrak{p}$ and $\mathfrak{p}'$ are not principal; there are no elements of norm 2 in the ring of integers of $\mathbb{Q}(\sqrt{-23})$. Thus $\mathfrak{p}$ is not trivial in the class group.

Since the class number of $\mathbb{Q}(\sqrt{-23})$ is 3, we see that $\mathfrak{p}^3$ is principal, and so $\mathfrak{p}$ has order 3 in the class group.

Let $\mathfrak{P}$ be a prime ideal of $\mathbb{Q}(\zeta_{23})$ lying above $\mathfrak{p}$. Write $N_{\mathbb{Q}(\zeta_{23})/\mathbb{Q}(\sqrt{-23})}(\mathfrak{P})$ as $\mathfrak{p}^f$ for some f. As $f|[\mathbb{Q}(\zeta_{23}) : \mathbb{Q}(\sqrt{-23})]$ (proven in the same way as Theorem 5.41), we see that $f|11$. It follows that $N_{\mathbb{Q}(\zeta_{23})/\mathbb{Q}(\sqrt{-23})}(\mathfrak{P}) = \mathfrak{p}$ or $\mathfrak{p}^{11}$.

But $\mathfrak{p}$ has order 3 in the class group, and so $\mathfrak{p}^f$ can only be principal if $3|f$. We conclude that $N_{\mathbb{Q}(\zeta_{23})/\mathbb{Q}(\sqrt{-23})}(\mathfrak{P})$ is not a principal ideal in $\mathbb{Q}(\sqrt{-23})$.

But if $\mathfrak{P}$ were a principal ideal in $\mathbb{Q}(\zeta_{23})$, the norm $N_{\mathbb{Q}(\zeta_{23})/\mathbb{Q}(\sqrt{-23})}(\mathfrak{P})$ would be a principal ideal in $\mathbb{Q}(\sqrt{-23})$, and we conclude that $\mathfrak{P}$ is not a principal ideal in $\mathbb{Q}(\zeta_{23})$.

It follows that $\mathbb{Q}(\zeta_{23})$ does not have unique factorisation.

Exercise 9.7 Verify explicitly that $\mathfrak{p}^3$ is principal in the above calculation by listing the generators 2^3, $2^2\rho$, $2\rho^2$ and ρ^3 for $\mathfrak{p}^3$, verifying that each is divisible by the element $\frac{3-\sqrt{-23}}{2}$, so that $\mathfrak{p}^3 \subseteq \langle \frac{3-\sqrt{-23}}{2} \rangle$, and checking that both sides have norm 8.

Remark 9.17 In fact, it was already known to Kummer that $\mathbb{Q}(\zeta_{23})$ fails to have unique factorisation. Kummer's argument was based on the product

$$\left(1 + \zeta + \zeta^5 + \zeta^6 + \zeta^7 + \zeta^9 + \zeta^{11}\right) \left(1 + \zeta^2 + \zeta^4 + \zeta^5 + \zeta^6 + \zeta^{10} + \zeta^{11}\right),$$

where we write ζ for ζ_{23}. This is easily seen to be divisible by 2 (see Exercise 9.8). It is also clear that neither factor is divisible by 2. However, Kummer showed that 2 is irreducible in $\mathbb{Z}[\zeta]$, the ring of integers in $\mathbb{Q}(\zeta_{23})$, and therefore the same methods as Chap. 4 show that 2 is not a prime element, and therefore that $\mathbb{Z}[\zeta]$ cannot be a UFD.

Exercise 9.8 Verify the assertion that the product in Remark 9.17 is divisible by 2, by multiplying out the brackets.

9.4 Remarks on Fermat's Last Theorem

The Fermat equation, $x^n + y^n = z^n$, has inspired more mathematical breakthroughs than any other. As is well-known, around 1638, Fermat wrote in the margin of his copy of a translation of the *Arithmetica* of Diophantus:

> Cubum autem in duos cubos, aut quadratoquadratum in duos quadratoquadratos, et generaliter nullam in infinitum ultra quadratum potestatem in duos eiusdem nominis fas est dividere cuius rei demonstrationem mirabilem sane detexi. Hanc marginis exiguitas non caperet.

In modern notation, this translates to the statement that when $n \geq 3$, the Fermat equation has no solution with x, y and z all nonzero integers. Fermat's comments that he had a marvellous proof but that the margin was too small led to this becoming the most notorious unsolved problem in mathematics for centuries. Several organisations offered prizes for its proof, but it was not until the work of Wiles [17], partly with Taylor [14], that the proof was finally completed. The work of Wiles and Taylor–Wiles is well beyond the scope of an undergraduate course, although Fermat's proof for $n = 4$, and Euler's proof for $n = 3$ would be more accessible.

The importance of the result is slight; however, the results which it inspired have been enormously influential (and include most of the contents of this book!). Thus the book would seem incomplete without a short sketch of how the Fermat equation has influenced the subject.

Following Fermat's proof of the case $n = 4$ (see [7], for example), it suffices to prove the result in the case that $n = p$ is an odd prime. It is usual to distinguish two cases, the *first case*, where none of x, y and z are divisible by p, and the *second case*, where exactly one of x, y and z is divisible by p. It tends to be easier to prove results using the elementary techniques available to us in the first case, and we shall restrict attention to that case in this sketch; however, with more work, the same results can generally be obtained in the second case also.

We now sketch an elementary (but false!) argument to prove Fermat's Last Theorem in the first case.

For $p = 3$, the equality $x^p + y^p = z^p$ can be regarded as a congruence modulo 9; as cubes are congruent to ± 1 or to 0, we can only recover the congruence if one of the cubes is 0 (mod 9). Suppose it is x^3, say. But this implies that x must be divisible by 3, and this contradicts the requirement of the first case, that $p \nmid x$.

Now consider the case where $p \geq 5$, and suppose that x, y and z are a nontrivial solution to the first case of the Fermat equation. Dividing throughout by a common factor, we can assume x, y and z have no common factor. We will try to establish a contradiction.

Let $\zeta = e^{\frac{2\pi i}{p}} \neq 1$ be a pth root of unity. Write $K = \mathbb{Q}(\zeta)$; we know that $[K : \mathbb{Q}] = p - 1$. From the Fermat equation, we deduce

$$z^p = x^p + y^p$$
$$= (x + y)(x + \zeta y)(x + \zeta^2 y) \cdots (x + \zeta^{p-1} y),$$

an equality in $\mathbb{Z}[\zeta]$.

The left-hand side of this equation is a pth power. We claim that the factors on the right-hand side are all coprime. Indeed, if $x + y$ and $x + \zeta y$, say, shared a common factor $\theta \in \mathbb{Z}[\zeta]$ which is not a unit in $\mathbb{Z}[\zeta]$, then $\theta | x + y$ and $\theta | x + \zeta y$ would imply that $\theta | (x + y) - (x + \zeta y)$, and so $\theta | (1 - \zeta) y$.

But we can't have $\theta | 1 - \zeta$, as $N_{K/\mathbb{Q}}(1 - \zeta) = p$, and if θ is not a unit, then we would need $N_{K/\mathbb{Q}}(\theta) = p$. But $\theta | x + y$, and $x + y$ is one of the factors of z. So $N_{K/\mathbb{Q}}(\theta)$ would divide $N_{K/\mathbb{Q}}(z) = z^{p-1}$. We would conclude that $p | z^{p-1}$, and so $p | z$, contradicting the assumption that we are in the first case.

So we must have $\theta | y$. But then $\theta | x + y$ and $\theta | y$ implies that $\theta | x$, and so θ is a common factor of x and y (and therefore z). Then $N_{K/\mathbb{Q}}(\theta) | N_{K/\mathbb{Q}}(x)$ and $N_{K/\mathbb{Q}}(\theta) | N_{K/\mathbb{Q}}(y)$, so that $N_{K/\mathbb{Q}}(\theta)$ is an integer dividing x^{p-1} and y^{p-1}. As x and y are coprime, we must have $N_{K/\mathbb{Q}}(\theta) = 1$, so that θ is a unit, again a contradiction.

Similar arguments apply (the reader might like to treat this as an exercise) if any pair $x + \zeta^i y$ and $x + \zeta^j y$ shared a common factor; this again leads to a contradiction.

So the factors on the right-hand side of the equality

$$z^p = (x + y)(x + \zeta y)(x + \zeta^2 y) \cdots (x + \zeta^{p-1} y)$$

are coprime; any divisor of z^p can therefore only divide one of the factors on the right-hand side—and since any divisor of z^p occurs to multiplicity p, we must have that each factor on the right-hand side is a pth power.

We have seen (Proposition 9.12) that an integral basis for $\mathbb{Z}[\zeta]$ is given by $\{1, \zeta, \ldots, \zeta^{p-2}\}$.

We can write $x + \zeta y = \alpha^p$ for some $\alpha \in \mathbb{Z}[\zeta]$, and can write $\alpha = a_0 + a_1\zeta + \cdots + a_{p-2}\zeta^{p-2}$. Then

$$\alpha^p \equiv a_0^p + a_1^p \zeta^p + \cdots + a_{p-2}^p (\zeta^{p-2})^p \pmod{p}$$
$$\equiv a_0^p + a_1^p + \cdots + a_{p-2}^p \pmod{p},$$

where we regard two elements of $\mathbb{Z}[\zeta]$ as congruent modulo p if each of the coefficients of $1, \zeta, \ldots, \zeta^{p-2}$ differs by a multiple of p.

Let $\overline{\alpha}$ denote the complex conjugate of α; if $\zeta = e^{\frac{2\pi i}{p}}$, say, then $\overline{\zeta} = e^{-\frac{2\pi i}{p}} = \zeta^{p-1}$. It follows that $\overline{\alpha} = a_0 + a_1\zeta^{p-1} + a_2\zeta^{p-2} + \cdots + a_{p-2}\zeta^2$, and the same argument shows that $\overline{\alpha}^p$ is also congruent modulo p to $a_0^p + a_1^p + \cdots + a_{p-2}^p \pmod{p}$.

On the other hand, $\overline{\alpha}^p = \overline{x + \zeta y} = x + \zeta^{p-1}y$. So we see that $\zeta y \equiv \zeta^{p-1}y \pmod{p}$. Then $p | \zeta y (\zeta^{p-2} - 1)$, and so $N_{K/\mathbb{Q}}(p) | N_{K/\mathbb{Q}}(y) N_{K/\mathbb{Q}}(\zeta^{p-2} - 1)$. But $p \nmid y$, as we are in the first case, and $N_{K/\mathbb{Q}}(p) = p^{p-1}$, and $N_{K/\mathbb{Q}}(\zeta^{p-2} - 1) = p$ by Lemma 9.10.

This is a contradiction.

Of course, things are really more complicated! We've made two false deductions in the course of this argument, and alert readers will probably have spotted both.

Firstly, we cannot conclude that $x + \zeta y = \alpha^p$; there may be units involved. It would have been more correct to conclude that $x + \zeta y = \epsilon \alpha^p$ for some unit ϵ and $\alpha \in \mathbb{Z}[\zeta]$. In fact, this is not such a serious issue; Kummer was able to study the structure of the units in $\mathbb{Z}[\zeta]$ carefully, and prove that the units are all of the form $r\zeta^k$, where r is a real number. The argument above can be modified to derive the same contradiction, with just a little more work.

More importantly, though, it may happen that $\mathbb{Z}[\zeta]$ does not have unique factorisation. This is a much more serious issue.

When $\mathbb{Z}[\zeta]$ does have unique factorisation, this does indeed complete the proof of Fermat's Last Theorem in the first case. However, this is rare!

As we have already seen, Kummer's solution was to work with ideals instead of elements. We use the above argument, and reach the equality

$$z^p = (x + y)(x + \zeta y)(x + \zeta^2 y) \cdots (x + \zeta^{p-1}y)$$

as before. We regard this now as an equality of principal ideals:

$$\langle z \rangle^p = \langle x + y \rangle \langle x + \zeta y \rangle \langle x + \zeta^2 y \rangle \cdots \langle x + \zeta^{p-1}y \rangle.$$

Since we have unique factorisation into ideals, we can now conclude that each ideal on the right-hand side is a pth power. That is, we have

$$\langle x + \zeta y \rangle = \mathfrak{a}^p$$

for some ideal $\mathfrak{a}$. The rest of the argument given above would work if we knew that $\mathfrak{a}$ was a principal ideal; then we could choose $\mathfrak{a} = \langle \alpha \rangle$, and use this α in the rest of the argument.

This is often possible. Indeed, it is clearly valid if every ideal is principal (which would mean that the class number is 1). But in fact we can see that it will hold in greater generality. We know that $\mathfrak{a}^p = \langle x + \zeta y \rangle$ is principal, and therefore the class of $\mathfrak{a}^p$ is trivial in the class group of K. If the size of this class group is coprime to p, we could conclude that the class of $\mathfrak{a}$ would have to be trivial also, as elements whose pth power is trivial in groups of order prime to p are necessarily themselves trivial. So as long as the class number of K is not divisible by p, we can derive the result.

Definition 9.18 The prime number p is *regular* if p does not divide the class number of $\mathbb{Q}(\zeta_p)$. If a prime p is not regular, it is *irregular*

In this way, Kummer was able to prove the first case of Fermat's Last Theorem for all regular primes p, and, as already remarked, he was also able to treat the second case using similar, but rather more complicated, ideas. The reader is referred to [1] or to [6] for complete details. Thus:

Theorem 9.19 (Kummer) *If p is a regular prime, then $x^p + y^p = z^p$ has no nontrivial solution.*

Many primes are regular; indeed, computational evidence suggests that, up to any given bound, there are more regular primes than irregular ones. Of the primes up to 100,000, for example, 3,789 are irregular, and 5,803 are regular. Unfortunately, it is not known that there are infinitely many regular primes, although it is known that there are infinitely many irregular primes. The first irregular primes are 37, 59, 67, 101 and 103.

In fact, there is a reasonably simple way to recognise regular primes, but again the details are sufficiently complicated that the reader is referred to [1]. The Bernoulli numbers appear in the evaluation of the Riemann zeta function at even integers, and are defined by the series expansion

$$\frac{x}{e^x - 1} = \sum_{n=0}^{\infty} \frac{B_n}{n!} x^n.$$

Then $B_0 = 1$, $B_1 = -\frac{1}{2}$, $B_2 = \frac{1}{6}$, $B_3 = 0$, $B_4 = -\frac{1}{30}$,..... (In fact, $B_{2k+1} = 0$ for all $k \geq 1$, as is easy to see from the observation that $\frac{x}{e^x-1} + \frac{x}{2}$ is an even function of x.) They can easily be computed recursively, given $B_0 = 1$, from the relation

$$\binom{k+1}{0} B_0 + \binom{k+1}{1} B_1 + \cdots + \binom{k+1}{k} B_k = 0$$

for all $k \geq 1$.

Then Kummer showed that

Proposition 9.20 *The prime p is regular if and only if it does not divide any numerator of $B_2, B_4, \ldots, B_{p-3}$.*

Exercise 9.9 Use this to compute B_k recursively for $k = 1, \ldots, 16$. Deduce that all primes up to 19 are regular, and use the proposition to give an irregular prime. Extend your calculations as far as you wish; you will need a computer algebra package to get far enough to see that 37 is irregular.

Chapter 10
Analytic Methods

Although this is a textbook on algebraic number theory, some interesting algebraic results can be obtained by incorporating some analytic techniques.

One well-known result is Dirichlet's Theorem on primes in arithmetic progressions. This states that for all a coprime to a modulus m, there are infinitely many primes which are congruent to $a \pmod{m}$. However, the proof is largely analytic in nature, and the use of algebraic objects (modular characters) in the analytic proof has a different nature from most of the topics previously developed in this book.

Instead, we will focus on another theorem of Dirichlet, the Analytic Class Number Formula. This formula expresses the residue of the zeta function of a number field (to be defined below) in terms of many of the invariants associated to the number fields which we have defined earlier in this text.

10.1 The Riemann Zeta Function

Recall that the zeta function was introduced by Euler, as a function on real numbers $s > 1$, and defined by

$$\zeta(s) = \sum_{n=1}^{\infty} \frac{1}{n^s}.$$

Euler computed the values of $\zeta(2k)$ (for $k \geq 1$), and was aware of the behaviour of the function as s gets closer to 1. Its importance lies in its relation to prime numbers; Euler also defined an alternative *Euler product* expression for $\zeta(s)$ as a product of terms for each prime (see Proposition 10.2) which begins to hint at the deep connections.

But it was Riemann who really developed the theory of the zeta function, to the extent that it is now named after him. He seems to have been the first to consider s as a complex variable, and to realise that questions about the distribution of prime numbers are inextricably linked with the complex behaviour of the zeta function,

F. Jarvis, *Algebraic Number Theory*, Springer Undergraduate
Mathematics Series, DOI: 10.1007/978-3-319-07545-7_10,
© Springer International Publishing Switzerland 2014

and, in particular, the values it takes to the left of the line $Re(s) = 1$, where the definition above no longer converges.

To do this, he explained that the zeta function had a meromorphic continuation to the entire complex plane, with a simple pole at $s = 1$ (with residue 1). Furthermore, this continuation has a symmetry, relating the values of ζ at s and at $1 - s$, known as the *functional equation*: if we write $\xi(s) = \pi^{-s/2} \Gamma(s/2)\zeta(s)$, where $\Gamma(z)$ denotes the usual Gamma function $\Gamma(z) = \int_0^\infty e^{-t} t^z \frac{dt}{t}$, then

$$\xi(s) = \xi(1 - s).$$

There are several proofs of this, and we will give one in the next section. Using classical formulae for the Gamma function, this formula can be rewritten

$$\zeta(s) = 2^s \pi^{s-1} \sin(\pi s/2)\Gamma(1 - s)\zeta(1 - s).$$

Let's look at how $\sum_{n=1}^\infty \frac{1}{n^s}$ behaves on the real line.

Proposition 10.1 $\sum_{n=1}^\infty \frac{1}{n^s}$ *converges absolutely for any real* $s > 1$, *and is not convergent for* $s \leq 1$.

Proof This is a standard result in analysis. We will prove the two statements using the comparison test.

First, consider the case $s \leq 1$. Then $\frac{1}{n^s} \geq \frac{1}{n}$. But the sum $\sum_{n=1}^\infty \frac{1}{n}$ diverges, as can be seen by grouping the terms:

$$\sum_{n=1}^\infty \frac{1}{n} = 1 + \frac{1}{2} + \left(\frac{1}{3} + \frac{1}{4}\right) + \left(\frac{1}{5} + \cdots + \frac{1}{8}\right) + \left(\frac{1}{9} + \cdots + \frac{1}{16}\right) + \cdots$$

$$> 1 + \frac{1}{2} + \left(\frac{1}{4} + \frac{1}{4}\right) + \left(\frac{1}{8} + \cdots + \frac{1}{8}\right) + \left(\frac{1}{16} + \cdots + \frac{1}{16}\right) + \cdots$$

$$= 1 + \frac{1}{2} + \frac{1}{2} + \frac{1}{2} + \frac{1}{2} + \cdots$$

which diverges.

Now let's treat the case $s > 1$. We can group the terms very slightly differently:

$$\sum_{n=1}^\infty \frac{1}{n^s} = 1 + \left(\frac{1}{2^s} + \frac{1}{3^s}\right) + \left(\frac{1}{4^s} + \cdots + \frac{1}{7^s}\right) + \left(\frac{1}{8^s} + \cdots + \frac{1}{15^s}\right) + \cdots$$

$$< 1 + \left(\frac{1}{2^s} + \frac{1}{2^s}\right) + \left(\frac{1}{4^s} + \cdots + \frac{1}{4^s}\right) + \left(\frac{1}{8^a} + \cdots + \frac{1}{8^s}\right) + \cdots$$

$$= 1 + 2.\frac{1}{2^s} + 4.\frac{1}{4^s} + 8.\frac{1}{8^s} + \cdots$$

$$= 1 + \frac{1}{2^{s-1}} + \frac{1}{4^{s-1}} + \frac{1}{8^{s-1}} + \cdots$$

$$= \frac{2^{s-1}}{2^{s-1} - 1},$$

using the formula for summing a geometric progression. This completes the proof.

□

We must therefore restrict attention to $s > 1$ when viewing $\zeta(s)$ as a real function.

In fact, the definition makes sense for s a complex number, although this won't be important for us, and we won't prove it; it turns out that the definition above of $\zeta(s)$ converges for $Re(s) > 1$.

One method to estimate the value of $\zeta(s)$ is to compare it with the integral $\int_1^\infty \frac{dx}{x^s}$. As

$$\int_n^{n+1} \frac{dx}{x^s} = \frac{n^{1-s} - (n+1)^{1-s}}{s-1},$$

and x^{-s} is a decreasing function, the area under the graph on this interval of length 1 lies in between the values of the function at n and at $n+1$:

$$(n+1)^{-s} < \frac{n^{1-s} - (n+1)^{1-s}}{s-1} < n^{-s};$$

summing this over $n = 1, 2, 3 \ldots$ gives

$$\zeta(s) - 1 < \frac{1}{s-1} < \zeta(s).$$

Writing this as two inequalities, and rearranging them gives

$$\frac{1}{s-1} < \zeta(s) < \frac{1}{s-1} + 1.$$

This gives the rate at which $\zeta(s) \to \infty$ as $s \to 1$. (The analytic class number formula, which forms the main result of the chapter, will be the corresponding formula for the zeta function of a number field.)

We have not yet explained the relationship of the ζ function to prime numbers. This is due to the *Euler product* for $\zeta(s)$.

Proposition 10.2 *If $Re(s) > 1$, then*

$$\zeta(s) = \sum_{n=1}^{\infty} \frac{1}{n^s} = \prod_p \left(1 - \frac{1}{p^s}\right)^{-1}.$$

Proof Simply observe that $|p^{-s}| = p^{-Re(s)} < 1$, so that

$$\left(1 - \frac{1}{p^s}\right)^{-1} = 1 + \frac{1}{p^s} + \frac{1}{p^{2s}} + \frac{1}{p^{3s}} + \cdots .$$

Now multiply all these together:

$$\prod_p \left(1 - \frac{1}{p^s}\right)^{-1} = \prod_p \left(1 + \frac{1}{p^s} + \frac{1}{p^{2s}} + \cdots\right)$$
$$= \zeta(s),$$

as every $\frac{1}{n^s}$ appears exactly once in the product (by uniqueness of prime factorisation). $\quad\square$

As a corollary, we note that this gives a proof that there are infinitely many prime numbers. Here are two ways to deduce this from the Euler product:

1. We know that $\zeta(s) \to \infty$ as $s \to 1$. This would not happen if there were only finitely many terms in the product (each term tends to the finite limit $\frac{p}{p-1}$).
2. Alternatively, you may know that

$$\zeta(2) = \sum_{n=1}^{\infty} \frac{1}{n^2} = \frac{\pi^2}{6}.$$

But π^2 is irrational, so we see that $\prod_p (1 - p^{-2})^{-1} = \prod_p \frac{p^2}{p^2-1} = \zeta(2) = \frac{\pi^2}{6}$ is irrational, which would not happen if there were only finitely many terms in the product.

Another implication of the Euler product is that

$$\log \zeta(s) = \sum_p -\log(1 - p^{-s}).$$

But recall that $\log(1 - z) = -(z + z^2/2 + z^3/3 + \cdots)$ if $|z| < 1$, so p^{-s} is a good approximation to $-\log(1 - p^{-s})$: the error is less than p^{-2s}. Indeed, for $s > 1$ real,

$$\log \zeta(s) = \sum_p p^{-s} + \sum_p \left(\frac{p^{-2s}}{2} + \frac{p^{-3s}}{3} + \frac{p^{-4s}}{4} + \cdots\right).$$

As

$$\frac{p^{-2s}}{2} + \frac{p^{-3s}}{3} + \frac{p^{-4s}}{4} + \cdots < \frac{p^{-2s}}{2} + \frac{p^{-3s}}{2} + \frac{p^{-4s}}{2} + \cdots = \frac{p^{-2s}}{2}\left(\frac{1}{1 - p^{-s}}\right) < p^{-2s}$$

for $s > 1$ (so $p^{-s} < 2^{-1} = \frac{1}{2}$),0

$$\log \zeta(s) \approx \sum_p p^{-s}$$

for $s > 1$; the error is at most $\sum_p p^{-2s} < \sum_n n^{-2s} = \zeta(2s) < \zeta(2)$. We deduce that, for $s > 1$ but near 1,

$$\sum_p p^{-s} \approx \log \frac{1}{s-1}.$$

Note that this proves that $\sum_p \frac{1}{p}$ diverges (and therefore yet another proof that there are infinitely many primes).

10.2 The Functional Equation of the Riemann Zeta Function

In fact, in this book we will only really need to understand the behaviour of the function for real values of s. However, in analytic number theory, one can prove some wonderful results about the distribution of the prime numbers by allowing s to be a complex variable. For completeness, although we will not need it elsewhere in the book, we sketch a proof of the functional equation for $\zeta(s)$. The reader should note that it will be valid for all $s \in \mathbb{C}$.

First set $\theta(t) = \sum_{n \in \mathbb{Z}} e^{-\pi n^2 t^2}$. We will need this only for $t \in \mathbb{R}$; for $t \neq 0$, the individual terms in the sum converge so fast to 0 that $\theta(t)$ converges for all $t \neq 0$.

Lemma 10.3 *For $t \neq 0$, we have $\theta(1/t) = t\theta(t)$.*

Proof Recall that $\int_{-\infty}^{\infty} e^{-\pi x^2} dx = 1$. Fix $t > 0$, and write $f(x) = e^{-\pi t^2 x^2}$. Define

$$F(x) = \sum_{n \in \mathbb{Z}} f(x+n) = \sum_{n \in \mathbb{Z}} e^{-\pi t^2 (x+n)^2},$$

which again converges because the terms tend to 0 very quickly.

By definition, $F(0) = \theta(t)$. Also, note that F is periodic with $F(x) = F(x+1)$, so it will have a Fourier series

$$F(x) = \sum_{m \in \mathbb{Z}} a_m e^{2\pi i m x},$$

where

$$a_m = \int_0^1 F(x) e^{-2\pi i m x} dx.$$

$$= \sum_{n \in \mathbb{Z}} \int_0^1 f(x+n) e^{-2\pi i m x} \, dx$$

$$= \sum_{n \in \mathbb{Z}} \int_0^1 f(x+n) e^{-2\pi i m (x+n)} \, dx$$

(as $e^{2\pi i m n} = 1$ for $m, n \in \mathbb{Z}$)

$$= \int_{-\infty}^{\infty} f(x) e^{-2\pi i m x} \, dx$$

$$= \int_{-\infty}^{\infty} e^{-\pi t^2 x^2 - 2\pi i m x} \, dx$$

$$= \int_{-\infty}^{\infty} e^{-\pi (tx + i \frac{m}{t})^2} e^{-\pi m^2 / t^2} \, dx$$

$$= e^{-\pi m^2 / t^2} \int_{-\infty}^{\infty} e^{-\pi (tx + i \frac{m}{t})^2} \, dx$$

$$= t^{-1} e^{-\pi m^2 / t^2}$$

by a change of variable (put $y = tx + i\frac{m}{t}$), and using Cauchy's Theorem to see that the integral along the real axis is the same as the integral along the line $Im(z) = \frac{m}{t}$. Now

$$\theta(t) = F(0) = \sum_{m \in \mathbb{Z}} a_m = \sum_{m \in \mathbb{Z}} t^{-1} e^{-\pi m^2 / t^2} = t^{-1} \theta(1/t),$$

so the result follows.   □

Next, we give a relationship between this function and the Riemann zeta function; the functional equation of the Riemann zeta function will follow from the lemma we've just proven.

Proposition 10.4 *For $Re(s) > 1$, we have*

$$\int_0^{\infty} (\theta(t) - 1) t^{s-1} \, dt = \pi^{-s/2} \Gamma(s/2) \zeta(s),$$

where $\Gamma(z)$ is the usual Gamma function, defined by $\Gamma(z) = \int_0^{\infty} e^{-t} t^z \frac{dt}{t}$.

Proof For $Re(s) > 1$, and from the definition of $\theta(t)$, the integral is

$$2 \int_0^\infty \sum_{n \geq 1} e^{-\pi n^2 t^2} t^{s-1} \, dt = 2 \sum_{n \geq 1} \int_0^\infty e^{-\pi n^2 t^2} t^{s-1} \, dt$$

$$= 2 \sum_{n \geq 1} n^{-s} \int_0^\infty e^{-\pi u^2} u^{s-1} \, du = 2\zeta(s) \int_0^\infty e^{-v} (v/\pi)^{s/2-1} (2\pi)^{-1} \, dv$$

$$= \pi^{-s/2} \Gamma(s/2) \zeta(s),$$

using the changes of variable $u = nt$ and $v = \pi u^2$, as required. $\square$

Now we prove the functional equation:

Theorem 10.5 *Suppose that $Re(s) > 1$. Write $\xi(s) = \pi^{-s/2} \Gamma(s/2) \zeta(s)$. Then*

$$\xi(s) = \xi(1 - s).$$

Proof We break up the integral defining $\xi(s)$ for $Re(s) > 1$:

$$\xi(s) = \int_1^\infty (\theta(t) - 1) t^{s-1} \, dt + \int_0^1 (\theta(t) - 1) t^{s-1} \, dt.$$

Now make the change of variable $u = 1/t$ in the second integral, and recall that $\theta(t) = \theta(1/t)/t$ to get

$$\xi(s) = \int_1^\infty (\theta(t) - 1) t^{s-1} \, dt + \int_0^1 (\theta(t) - 1) t^{s-1} \, dt$$

$$= \int_1^\infty (\theta(t) - 1) t^{s-1} \, dt + \int_1^\infty (u\theta(u) - 1) u^{-s-1} \, du$$

$$= \int_1^\infty (\theta(t) - 1) t^{s-1} \, dt + \int_1^\infty \theta(u) u^{-s} \, du - \int_1^\infty u^{-s-1} \, du$$

$$= \int_1^\infty (\theta(t) - 1) t^{s-1} \, dt + \int_1^\infty \theta(u) u^{-s} \, du - \frac{1}{s}$$

$$= \int_1^\infty (\theta(t) - 1) t^{s-1} \, dt + \int_1^\infty (\theta(u) - 1) u^{-s} \, du - \frac{1}{s} - \frac{1}{1-s}$$

$$= \int\limits_{1}^{\infty} (\theta(t) - 1) \left[t^{s-1} + t^{-s} \right] dt - \frac{1}{s} - \frac{1}{1-s}.$$

This integral converges for all $s \in \mathbb{C}$ to a holomorphic function, and the final expression is clearly unchanged when s is replaced by $1 - s$. So we see that $\xi(s) = \xi(1 - s)$ (and also that ξ has simple poles at $s = 0$ and $s = 1$ with residues -1 and $+1$ respectively).   □

Since $\xi(s)$ is defined for all $s \neq 0, 1$ from the proof above, and $\xi(s) = \pi^{-s/2} \Gamma(s/2) \zeta(s)$, we can define $\zeta(s)$ for all $s \neq 0, 1$. The theorem then gives a relation, known as the *functional equation*, between $\zeta(s)$ and $\zeta(1 - s)$. In fact, standard properties of the Gamma function allow one to rewrite this as

$$\zeta(s) = 2^s \pi^{s-1} \sin\left(\frac{\pi s}{2}\right) \Gamma(1 - s) \zeta(1 - s).$$

10.3 Zeta Functions of Number Fields

Around the same time that Riemann was doing his work on the zeta function, others (notably Kummer, Dedekind and Dirichlet) were developing the theory of number fields. We know that we do not have unique factorisation in general number fields, so if we were to try to generalise the definition of the Riemann zeta function to other number fields using elements in the ring of integers, we would not have an Euler product. However, as we have seen, ideals in the ring of integers do factorise uniquely into prime ideals, and we will therefore generalise the definition of the zeta function by using ideals.

For this reason, it is natural to make the definition:

Definition 10.6 The *Dedekind zeta function* of K is given by

$$\zeta_K(s) = \sum_{\mathfrak{a}} \frac{1}{N_{K/\mathbb{Q}}(\mathfrak{a})^s},$$

where $\mathfrak{a}$ runs through all distinct non-zero integral ideals of the field K (i.e., the ideals of $\mathbb{Z}_K$).

Note that when $K = \mathbb{Q}$, we get exactly the Riemann zeta function. We will see later that the Dedekind zeta function is also convergent for $Re(s) > 1$.

Remark 10.7 $\zeta_K(s)$ has an Euler product (valid for $Re(s) > 1$):

$$\zeta_K(s) = \prod_{\mathfrak{p}} \frac{1}{1 - N_{K/\mathbb{Q}}(\mathfrak{p})^{-s}},$$

where the product is taken over all of the prime ideals of $\mathbb{Z}_K$. The proof is identical to that for the Riemann zeta function, and is equivalent to unique factorisation of ideals.

It is natural to ask for generalisations of Riemann's results for $\zeta(s)$ to $\zeta_K(s)$. The first important result is due to Dirichlet, who proved the *analytic class number formula*, showing that $\zeta_K(s)$ has a singularity at $s = 1$, and computing the limit of $(s - 1)\zeta_K(s)$ as $s \to 1$. (In complex variable language, $\zeta_K(s)$ has a simple pole at $s = 1$, and Dirichlet's formula gives the residue.) The formula for the residue involves many of the arithmetic quantities related to K, such as the class number, the discriminant, the numbers of real and complex embeddings, and so on.

It is also true that $\zeta_K(s)$ has a meromorphic continuation to the whole complex plane, and satisfies a functional equation. This proof is just beyond the scope of this book. It was first done by Hecke in the 1920s, but nowadays it is usual to follow Tate's argument from his thesis of 1950. A modern treatment of Hecke's proof can be found in [10]; Tate's thesis was republished in [2].

10.4 The Analytic Class Number Formula

The geometrical methods of Chap. 7, together with a little analysis, allow us to prove the beautiful analytic class number formula.

First we need to recall some definitions from Chap. 7, and especially concerning Dirichlet's Unit Theorem Theorem 7.31). There, we showed that if K is a number field with r_1 real embeddings $\{\rho_1, \ldots, \rho_{r_1}\}$ and r_2 pairs of complex embeddings $\{\sigma_1, \overline{\sigma}_1, \ldots, \sigma_{r_2}, \overline{\sigma}_{r_2}\}$, then there are $r = r_1 + r_2 - 1$ fundamental units $\epsilon_1, \ldots, \epsilon_r$ such that every unit ϵ can be written

$$\epsilon = \zeta \epsilon_1^{\nu_1} \ldots \epsilon_r^{\nu_r}$$

with $\zeta \in \mu(K)$, the roots of unity in K, and $\nu_i \in \mathbb{Z}$.

The proof of this result used lattice-theoretic methods, and Minkowski's Theorem in particular. Recall that we had a commutative diagram:

$$
\begin{array}{ccccc}
K^\times & \xrightarrow{\ i\ } & K_{\mathbb{R}}^\times & \xrightarrow{\ \ell\ } & \mathbb{R}^{r_1+r_2} \\
{\scriptstyle N_{K/\mathbb{Q}}}\downarrow & & {\scriptstyle N}\downarrow & & {\scriptstyle \mathrm{tr}}\downarrow \\
\mathbb{Q}^\times & \longrightarrow & \mathbb{R}^\times & \xrightarrow{\ \ell\ } & \mathbb{R}
\end{array}
$$

where $K_{\mathbb{R}} = \mathbb{R}^{r_1} \times \mathbb{C}^{r_2}$. The proof of Dirichlet's Unit Theorem worked by showing that the units $\mathbb{Z}_K^\times \subset K^\times$ mapped to a complete lattice in the r-dimensional subspace $H \subset \mathbb{R}^{r_1+r_2}$ defined by $H = \{x \in \mathbb{R}^{r_1+r_2} \mid \mathrm{tr}(x) = 0\}$. In particular, the image of $\mathbb{Z}_K^\times$ is an r-dimensional lattice in $\mathbb{R}^{r_1+r_2}$, and if we write $\lambda = \ell \circ i$, the vectors $\lambda(\epsilon_1), \ldots, \lambda(\epsilon_r)$ are a basis for the lattice, and so span H.

The analytic class number formula will also involve a term describing how "widely spaced" the units are in H (in a similar way to how the discriminant describes how widely spaced the integers $\mathbb{Z}_K$ are). Recall that if $x \in \mathbb{Z}_K$,

$$\lambda(x) = (\log|\rho_1(x)|, \ldots, \log|\rho_{r_1}(x)|, \log|\sigma_1(x)|^2, \ldots, \log|\sigma_{r_2}(x)|^2).$$

We can measure how widely spaced the units are in H, by applying λ to the fundamental units, and taking a determinant.

Definition 10.8 Let $\epsilon_1, \ldots, \epsilon_r$ denote a set of fundamental units, where $r = r_1 + r_2 - 1$. Consider the map $\lambda : K \longrightarrow \mathbb{R}^{r_1+r_2}$ above, and write $\lambda(x) = (\lambda_1(x), \ldots, \lambda_{r_1+r_2}(x))$, so that $\lambda_i(x) = \log|\rho_i(x)|$ if $1 \leq i \leq r_1$ and is $\log|\sigma_{i-r_1}(x)|^2$ if $i > r_1$. Consider the $(r+1) \times r$-matrix whose entries are $\lambda_i(\epsilon_j)$, and define the *regulator* R_K to be the absolute value of the determinant of any $r \times r$-minor of this matrix.

Exercise 10.1 Show that the value of the regulator does not depend on the choice of the $r \times r$-minor.

Exercise 10.2 If K is a real quadratic field, show that $R_K = \log \eta$, where $\eta > 1$ is a fundamental unit for K. Hence give the regulator of $\mathbb{Q}(\sqrt{2})$.

Now we turn to the statement and proof of the analytic class number formula.

Theorem 10.9 (Analytic Class Number Formula) $\zeta_K(s)$ *converges for all* $Re(s) > 1$. *It has a simple pole at* $s = 1$, *and*

$$\lim_{s \to 1}(s - 1)\zeta_K(s) = \frac{2^{r_1+r_2}\pi^{r_2}R_K}{m|D_K|^{\frac{1}{2}}}h_K,$$

where R_K *is the regulator of* K, h_K *is the class number of* K, *and* $m = |\mu(K)|$, *the number of roots of unity in* K.

We are going to translate the result into a calculation of volumes of certain regions in $K_{\mathbb{R}} \cong \mathbb{R}^n$.

Definition 10.10 We say that a *cone* in $\mathbb{R}^n$ is a subset $X \subset \mathbb{R}^n$ such that if $x \in X$ and $\lambda \in \mathbb{R}_{>0}$, then $\lambda x \in X$.

Note that the same definition applies to any real vector space, such as $K_{\mathbb{R}}$.

Proposition 10.11 *Let X be a cone in $\mathbb{R}^n$, and let*

$$F : X \longrightarrow \mathbb{R}_{>0}$$

be a function satisfying $F(\xi x) = \xi^n F(x)$ for $x \in X$, $\xi \in \mathbb{R}_{>0}$. Suppose that the set $T = \{x \in X \mid F(x) \leq 1\}$ is bounded, with non-zero volume $v = \mathrm{vol}(T)$, and let Γ be a lattice in $\mathbb{R}^n$, with $\Delta = \mathrm{vol}(\Gamma)$. Then the function

$$Z(s) = \sum_{\Gamma \cap X} \frac{1}{F(x)^s}$$

converges for Re(s) > 1 and $\lim_{s \to 1} (s-1)Z(s) = \frac{v}{\Delta}$.

Proof Note that, for all $r \in \mathbb{R}_{>0}$, $\mathrm{vol}(\frac{1}{r}\Gamma) = \frac{\Delta}{r^n}$. If $N(r)$ denotes the number of points in $\frac{1}{r}\Gamma \cap T$, then

$$v = \mathrm{vol}(T) = \lim_{r \to \infty} N(r)\frac{\Delta}{r^n} = \Delta \lim_{r \to \infty} \frac{N(r)}{r^n}.$$

But $N(r)$ is also the number of points in $\{x \in \Gamma \cap X \mid F(x) \le r^n\}$, at least for the nice F we consider. Order the points of $\Gamma \cap X$ so that

$$0 < F(x_1) \le F(x_2) \le \dots,$$

and let $r_k = F(x_k)^{\frac{1}{n}}$. Then $N(r_k - \epsilon) < k \le N(r_k)$ for all $\epsilon > 0$. It follows that

$$\frac{N(r_k - \epsilon)}{(r_k - \epsilon)^n} \left(\frac{r_k - \epsilon}{r_k}\right)^n < \frac{k}{r_k^n} \le \frac{N(r_k)}{r_k^n},$$

and thus

$$\lim_{r_k \to \infty} \frac{k}{r_k^n} = \lim_{k \to \infty} \frac{k}{F(x_k)} = \frac{v}{\Delta},$$

as the two outer terms have the same limit. We use this to approximate the terms in the sum $Z(s)$. From the above, given $\epsilon > 0$, there exists k_0 such that for all $k \ge k_0$, one has

$$\left(\frac{v}{\Delta} - \epsilon\right)\frac{1}{k} < \frac{1}{F(x_k)} < \left(\frac{v}{\Delta} + \epsilon\right)\frac{1}{k}.$$

Summing,

$$\left(\frac{v}{\Delta} - \epsilon\right)^s \sum_{k=k_0}^{\infty} \frac{1}{k^s} < \sum_{k=k_0}^{\infty} \frac{1}{F(x_k)^s} < \left(\frac{v}{\Delta} + \epsilon\right)^s \sum_{k=k_0}^{\infty} \frac{1}{k^s}.$$

Note also that as the Riemann zeta function converges for $Re(s) > 1$, the same holds for $Z(s)$. We also know the residue of $\zeta(s)$ at $s = 1$, so we multiply through by $(s-1)$, and let s tend to 1 from above.

As

$$\lim_{s \to 1} (s-1)\zeta(s) = 1,$$

and we observe that

$$\lim_{s \to 1} (s-1)[\text{a finite sum}] = 0,$$

we conclude that

$$\frac{v}{\Delta} - \epsilon \le \lim_{s \to 1}(s-1)Z(s) \le \frac{v}{\Delta} + \epsilon.$$

As this holds for all $\epsilon > 0$, the result follows. $\square$

This will be useful both to see the convergence of $\zeta_K(s)$ for $Re(s) > 1$ and in studying the behaviour of $\zeta_K(s)$ as $s \to 1$ from above.

We begin by making a few simple manipulations. Write

$$\zeta_K(s) = \sum_{C \in C_K} f_C(s),$$

the sum running over ideal classes in the class group, and where

$$f_C(s) = \sum_{\mathfrak{a} \in C} \frac{1}{N_{K/\mathbb{Q}}(\mathfrak{a})^s}.$$

We will compute $\lim_{s \to 1}(s-1)f_C(s)$ for each ideal class C, and will observe that the result is independent of C. This will account for the factor h_K in the formula.

Choose any integral $\mathfrak{b}$ in the class C^{-1}, the inverse class; then for all $\mathfrak{a} \in C$, $\mathfrak{ab}$ is therefore principal, $\langle \alpha \rangle$, say. The association $\mathfrak{a} \mapsto \langle \alpha \rangle$ gives a bijection between integral ideals $\mathfrak{a} \in C$, and principal ideals $\langle \alpha \rangle$ divisible by $\mathfrak{b}$ (i.e., elements $\alpha \in \mathfrak{b}$).

It follows that

$$f_C(s) = N_{K/\mathbb{Q}}(\mathfrak{b})^s \sum_{\mathfrak{b}|\langle \alpha \rangle} \frac{1}{|N_{K/\mathbb{Q}}(\alpha)|^s}.$$

Note that $\langle \alpha \rangle = \langle \alpha' \rangle$ if and only if α and α' are associate. We may therefore assume that α runs over a complete set $\mathcal{B}$ of non-associate members of $\mathfrak{b}$.

Let

$$\Gamma = i(\mathfrak{b}) = \{x \in K_{\mathbb{R}} \mid x = i(\alpha) \text{ for some } \alpha \in \mathfrak{b}\},$$

and

$$\Theta = \{x \in K_{\mathbb{R}} \mid x = i(\alpha) \text{ for some } \alpha \in \mathcal{B}\}.$$

Thus

$$f_C(s) = N_{K/\mathbb{Q}}(\mathfrak{b})^s \sum_{x \in \Theta} \frac{1}{|N(x)|^s}.$$

We will find a cone $X \subset K_{\mathbb{R}}$ such that every $\alpha \in \mathcal{B}$ has $i(\alpha)$ associate to precisely one member of X. Then $\Theta = \Gamma \cap X$, and we will be able to apply Proposition 10.11 with $F(x) = |N(x)|$.

First, we define the cone X. Let $\epsilon_1, \ldots, \epsilon_r$ be fundamental units (where $r = r_1 + r_2 - 1$). Write $\lambda = (1, \ldots, 1, 2, \ldots, 2)$ be the vector in $\mathbb{R}^{r_1+r_2}$ whose components are $\lambda_i = 1$ if $i \le r_1$ (corresponding to the real components in $K_{\mathbb{R}}$) and $\lambda_i = 2$ if $i > r_1$ (corresponding to the complex embeddings).

The vectors $\lambda(\epsilon_1), \ldots, \lambda(\epsilon_r)$ span H, as we saw in Corollary 7.30; thus the set

$$\{\lambda, \lambda(\epsilon_1), \ldots, \lambda(\epsilon_r)\}$$

are a basis for $\mathbb{R}^{r_1+r_2}$. So for all $\ell(x) \in \mathbb{R}^{r_1+r_2}$, we can write

$$\ell(x) = \xi\lambda + \xi_1\lambda(\epsilon_1) + \cdots + \xi_r\lambda(\epsilon_r) \tag{10.1}$$

for some coefficients $\xi, \xi_i \in \mathbb{R}$.

Observe that $\operatorname{tr}\lambda(\epsilon_i) = 0$ (as $\lambda(\epsilon_i) \in H$), and so $\operatorname{tr}\ell(x) = \xi.\operatorname{tr}\lambda = \xi n$. But $\operatorname{tr}\ell(x) = \log|N(x)|$, and so $\xi = \frac{1}{n}\log|N(x)|$.

Definition 10.12 The cone $X \subset K_{\mathbb{R}}$ will be defined to consist of all x such that

1. $N(x) \ne 0$
2. the coefficients ξ_i $(i = 1, \ldots, r)$ of $\ell(x)$ satisfy $0 \le \xi_i < 1$
3. $0 \le \arg(x_1) < \frac{2\pi}{m}$, where x_1 is the first component of x.

Let $\xi \in \mathbb{R}_{>0}$. Then $N(\xi x) = \xi^n N(x) \ne 0$. Next, $\ell(\xi x) = (\log\xi)\lambda + \ell(x)$, so the coefficients of $\lambda(\epsilon_i)$ are unchanged. Finally, $\arg(\xi x_1) = \arg(x_1)$. Thus if $x \in X$, and $\xi \in \mathbb{R}_{>0}$, then $\xi x \in X$, and we deduce that X is a cone in $K_{\mathbb{R}}$.

Lemma 10.13 *Let $y \in \mathbb{R}^n$, with $N(y) \ne 0$. Then y is uniquely of the form $x.i(\epsilon)$, where $x \in X$ and $\epsilon \in \mathbb{Z}_K^\times$.*

Proof One has

$$\ell(y) = \gamma\lambda + \gamma_1\lambda(\epsilon_1) + \cdots + \gamma_r\lambda(\epsilon_r).$$

Write $\gamma_i = k_i + \xi_i$, with $k_i \in \mathbb{Z}$, $\xi_i \in [0, 1)$. Let

$$\eta = \epsilon_1^{k_1} \ldots \epsilon_r^{k_r},$$

and put $z = y.i(\eta^{-1})$. Suppose $\arg(z_1) = \phi$, and write

$$0 \le \phi - \frac{2\pi k}{m} < \frac{2\pi}{m}$$

for some integer k. Choose $\zeta \in \mu(K)$ such that $\tau_1(\zeta) = e^{\frac{2\pi i}{m}}$, where τ_1 gives the first component of the map $K \longrightarrow K_{\mathbb{R}}$, and then $x = y.i(\eta^{-1}).i(\zeta^{-k}) \in X$. Clearly then $y = x.i(\epsilon)$ for a unit ϵ, and this decomposition is clearly unique from the construction. $\square$

It follows that in every class of associate members of $\mathbb{Z}_K$, there is a unique one whose image in $\mathbb{R}^n$ lies in X.

Then we have

$$f_C(s) = N_{K/\mathbb{Q}}(\mathfrak{b})^s \sum_{x \in \Gamma \cap X} \frac{1}{|N(x)|^s}$$

and we can evaluate the sum exactly as in Proposition 10.11.

Recall that we needed to calculate $v = \operatorname{vol}(T)$, where T was the set

$$\{x \in X \mid |N(x)| \leq 1\}$$

and $\Delta = \operatorname{vol}(\Gamma)$. But we computed the latter in Proposition 7.13; $\Delta = N_{K/\mathbb{Q}}(\mathfrak{b})$ $|D_K|^{\frac{1}{2}}$. It merely remains to calculate v.

Proposition 10.14 $\operatorname{vol}(T)$ *is given by*

$$v = \frac{2^{r_1+r_2} \pi^{r_2} R_K}{m}.$$

Proof First note that if $\epsilon \in \mathbb{Z}_K^\times$, then multiplication by ϵ is volume-preserving. This is simply because the volume form is multiplied by value of the determinant of the transformation $x \mapsto x.i(\epsilon)$, which is $|N_{K/\mathbb{Q}}(\epsilon)| = 1$.

Put

$$\widetilde{T} = \bigcup_{k=0}^{m-1} T.i(\zeta^k).$$

Then $\widetilde{T}$ corresponds to the cone X defined only by conditions (1) and (2) of Definition 10.12. It follows that $\operatorname{vol}(\widetilde{T}) = m.\operatorname{vol}(T)$.

Now let $\overline{T}$ denote the set

$$\{x \in \widetilde{T} \mid x_i > 0 \text{ for all } i = 1, \ldots, r_1\}.$$

It follows that

$$\operatorname{vol}(T) = \frac{2^{r_1}}{m} \operatorname{vol}(\overline{T}).$$

Thus it suffices to calculate $\operatorname{vol}(\overline{T})$. We make several changes of variables, before computing the volume. Firstly, we consider the isomorphism

$$K_{\mathbb{R}} \longrightarrow \mathbb{R}^n$$
$$(x_1, \ldots, x_{r_1}, z_1, \ldots, z_{r_2}) \mapsto (x_1, \ldots, x_{r_1}, R_1, \phi_1, \ldots, R_{r_2}, \phi_{r_2})$$

where $z_k = R_k e^{i\phi_k}$, so that

$$\ell(x_1, \ldots, x_{r_1}, z_1, \ldots, z_{r_2}) = (\log x_1, \ldots, \log x_{r_1}, \log R_1^2, \ldots, \log R_{r_2}^2).$$

The Jacobian of this change of variables is easily computed to be $R_1 \ldots R_{r_2}$.

Then $\overline{T}$ is given by

1. $x_1 > 0, \ldots, x_{r_1} > 0, R_1 > 0, \ldots, R_{r_2} > 0$ and $x_1 \cdots x_{r_1}(R_1 \cdots R_{r_2})^2 \le 1$; note that this quantity is $N(x)$.
2. in the formula (10.1) giving the jth component of $\ell(x)$,

$$\ell(x) = \xi\lambda + \xi_1\lambda(\epsilon_1) + \cdots + \xi_r\lambda(\epsilon_r),$$

one has $0 \le \xi_k < 1$.

The $\phi_{r_1+1}, \ldots, \phi_{r_1+r_2}$ independently take values in $[0, 2\pi)$. We replace the variables $x_1, \ldots, x_{r_1}, R_1, \ldots, R_{r_2}$ by the variables $\xi, \xi_1, \ldots, \xi_r$, got from the formula

$$\ell(x) = \xi\lambda + \xi_1\lambda(\epsilon_1) + \cdots + \xi_r\lambda(\epsilon_r),$$

so that $\xi = N(x)$. Now the image of $\overline{T}$ is given simply by $0 < \xi \le 1, 0 \le \xi_k < 1$ for all k. We need to compute the Jacobian of this change of variable. Considering the jth components, we get

$$\log x_j = \frac{1}{n}\log\xi + \sum_{k=1}^{r}\xi_k\lambda_j(\epsilon_k)$$

$$\log R_j^2 = \frac{2}{n}\log\xi + \sum_{k=1}^{r}\xi_k\lambda_{r_1+j}(\epsilon_k)$$

and we can read off

$$\frac{\partial x_j}{\partial\xi} = \frac{x_j}{n\xi}; \quad \frac{\partial x_j}{\partial\xi_k} = x_j\lambda_j(\epsilon_k); \quad \frac{\partial R_j}{\partial\xi} = \frac{R_j}{n\xi}; \quad \frac{\partial R_j}{\partial\xi_k} = \frac{R_j}{2}\lambda_{r_1+j}(\epsilon_k).$$

Thus the Jacobian of this change of variables is given by

$$J = \begin{vmatrix} \frac{x_1}{n\xi} & x_1\lambda_1(\epsilon_1) & \cdots & x_1\lambda_1(\epsilon_r) \\ \vdots & \vdots & \ddots & \vdots \\ \frac{x_{r_1}}{n\xi} & x_{r_1}\lambda_{r_1}(\epsilon_1) & \cdots & x_1\lambda_{r_1}(\epsilon_r) \\ \frac{R_1}{n\xi} & \frac{R_1}{2}\lambda_{r_1+1}(\epsilon_1) & \cdots & \frac{R_1}{2}\lambda_{r_1+1}(\epsilon_r) \\ \vdots & \vdots & \ddots & \vdots \\ \frac{R_{r_2}}{n\xi} & \frac{R_{r_2}}{2}\lambda_{r_1+r_2}(\epsilon_1) & \cdots & \frac{R_{r_2}}{2}\lambda_{r_1+r_2}(\epsilon_r) \end{vmatrix}$$

$$= \frac{x_1 \ldots x_{r_1} R_1 \ldots R_{r_2}}{2^{r_2} n \xi} \begin{vmatrix} 1 & \lambda_1(\epsilon_1) & \ldots & \lambda_1(\epsilon_r) \\ \vdots & \vdots & \ddots & \vdots \\ 1 & \lambda_{r_1}(\epsilon_1) & \ldots & \lambda_{r_1}(\epsilon_r) \\ 2 & \lambda_{r_1+1}(\epsilon_1) & \ldots & \lambda_{r_1+1}(\epsilon_r) \\ \vdots & \vdots & \ddots & \vdots \\ 2 & \lambda_{r_1+r_2}(\epsilon_1) & \ldots & \lambda_{r_1+r_2}(\epsilon_r) \end{vmatrix}$$

One adds all rows to the top one to get

$$J = \frac{x_1 \ldots x_{r_1} R_1 \ldots R_{r_2}}{2^{r_2} n \xi} \begin{vmatrix} n & 0 & \cdots & 0 \\ \cdot & \lambda_2(\epsilon_1) & \cdots & \lambda_2(\epsilon_r) \\ \vdots & \vdots & \ddots & \vdots \\ \cdot & \lambda_{r_1+r_2}(\epsilon_1) & \cdots & \lambda_{r_1+r_2}(\epsilon_r) \end{vmatrix}$$

Expanding along the top row, we see that this determinant is exactly $n R_K$, where R_K denotes the regulator. Recalling that $\xi = x_1 \ldots x_{r_1}(R_1 \ldots R_{r_2})^2$, it follows that

$$|J| = \frac{R_K}{2^{r_2} R_1 \ldots R_{r_2}}.$$

Finally, we can deduce the result:

$$\text{vol}(\overline{T}) = 2^{r_2} \text{vol}_{\mathbb{R}}(\overline{T})$$

$$= 2^{r_2} \int_{\overline{T}} dx_1 \ldots dx_{r_1} dy_{r_1+1} dz_{r_1+1} \ldots dy_{r_1+r_2} dz_{r_1+r_2}$$

$$= 2^{r_2} \int_{\overline{T}} R_1 \ldots R_{r_2} dx_1 \ldots dx_{r_1} dR_1 \ldots dR_{r_2} d\phi_1 \ldots d\phi_{r_2}$$

$$= 2^{r_2} (2\pi)^{r_2} \int R_1 \ldots R_{r_2} dx_1 \ldots dx_{r_1} dR_1 \ldots dR_{r_2}$$

$$= 2^{r_2} (2\pi)^{r_2} \int |J| R_1 \ldots R_{r_2} d\xi d\xi_1 \ldots d\xi_r$$

$$= 2^{r_2} \pi^{r_2} R_K.$$

Thus

$$\text{vol}(T) = \frac{2^{r_1+r_2} \pi^{r_2} R_K}{m}.$$

□

Thus

$$\lim_{s \to 1} (s - 1) f_C(s) = N_{K/\mathbb{Q}}(\mathfrak{b}) \frac{v}{\Delta} = N_{K/\mathbb{Q}}(\mathfrak{b}) \frac{2^{r_1+r_2} \pi^{r_2} R_K}{m} \bigg/ N_{K/\mathbb{Q}}(\mathfrak{b}) |D_K|^{\frac{1}{2}}.$$

So

$$\lim_{s \to 1} (s - 1) f_C(s) = \frac{2^{r_1+r_2} \pi^{r_2} R_K}{m |D_K|^{\frac{1}{2}}},$$

which is independent of C. Summing over the ideal classes gives

$$\lim_{s \to 1} (s - 1) \zeta_K(s) = \frac{2^{r_1+r_2} \pi^{r_2} R_K}{m |D_K|^{\frac{1}{2}}} h_K,$$

which is the analytic class number formula.

10.5 Explicit Class Number Formulae

The main result of the last section, that

$$\lim_{s \to 1} (s - 1) \zeta_K(s) = \frac{2^{r_1} (2\pi)^{r_2} h_K R_K}{m |D_K|^{\frac{1}{2}}},$$

can be used to give explicit expressions for class numbers. Indeed, apart from the class number, the other terms in the formula, as well as the residue of the zeta function, can often be calculated numerically with relative ease, and so the class number can be recovered from these calculations.

We will present explicit formulae for quadratic fields. Thus we consider $K = \mathbb{Q}(\sqrt{d})$, with d squarefree. As already noted, the principal ideal $\langle p \rangle$ for a prime p of $\mathbb{Z}$ can factorise in $\mathbb{Z}_K$ in three different ways:

1. p can *split*, so that $\langle p \rangle = \mathfrak{p}_1 \mathfrak{p}_2$ with $\mathfrak{p}_1 \neq \mathfrak{p}_2$, and $N_{K/\mathbb{Q}}(\mathfrak{p}_1) = N_{K/\mathbb{Q}}(\mathfrak{p}_2) = p$;
2. p can be *inert*, so that $\langle p \rangle$ remains a prime ideal in $\mathbb{Z}_K$, with norm p^2;
3. p can *ramify*, so that $\langle p \rangle = \mathfrak{p}^2$ for some prime ideal $\mathfrak{p}$ of norm p.

We define

$$\chi(p) = \begin{cases} 1, & p \text{ splits} \\ -1, & p \text{ is inert} \\ 0, & p \text{ ramifies} \end{cases}$$

As we remarked in Sect. 5.9, χ is actually a Dirichlet character modulo D_K, that is, $\chi(p)$ depends only on the value of $p \pmod{D_K}$, and it can be extended to all integers n, in such a way that if m and n are coprime, then $\chi(mn) = \chi(m)\chi(n)$.

Consider the factors corresponding to the primes dividing $\langle p \rangle$ in the Euler product

$$\zeta_K(s) = \prod_{\mathfrak{p}} \left(1 - \frac{1}{N_{K/\mathbb{Q}}(\mathfrak{p})^s} \right)^{-1}.$$

By the remark above, these are

$$\left(1 - \frac{1}{p^s} \right)^{-2}, \quad \left(1 - \frac{1}{p^{2s}} \right)^{-1}, \quad \left(1 - \frac{1}{p^s} \right)^{-1}$$

in the split, inert and ramified cases respectively. In each case, there is a factor $\left(1 - \frac{1}{p^s} \right)^{-1}$, which is the Euler factor of the Riemann zeta function at p, and the other factor is given by $\left(1 - \frac{\chi(p)}{p^s} \right)^{-1}$. We define the *Dirichlet L-function*

$$L(s, \chi) = \prod_{p} \left(1 - \frac{\chi(p)}{p^s} \right)^{-1},$$

and we conclude that

$$\zeta_K(s) = \zeta(s) L(s, \chi).$$

(Note in passing that the multiplicativity of χ and unique factorisation in $\mathbb{Z}$ give the alternative expression $L(s, \chi) = \sum_{n=1}^{\infty} \frac{\chi(n)}{n^s}$.)

We can multiply through by $(s-1)$ and let $s \to 1$ in this expression. The Riemann zeta-function has a simple pole at $s = 1$ with residue 1, and the residue of $\zeta_K(s)$ is given by the analytic class number formula. We get

$$\frac{2^{r_1}(2\pi)^{r_2} h_K R_K}{m |D_K|^{\frac{1}{2}}} = L(1, \chi).$$

In the case of a real quadratic field, we have $r_1 = 2$, $r_2 = 0$, $m = 2$, and $R_K = \log \epsilon$, where $\epsilon > 1$ is a fundamental unit, and we conclude that

$$h_K = \frac{\sqrt{|D_K|}}{2 \log \epsilon} L(1, \chi).$$

If K is imaginary quadratic, then $r_1 = 0$, $r_2 = 1$, $R_K = 1$, and so

$$h_K = \frac{m \sqrt{|D_K|}}{2\pi} L(1, \chi).$$

(Recall that $m = 2$ except for the fields $\mathbb{Q}(i)$ and $\mathbb{Q}(\sqrt{-3})$, both of which have class number one.)

The quantities $L(1, \chi)$ are not so easy to compute exactly (see [1] for more details); however, it is sometimes relatively easy to compute enough terms in the sum $L(1, \chi) = \sum_{n=1}^{\infty} \frac{\chi(n)}{n}$ to get an idea of the value of h_K (especially in the imaginary quadratic case), recalling that it must be integral.

Example 10.15 *Consider $K = \mathbb{Q}(\sqrt{2})$. Then the fundamental unit is $\epsilon = 1 + \sqrt{2}$, and the discriminant is $D_K = 8$. Then*

$$h_K = \frac{\sqrt{8}}{2 \log(1 + \sqrt{2})} L(1, \chi) \approx 1.605 L(1, \chi).$$

But

$$L(1, \chi) = 1 - \frac{1}{3} - \frac{1}{5} + \frac{1}{7} + \cdots < 1,$$

as is easily seen by grouping terms suitably, and so h_K is a positive integer less than 1.605; we conclude that $h_K = 1$.

Exercise 10.3 Show that the class number of $\mathbb{Q}(\sqrt{3})$ is 1.

As a final comment, we remark that there are closed form expressions for $L(1, \chi)$ in both the real and imaginary quadratic cases, but their evaluation is beyond the scope of this book, and we merely state the result, referring the reader to [1] for details.

Theorem 10.16 *Let $K = \mathbb{Q}(\sqrt{d})$ be a quadratic field with discriminant D_K and character χ.*
If $d > 0$, let $\epsilon > 1$ denote its fundamental unit; the product

$$\eta = \prod_{0 < a < D_K/2} \sin\left(\frac{\pi a}{D_K}\right)^{\chi(a)}$$

is a unit in K, and h_K can be computed from the equality $\eta = \epsilon^{-h_K}$.
If $d < 0$ and $D_K < -4$, then

$$h_K = \frac{1}{2 - \chi(2)} \sum_{0 < a < |D_K|/2} \chi(a),$$

where $\chi(a) = 0$ if a and D_K are not coprime.

Exercise 10.4 Use these closed form expressions to compute the class numbers of $\mathbb{Q}(\sqrt{17})$ and $\mathbb{Q}(\sqrt{-23})$.

For more on the numerical computation of class numbers, see [3].

10.6 Other Embeddings

In Chap. 7 and again in this chapter, we have proven results about number fields by using their embeddings into $\mathbb{R}$ and $\mathbb{C}$. The aim of this short section is to orient the reader towards further topics, although lack of space means that we will merely hint at some consequences of the theory we begin to develop.

These embeddings arise as *completions* of K with respect to *valuations* defined with the aid of embeddings. For example, a real embedding $K \longrightarrow \mathbb{R}$ enables us to give some notion of "size" to elements of K, as the real numbers come equipped with a notion of absolute value. Similarly, complex numbers have a modulus, and this enables us to assign a notion of size with respect to (conjugate pairs of) embeddings.

Any notion of "size" ought to satisfy the following:

Definition 10.17 A *norm* on a number field K is a function $| \ | : K \longrightarrow \mathbb{R}_{\geq 0}$ such that

- $|a| = 0$ if and only if $a = 0$;
- $|ab| = |a|.|b|$ for all $a, b \in K$;
- $|a + b| \leq |a| + |b|$ for all $a, b \in K$.

This should look a little like the definition of a metric, but we are using both the multiplicative and additive properties of the field here. The third condition is known as the *triangle inequality*. These properties encapsulate what we know about the absolute value on $\mathbb{R}$, and the modulus on $\mathbb{C}$, and it follows that each embedding τ from K into $\mathbb{R}$ or $\mathbb{C}$ then gives us a norm $| \ |_\tau$, defined by

$$|\alpha|_\tau = |\tau(\alpha)|,$$

where the right-hand side is the absolute value if τ is a real embedding, and is the modulus if τ is a complex embedding. Notice that τ and $\overline{\tau}$ give the same norm in the latter case.

Exercise 10.5 Verify that $|1| = 1$ and that $|-1| = 1$.

Here is another (not very interesting) example of a norm:

Definition 10.18 The *trivial norm* $| \ |_0$ is defined by

$$|a|_0 = \begin{cases} 0 & \text{if } a = 0, \\ 1 & \text{if } a \neq 0. \end{cases}$$

Exercise 10.6 Show that if $| \ |$ is a norm on K such that $|x| = 1$ for all nonzero $x \in \mathbb{Z}_K$ then $| \ |$ is the trivial norm.

There are more interesting norms on $\mathbb{Q}$! Another interesting family is given by the *p-adic norms*, defined in the following exercise:

Exercise 10.7 Suppose that p is a prime number. If $\frac{a}{b} \in \mathbb{Q}$, write $\frac{a}{b} = p^r \frac{m}{n}$, where neither m nor n is divisible by p. Show that the function $|\ |_p$, defined by

$$\left|\frac{a}{b}\right|_p = p^{-r}$$

is a norm on $\mathbb{Q}$.

The problem of classifying all norms on $\mathbb{Q}$ was solved by Ostrowski. Let us write $|\ |_\infty$ for the usual absolute value on $\mathbb{Q}$.

Theorem 10.19 (Ostrowski's Lemma) *Every non-trivial norm on $\mathbb{Q}$ is either equal to a power $|a|_\infty^t$ of the ordinary absolute value, where $0 < t \leq 1$, or is a power $|a|_p^t$ of the p-adic norm for some prime p, and some $t > 0$.*

Proof Suppose that $|\ |$ is a non-trivial norm. We will distinguish two cases.

Firstly, suppose that for every integer $m \geq 1$, we have $|m| \leq 1$. If $|m| = 1$ for all $m \in \mathbb{Z}$, then $|\ |$ is trivial. So there must be some $m > 1$ with $|m| < 1$. Let p be the smallest such integer; we'll show that it is prime. If not, we could write $p = rs$ with $r > 1, s > 1$, and as $|r||s| = |p| < 1$, either $|r| < 1$ or $|s| < 1$, contradicting the minimality of p.

If $m \in \mathbb{Z}$, with $p \nmid m$, write $m = qp + r$ with $0 < r < p$, and $q \in \mathbb{Z}$. As $q \in \mathbb{Z}$, $|q| \leq 1$, and so $|qp| = |q||p| < 1$. Also $|r| = 1$, as p was the smallest positive integer with $|p| < 1$, and $r < p$. So $|m| = 1$.

That is, if $m \in \mathbb{Z}$ is nonzero, we have $|m| = 1$ if $p \nmid m$. Write $c = |p| < 1$. Given any integer n, take its prime factorisation

$$n = \prod_q q^{n_q},$$

and then the multiplicativity of the norm shows that $|n| = c^{n_p}$. The same conclusion applies to quotients of two integers $\frac{m}{n}$, and we conclude that $|\ |$ is a power of the p-adic norm as in the statement.

The second case occurs when $|x| > 1$ for some $x \in \mathbb{Z}$. We can assume that $x > 1$ (using the fact that $|-1| = 1$). We need to show that $|\ |$ is then a power of the absolute value.

We will consider powers x^n of x, and let n vary. Let $y > 1$ be an integer, and write x^n in base y:
$$x^n = c_m y^m + c_{m-1} y^{m-1} + \cdots + c_0,$$

where $c_m, \ldots, c_0 \in \{0, 1, \ldots, y - 1\}$ and $c_m \neq 0$. Of course y and the coefficients c_i depend on n. Then

$$m \leq \frac{\log x^n}{\log y} = \frac{n \log x}{\log y}.$$

By the triangle inequality,

$$|x|^n = |x^n| \le (m+1)M \max\{|y|^m, 1\},$$

where $M = \max\{|1|, \ldots, |m-1|\}$ is independent of n. So

$$|x|^n = |x^n| \le \left(\frac{n \log x}{\log y} + 1\right) M \max\{|y|^{n \log x / \log y}, 1\};$$

taking nth roots and letting $n \to \infty$ gives

$$|x| \le \max\{|y|^{\log x / \log y}, 1\}.$$

But x was chosen so that $|x| > 1$, and we conclude that $|y| > 1$ also.

So if $|x| > 1$ for some integer $x > 1$, we see that $|y| > 1$ for *all* integers $y > 1$, and we conclude that

$$|x| \le |y|^{\log x / \log y}$$

or

$$|x|^{1/\log x} \le |y|^{1/\log y}.$$

Now we know that $|y| > 1$, we see that we can apply the same argument with x and y interchanged, to get the opposite inequality, and therefore an equality

$$|x|^{1/\log x} = |y|^{1/\log y}.$$

If $|x|^{1/\log x} = c$, we conclude that $|y|^{1/\log y} = c$ for all integers $y > 1$. Then $|y| = c^{\log y} = y^t$ for some t. As $|-1| = 1$, we see that $|y| = |y|_\infty^t$ for all $y \in \mathbb{Z}$, and then, again using the multiplicativity of norms, that $|y| = |y|_\infty^t$ for all $y \in \mathbb{Q}$. Note that we need $0 < t \le 1$ here so that the triangle inequality is satisfied. $\qquad \square$

Definition 10.20 Two norms, $|\ |$ and $|\ |'$ are *equivalent* if one is a power of the other.

Two equivalent norms have the same open sets, and therefore the same topology; we may wish to identify them for this reason. Notice that the absolute value is not equivalent to any p-adic norm, and nor is any p-adic norm equivalent to any q-adic norm if p and q are different primes.

Remarkably, then, the usual absolute value and the p-adic norms introduced in Exercise 10.7 are essentially all the possible norms on $\mathbb{Q}$—every non-trivial norm is equivalent to one of these.

So norms are (essentially) in bijection with the set of all primes, plus one other (we often write $|\ |_\infty$ for the absolute value, and then write ∞ for the other member of the set). We say that a norm is *nonarchimedean* if we have a stronger version of the triangle inequality: $|a + b| \le \max(|a|, |b|)$ (this is known as the *ultrametric inequality*); the standard p-adic norms are all nonarchimedean, whereas the usual absolute value is not (so it is *archimedean*).

We remark that there is a similar result in the case of a general number field; the non-trivial nonarchimedean norms are in bijection with the prime ideals in the ring of integers, while the archimedean norms can be derived after taking the usual real or complex absolute values after some embedding $K \hookrightarrow \mathbb{C}$. Let's simply state the nonarchimedean result:

Theorem 10.21 *Every nontrivial nonarchimedean norm on a number field K is equivalent to one of the form*

$$|x|_{\mathfrak{p}} = c^{v_{\mathfrak{p}}(x)},$$

where $\mathfrak{p}$ is some nonzero prime ideal of $\mathbb{Z}_K$, and $v_{\mathfrak{p}}(x)$ is defined for $x \in \mathbb{Z}_K$ to be the largest integer m such that $x \in \mathfrak{p}^m$, and extended multiplicatively to nonzero elements of K.

Remark 10.22 Fix any $0 < c < 1$. From the proof of Ostrowski's Lemma, defining $|\frac{a}{b}| = c^r$, where $\frac{a}{b} = p^r \frac{s}{t}$, gives us a p-adic norm. One might wonder why we have chosen to fix $c = p^{-1}$. The answer is that it is sometimes useful to have the *product formula*: for all $x \in \mathbb{Q}^{\times}$,

$$|x|_{\infty} \times \prod_p |x|_p = 1,$$

and it is an easy exercise to convince yourself that the normalisation $c = p^{-1}$ in the p-adic norm makes this hold.

Exercise 10.8 Put $x = -\frac{360}{91}$. Verify the product formula in the remark.

Recall that we can complete a metric space with respect to a metric; the *completion* consists of all Cauchy sequences in the spaces modulo the set of sequences with limit 0.

A norm $|\ |$ on a number field gives a metric via $d(x, y) = |x - y|$; the metric given by the usual absolute value, for example, is just the usual distance apart of two rationals on the number line. As any real number can be the limit of a Cauchy sequence in $\mathbb{Q}$, we can identify the completion of $\mathbb{Q}$ with respect to the usual absolute value $|\ |_{\infty}$ with $\mathbb{R}$.

We can consider the completion of $\mathbb{Q}$ with respect to the metrics coming from the other norms, and we write $\mathbb{Q}_p$ for the completion of $\mathbb{Q}$ with respect to the p-adic metric induced by $|\ |_p$. In fact, $\mathbb{Z}$ is not complete with respect to the p-adic metrics, and we also write $\mathbb{Z}_p$ for the completion of $\mathbb{Z}$ with respect to the p-adic metric. There is a sense in which $\mathbb{Z}_p$ is the ring of integers of $\mathbb{Q}_p$. Similarly, every nonzero prime ideal $\mathfrak{p}$ of $\mathbb{Z}_K$ gives a $\mathfrak{p}$-adic norm $|\ |_{\mathfrak{p}}$, which gives a $\mathfrak{p}$-adic metric, and completions $K_{\mathfrak{p}}$ and $\mathbb{Z}_{K,\mathfrak{p}}$.

These completions are very important in modern algebraic number theory. The rings $\mathbb{Z}_p$ are *local*; they have a unique maximal ideal consisting of all multiples of p. Very often problems relating to a number field as a whole can be tackled by passing to the completions at all finite primes. This has the effect of isolating the part of the problem relating to one prime at a time, and one can sometimes deal with the primes individually.

In particular, we studied Dirichlet's Analytic Class Number Formula earlier in this chapter, which generalised the well-known result on the pole and residue of the Riemann zeta function at $s = 1$ to the Dedekind zeta function of a number field. One might reasonably expect other properties of the zeta function to generalise also; in particular, one might expect the Dedekind zeta function to have a meromorphic continuation to the whole complex plane, and satisfy some sort of functional equation. This was first proven by Hecke, around 1920, but the proof is cumbersome (see [10] for a modern treatment). Tate's thesis (1950, but published in [2]) gave a new way to view zeta functions, with the term at $\mathfrak{p}$ in the Euler product for the Dedekind zeta function coming from an integral on $\mathbb{Z}_{K,\mathfrak{p}}$. Then one can get the functional equation by developing a theory of integration (and particularly Fourier theory) for these completions.

Chapter 11
The Number Field Sieve

Like most pure mathematics, number theory developed with little thought to applications. The advent of the computer, however, has led to increasing importance of pure mathematics within cryptography and data-transmission. The problem of factorising large numbers, as we shall see in a moment, is of great importance in cryptography, and the fastest factorisation method currently known uses algebraic number theory in a crucial way. In this chapter, we will introduce this method, the *number field sieve*, in some detail.

We'll begin by reminding the reader about the RSA algorithm, which motivates the problem of factorisation, and also the *quadratic sieve*, one of the previous factoring methods of choice, many elements of which continue to be useful in the number field sieve.

11.1 The RSA Cryptosystem and the Problem of Factorisation

A very well-known protocol is the RSA system for encryption and decryption of messages. At the heart of this system lies the observation that while multiplication of two numbers is easy, recovering the factors of a product is difficult. Sometimes the analogy is made with mixing pots of paint of different colours—mixing a pot of red paint and a pot of white paint to get a suitable shade of pink is easy, but given a mixed pot of pink paint, separating out the white and red paints is very difficult!

Let's briefly recall the RSA system.

We suppose that Bob wishes to send a message to Alice.

Alice chooses two primes, p and q, which she keeps secret. She computes $N = pq$, and tells this to Bob. Alice also publishes an *encryption key*, a number e (as we will see below, e should be coprime to $\phi(N) = (p-1)(q-1)$).

They agree on a system to convert any message into a sequence of numbers less than N.

F. Jarvis, *Algebraic Number Theory*, Springer Undergraduate
Mathematics Series, DOI: 10.1007/978-3-319-07545-7_11,
© Springer International Publishing Switzerland 2014

In order to transmit a number M to Alice, Bob computes M^e (mod N) and sends it to Alice.

Unless Bob is so unlucky that M happens not to be coprime to N, the Fermat-Euler theorem tells us that $M^{\phi(N)} \equiv 1$ (mod N).

Since Alice knows p and q, she knows $\phi(N) = (p-1)(q-1)$, and can compute the *decryption key* d so that $de \equiv 1$ (mod $\phi(N)$), and she can therefore compute $(M^e)^d \equiv M$ (mod N) to recover the original transmitted message M.

Anyone who can factor N to discover the primes p and q can decrypt the message; the primes p and q are therefore chosen very large. No-one has yet discovered a fast way to decrypt the message without factoring N.

This system has meant that the problem of factorisation is increasingly important.

For fairly small numbers, the simplest factoring idea is simply to test N for divisibility by primes not exceeding $\sqrt{N}$. After all, if $N = ab$ with $a, b > 1$, then we can't have both of a and b larger than $\sqrt{N}$, so N must have some non-trivial divisor no larger than $\sqrt{N}$—and then any prime divisor of this will again be no larger than $\sqrt{N}$. Therefore if N is composite, it will have some prime factor at most $\sqrt{N}$. For example, to factor 10001, it suffices to check just the 25 primes up to $\lfloor\sqrt{10001}\rfloor = 100$.

However, if N gets bigger, this trial division method quickly becomes impractical. Even for 20-digit numbers it would involve testing by all primes up to 10^{10}; this is feasible, but still lengthy. For numbers of 30 digits, it is not really feasible.

But here is another way to factor 10001: try to write it as $x^2 - y^2$, the difference of two squares. Then the factorisation $x^2 - y^2 = (x+y)(x-y)$ will give us a factorisation of 10001.

We therefore list the squares just larger than 10001:

$$101^2 = 10201, \quad 102^2 = 10404, \ldots$$

and check the differences between these squares and 10001; if we find that the difference is a square, we can then view 10001 as the desired difference of two squares. We soon find that $105^2 = 11025$ differs from 10001 by $1024 = 32^2$. Thus

$$10001 = 105^2 - 32^2 = (105 + 32)(105 - 32) = 137 \times 73,$$

and we have factored 10001.

Exercise 11.1   Factor $N = 12707$ using this method.

This approach works well for numbers which are the product of two similar-sized numbers. But if we had a number $N = pq$ which was the product of two primes, say of 40 digits and 50 digits, this approach would also take too long. The two primes might seem "close" as 40 and 50 are similar—but the second is of course about 10^{10} times larger than the first, so this method is no better than trial division.

The next observation is that solving $N = x^2 - y^2$ is stronger than needed—it may be enough to solve $x^2 - y^2 \equiv 0$ (mod N). This means that N divides $(x+y)(x-y)$, and unless $x \equiv \pm y \pmod{N}$, neither $x + y$ nor $x - y$ is divisible by N; it follows that N breaks up as a product, part of which divides $x + y$ and the other part divides $x - y$. For example, given that

$$107^2 \equiv 38^2 \pmod{667},$$

we could deduce that $667|(107 + 38)(107 - 38) = 145 \times 69$. Since $667 \nmid 145$ and $667 \nmid 69$, we deduce that 667 is not prime, and that we can write $667 = ab$ with $a|145$ and $b|69$. We can compute a and b since a is a common factor of 667 and 145, and b is a common factor of 667 and 69. Euclid's algorithm gives $(667, 145) = 29$ and $(667, 69) = 23$, and we can check that $667 = 23 \times 29$.

The idea of finding a pair of integers x and y with $x^2 \equiv y^2 \pmod{N}$ is at the heart of most of today's factoring algorithms, and this is the approach taken by the number field sieve, and its precursor, the quadratic sieve, which we will discuss first.

11.2 The Quadratic Sieve

The quadratic sieve was invented in the early 1980s by Carl Pomerance. It is still the fastest factorisation method for numbers up to 100 digits or so, but the more complicated number field sieve is asymptotically faster, as we shall explain below.

As mentioned above, the approach to factorisation taken by the quadratic sieve and the number field sieve is to find x and y with $x^2 \equiv y^2 \pmod{N}$.

Since we include the quadratic sieve only as motivation for the number field sieve, we shall not go into any practical implementation issues here.

The quadratic sieve is a refinement of the naive idea:

Choose x so that x^2 is "small" modulo N. If we are lucky, we will find x so that $x^2 \equiv y^2 \pmod{N}$. Choosing x near to the square root of N is likely to give small values of $x^2 \pmod{N}$. If we do not get a square, try another value of x, and keep going until a square is found.

Example 11.1 Let's factor $N = 24511$. First compute its square root: $\sqrt{N} = 156.559\ldots$. Then select values of x near this square root, and compute $x^2 \pmod{N}$.

x	$x^2 \pmod{N}$	Square?
157	138	No
158	453	No
159	770	No
160	1089	Yes (33^2)

So $160^2 \equiv 33^2 \pmod{24511}$, and this gives

$$24511|160^2 - 33^2 = (160 + 33)(160 - 33) = 193 \times 127.$$

However, finding squares like this is rather unusual. Our refinement makes use of *index calculus*.

Even though we may not easily find a square, all is not lost. The idea is to multiply together several of these congruences so that the right-hand side becomes a square. Since each left-hand side is already a square, the product of the left-hand sides is automatically square.

We select a *smoothness bound B*, which will depend on the number N to some extent. We say that a number is *B-smooth* if all of its prime factors are no larger than B, and that the *factorbase* consists of primes below B.

The idea is to find squares which are B-smooth modulo N. Since numbers are more likely to factorise into small primes if they themselves are not too big, we run through values of x with x^2 "small" modulo N, and then try to factorise these values. Ignore the results if they are not B-smooth, but if they are, store them for the next stage of the algorithm, when we combine some subset of those obtained to give a congruence of the desired form. Let's explain this with an example.

Example 11.2 Let's select $N = 227179$. Then $\sqrt{N} = 476.732\ldots$; consider the numbers either side of $\sqrt{N}$.

$$476^2 \equiv -603 = -3^2 \times 67 \ (\text{mod } N)$$

and

$$477^2 \equiv 350 = 2 \times 5^2 \times 7 \ (\text{mod } N).$$

If our smoothness bound is $B = 25$, say, ignore the first congruence, but keep the second; the largest factor involved in 350 is 7, so 477^2 is B-smooth for $B = 25$.

Generally, one looks at all values of x in some interval either side of $\sqrt{N}$. Because we will allow negative small values of x^2 (mod N), we will factorise $x^2 - N$. Let's tabulate the 25-smooth values of $x^2 - N$, where we allow x to be in the range $460 \leq x \leq 495$, within about 15 of $\sqrt{N}$.

x	$x^2 - N$	Factorisation
470	$-\ 6729$	$-3 \times 7 \times 13 \times 23$
473	$-\ 3450$	$-2 \times 3 \times 5^2 \times 23$
477	350	$2 \times 5^2 \times 7$
482	5145	$3 \times 5 \times 7^3$
493	15870	$2 \times 3 \times 5 \times 23^2$

Now we combine these in such a way to form squares. In this example, it is easy to do this by inspection.

Notice that the last three values of x combine so that the right-hand side is a square:

$$(477 \times 482 \times 493)^2 = 477^2 \times 482^2 \times 493^2$$
$$\equiv (2 \times 5^2 \times 7) \times (3 \times 5 \times 7^3) \times (2 \times 3 \times 5 \times 23^2)$$
$$\equiv 2^2 \times 3^2 \times 5^4 \times 7^4 \times 23^2$$
$$\equiv (2 \times 3 \times 5^2 \times 7^2 \times 23)^2 \pmod{227179}.$$

We also compute

$$477 \times 482 \times 493 \equiv 212460 \pmod{227179}$$
$$2 \times 3 \times 5^2 \times 7^2 \times 23 \equiv 169050 \pmod{227179}.$$

It follows that
$$212460^2 \equiv 169050^2 \pmod{227179},$$

and so $227179|(212460^2 - 169050^2)$, and we compute

$$(227179, 212460 + 169050) = 157, \qquad (227179, 212460 - 169050) = 1447,$$

giving the factorisation
$$227179 = 157 \times 1447.$$

So the outcome of the first stage of the algorithm is a matrix consisting of values of x such that $x^2 \pmod{N}$ is B-smooth, and we need to find some combination whose product is a square.

In general, in order to find combinations giving squares systematically, we need to do some linear algebra modulo 2 on the exponents. Squares which are B-smooth should be converted into vectors modulo 2, with components corresponding to -1 and each prime in the factorbase.

In the example just seen, the factorbase consists of those primes up to 25; there are 9 such primes, and so our vectors are of the form

$$(x_{-1}, x_2, x_3, x_5, x_7, x_{11}, x_{13}, x_{17}, x_{19}, x_{23}),$$

where x_p is 1 if the exponent of p in the factorisation of $x^2 \pmod{N}$ is odd, and 0 if it is even (and $x_{-1} = 1$ if $x^2 - N < 0$ and is 0 if $x^2 - N > 0$). The five B-smooth values in Example 11.2 become:

x	Factorisation	Vector
470	$-3 \times 7 \times 13 \times 23$	(1, 0, 1, 0, 1, 0, 1, 0, 0, 1)
473	$-2 \times 3 \times 5^2 \times 23$	(1, 1, 1, 0, 0, 0, 0, 0, 0, 1)
477	$2 \times 5^2 \times 7$	(0, 1, 0, 0, 1, 0, 0, 0, 0, 0)
482	$3 \times 5 \times 7^3$	(0, 0, 1, 1, 1, 0, 0, 0, 0, 0)
493	$2 \times 3 \times 5 \times 23^2$	(0, 1, 1, 1, 0, 0, 0, 0, 0, 0)

and the congruence of squares that we found is equivalent to the sum of the last three vectors being $(0, 0, 0, 0, 0, 0, 0, 0, 0, 0)$ modulo 2.

To guarantee that we can find nontrivial linear relations between the exponent vectors, elementary linear algebra tells us that we need to have more B-smooth squares than primes in the factorbase. With a factorbase of size 10, we would generally need at least 11 B-smooth values of $x^2 - N$ to guarantee a nontrivial relation.

It is worth remarking that even for nontrivial linear relations, it happens quite often that in the resulting expression $x^2 \equiv y^2 \pmod{N}$, we find that $x \equiv \pm y \pmod{N}$. In this case, all the factors of N divide $x + y$ and none divide $x - y$, or vice versa. When this happens, we need to find alternative combinations, and try again.

Doing mod 2 linear algebra on the exponent vectors is known as *index calculus*.

Note that the size of the smoothness bound is rather critical. If B is chosen too small, it will be very difficult to find many squares which are B-smooth. But if B is too large, we will need to find many B-smooth squares—the linear algebra involves a matrix with a row for each B-smooth square and a column for each prime in the factorbase; cutting the size of the matrix can save a lot of time in the linear algebra section.

Exercise 11.2 Factor 21311 by this method using the values $x = 132$ and $x = 144$, and using index calculus.

Remark 11.3 Notice that a prime p can only divide $x^2 - N$ if N is a quadratic residue modulo p, and so we need only include these primes in our factorbase. Again in the setting of Example 11.2, with $B = 25$, this rules out the possibility of 11 or 19 ever dividing any of the values of $x^2 - N$. We could therefore have chosen the factorbase $\{-1, 2, 3, 5, 7, 13, 17, 23\}$ (of size 8), and we can guarantee a nontrivial relationship between exponent vectors if we had at least 9 B-smooth values of $x^2 - N$.

This process has been automated to factorise numbers in excess of 100 digits.

Remark 11.4 We should remark that there is a simple way to find candidate values of x for which $x^2 - N$ is smooth, which justifies the inclusion of the word "sieve" in the name of the algorithm.

Having chosen our interval, we then proceed as follows. For every prime p in our factorbase, we work out the solutions to $x^2 \equiv N \pmod{p}$. In general, this will have two solutions, $x \equiv a_p \pmod{p}$ or $x \equiv b_p \pmod{p}$. We observe that this means that the values of x where $x^2 - N$ has a factor of p are exactly those satisfying one of the congruences $x \equiv a_p \pmod{p}$ or $x \equiv b_p \pmod{p}$. Rather than compute each $x^2 - N$, and factor the result, it is much faster to find the first solutions in our interval to the congruences $x \equiv a_p \pmod{p}$ or $x \equiv b_p \pmod{p}$. We record that these values of x give values of $x^2 - N$ with factors of p. Then we step through our interval, exactly as in the *sieve of Eratosthenes* (see [7]), successively adding p to the values of solutions, and recording that these x-values also satisfy the congruences, and therefore that the corresponding values of $x^2 - N$ also have factors of p.

After sieving for each prime in the factorbase, we have now recorded the primes in the factorbase dividing each value of $x^2 - N$. We can even adapt this method to deal with powers of primes in the factorbase, so that for every x in our interval, we can easily compute whether or not every prime dividing $x^2 - N$ (mod N) lies in our factorbase.

(This is not exactly how sieving is performed in practice, although it is essentially similar. For more details, see Crandall and Pomerance [4], for example.)

Remark 11.5 In practice, often a "large prime" variant is used. This means that relations are retained if $x^2 - N$ factors completely over the factorbase *except possibly for one large prime between B and B^2*. It is easy to incorporate this into the sieving process, and it clearly may give extra relations. If two relations both involve the same large prime, they can be multiplied, and this gives another relation purely involving primes in the factorbase.

11.3 The Number Field Sieve: A First Example

The number field sieve may be viewed as a much more sophisticated variant of the quadratic sieve, but working in an algebraic number field.

We will build up towards a fairly full description of the algorithm gradually, starting with a very simple example. However, we will be extremely fortunate in this example that everything works so simply!

It will be extremely rare that we can find everything working so well—we will successively improve our algorithm so that it works for more and more situations, but (at least before the final description) we will explain that there are still reasons why our procedure will not work in general.

Example 11.6 Let us begin by factoring $N = 119$, using a different method to find a congruence of the form $x^2 \equiv y^2$ (mod 119).

We observe that $119 = 11^2 - 2$, i.e., that 11 is a root of $x^2 - 2$ modulo 119. This means that there is a ring homomorphism

$$\phi : \mathbb{Z}[\sqrt{2}] \longrightarrow \mathbb{Z}/119\mathbb{Z}$$
$$a + b\sqrt{2} \mapsto a + 11b \text{ (mod 119)}$$

from the ring $\mathbb{Z}[\sqrt{2}]$ to the integers modulo 119.

Now we observe that $3 + 2\sqrt{2}$ maps to $3 + 11 \cdot 2 = 25 = 5^2$.

However, we also observe that $3 + 2\sqrt{2} = (1 + \sqrt{2})^2$, and that $1 + \sqrt{2}$ maps to $1 + 11 \cdot 1 = 12$. As

$$3 + 2\sqrt{2} = (1 + \sqrt{2})^2,$$

and ϕ is a ring homomorphism, it must be true that

$$\phi(3 + 2\sqrt{2}) = \phi(1 + \sqrt{2})^2.$$

This means that $25 = 12^2$ in $\mathbb{Z}/119\mathbb{Z}$, or that $5^2 \equiv 12^2 \pmod{119}$. As above, we compute $(119, 12 + 5) = 17$ and $(119, 12 - 5) = 7$, and find the factorisation $119 = 17 \times 7$.

We have therefore recovered our factorisation partly by using arithmetic in $\mathbb{Z}[\sqrt{2}]$.

We were very lucky for several reasons. Firstly, it happened that we could write 119 in the form $x^2 - 2$. This enabled us to find a ring homomorphism ϕ of a convenient form. More importantly, we found a square $3 + 2\sqrt{2} = (1 + \sqrt{2})^2$ in $\mathbb{Z}[\sqrt{2}]$ which itself mapped to a square under this homomorphism. This allowed us to recover a congruence of squares modulo 119. As we try to extend this calculation to cover more and more cases, we'll see that, in fact, there were even more lucky features than those mentioned so far: for example, that $\mathbb{Z}[\sqrt{2}]$ is the full ring of integers in its field of fractions $\mathbb{Q}(\sqrt{2})$, and that it has unique factorisation.

Exercise 11.3 Factorise $527 = 23^2 - 2$ using this method, again considering $3 + 2\sqrt{2}$.

Our goal will be to extend this to cover more and more examples. Since most of the difficulties are already apparent in the case of quadratic fields, we will restrict attention to this case for all of the chapter. Our rings will (usually) be $\mathbb{Z}[\sqrt{d}]$ for some d, and we will write K for the field of fractions of $\mathbb{Z}[\sqrt{d}]$, which will be the field in which some of the calculations take place. (There will be different d for each example, but we will just write K for simplicity in each case.)

11.4 Index Calculus

One lucky point above is that we chose a square that happened to map to a square. This is lucky in the same way as Example 11.1, and generally we need to use some form of index calculus to combine smooth relations into those where squares map to squares.

In this first example, with $N = 119 = 11^2 - 2$, we plucked $3 + 2\sqrt{2}$ from nowhere, and observed that it could be used to give the factorisation. In general, if we hadn't been so lucky, we would have had to find many pairs (a, b), and consider the corresponding elements $a + b\sqrt{2}$ and their images $a + 11b \pmod{119}$, and do some linear algebra to determine a collection of pairs (a, b) such that the product of the corresponding $a + b\sqrt{2}$ and also the product of the images $a + 11b \pmod{119}$ are squares.

We need to find a set S of pairs (a, b), corresponding to $a + b\alpha$ and $a + bm \pmod{N}$, where *both* $\prod_{(a,b)\in S}(a + b\alpha) \in \mathbb{Z}[\alpha]$, and $\prod_{(a,b)\in S}(a + bm)$ $\pmod{N}$ are squares.

We will again follow the idea of the quadratic sieve, and choose a smoothness bound B. We will look at many pairs (a, b) and hope that *both $a + b\alpha \in \mathbb{Z}[\alpha]$ and $a + bm \in \mathbb{Z}/N\mathbb{Z}$ are B-smooth*. We will then hope that we can combine pairs by doing some mod 2 linear algebra so that the products are squares.

We have already defined smoothness for integers—recall that an integer is *B*-smooth if all of its prime factors are no greater than *B*. But we also need a similar notion for elements in $\mathbb{Z}[\alpha]$.

Definition 11.7 Let *B* be a positive number. An algebraic number is called *B-smooth* if every prime number dividing its norm is at most *B*.

Recall that algebraic number fields have unique factorisation of ideals into prime ideals, and that every prime ideal has norm which is the power of a prime number in $\mathbb{Z}$. Given an algebraic integer β in some number field *K*, we can consider the ideal $\langle \beta \rangle$, and factor it as a product of prime ideals of $\mathbb{Z}_K$. Then β will be *B*-smooth if all the prime ideals in its factorisation are *B*-smooth (since the norm of $\langle \beta \rangle$ is the product of the norms of the ideals in the prime factorisation). The norm of any prime ideal in $\mathbb{Z}_K$ must be the power of a prime number, and so for β to be smooth, we really require that all prime ideals in the factorisation of $\langle \beta \rangle$ lie above a prime number which is at most *B*.

Example 11.8 Let $N = 115$, and observe that $115 = 11^2 - 6$. So 11 is a root of $x^2 - 6$ modulo 115, and so there is a ring homomorphism $\mathbb{Z}[\sqrt{6}] \longrightarrow \mathbb{Z}/115\mathbb{Z}$ given by $a + b\sqrt{6} \mapsto a + 11b \pmod{115}$.

We will tabulate pairs (a, b), and consider when $a + b\sqrt{6}$ is *B*-smooth, and when $a + 11b \pmod{115}$ is *B*-smooth (one could use different smoothness bounds on the algebraic side from the mod 115 side, but for simplicity we will use the same bound). A moment's thought should convince you that we needn't bother with those where *a* and *b* share a non-trivial common factor; it will essentially just give a duplicate row in the matrix.

Choose $B = 7$.

By definition, $a + b\sqrt{6}$ is smooth when $N(a + b\sqrt{6})$ is smooth, i.e., when $a^2 - 6b^2$ is smooth.

(a, b)	$a^2 - 6b^2$	-1	2	3	5	7	$a + 11b$	2	3	5	7
$(3, 1)$	3	0	0	1	0	0	14	1	0	0	1
$(4, 1)$	10	0	1	0	1	0	15	0	1	1	0
$(9, 1)$	75	0	0	1	0	0	20	0	0	1	0
$(-1, 1)$	-5	1	0	0	1	0	10	1	0	1	0
$(-2, 1)$	-2	1	1	0	0	0	9	0	0	0	0
$(-3, 1)$	3	0	0	1	0	0	8	1	0	0	0
$(-4, 1)$	10	0	1	0	1	0	7	0	0	0	1
$(-6, 1)$	30	0	1	1	1	0	5	0	0	1	0
$(-9, 1)$	75	0	0	1	0	0	2	1	0	0	0
$(3, 2)$	-15	1	0	1	1	0	25	0	0	0	0
$(2, 3)$	-50	1	1	0	0	0	35	0	0	1	1
$(7, 3)$	-5	1	0	0	1	0	40	1	0	1	0

We will list pairs (a, b) with $-10 \leq a \leq 10$, $1 \leq b \leq 3$ where both $a^2 - 6b^2$ and $a + 11b \pmod{115}$ are 7-smooth. As in the quadratic sieve case, it may be that $a^2 - 6b^2 < 0$, and so we will have an additional column to deal with a possible minus sign.

We have 9 columns corresponding to our factorbase, and 12 rows. Since we have more rows in the matrix than we have columns, there will be some combination of pairs for which the products give squares on both sides. (Again, we should note that the choice of the smoothness bound is crucial—if B is too small, it will be hard to find rows at all, whereas if B is too big, we will have many columns, and although it will be easier to find rows, the linear algebra required to find the desired relations may become difficult.)

In this example, we can see that the pairs $(-3, 1)$ and $(-9, 1)$ are the same, and so multiplying them (which corresponds to adding rows) should give what we want. Indeed,

$$(-3 + \sqrt{6})(-9 + \sqrt{6}) = 33 - 12\sqrt{6} = (-3 + 2\sqrt{6})^2.$$

We also see that $\phi(33 - 12\sqrt{6}) = 16 = 4^2$, and can easily compute

$$\phi(-3 + 2\sqrt{6}) = -3 + 2 \cdot 11 = 19,$$

so we get that

$$4^2 = \phi(33 - 12\sqrt{6}) = \phi((-3 + 2\sqrt{6})^2) = \phi(-3 + 2\sqrt{6})^2 = 19^2$$

in $\mathbb{Z}/115\mathbb{Z}$, and therefore get the congruence $19^2 \equiv 4^2 \pmod{115}$. We compute the highest common factors $(115, 19 + 4) = 23$ and $(115, 19 - 4) = 5$, and get the factorisation $115 = 23 \times 5$.

We have therefore incorporated index calculus into the method we gave earlier. Once again, we could actually compute all the rows in the matrix by means of a sieving process, but we will not go into implementation details here, referring instead to [4].

Exercise 11.4 Factorise 1679 by writing it as $41^2 - 2$, and using the pairs $(a, b) = (-1, 2)$ and $(5, 4)$.

Before we go on, it is perhaps worth noting that there are no entries in the final column corresponding to the prime 7 on the algebraic side. It is easy to see why; we would need $a^2 - 6b^2$ to be divisible by 7, and this would require 6 to be a square modulo 7. The same argument suggests that primes which occur on the algebraic side must have 6 as a quadratic residue. We will return to this observation in the next section, when we discuss the algebraic factorbase more precisely.

In the quadratic sieve, one side of the congruence is automatically a square, by the way we try to construct our congruences. Indeed, our aim there is to find squares which are fairly small, and to hope that this means that they are smooth. So our congruences are "squares $\equiv$ smooth". In the Number Field Sieve, we will not necessarily have a square on either side of the congruence, but will instead look for smooth algebraic numbers which map to smooth numbers modulo N, and then the linear algebra will try to combine "smooth $\equiv$ smooth" relations to get the "square $\equiv$ square" congruence we want.

To do index calculus, it is necessary to have a factorbase. If B is the smoothness bound, then the factorbase on the modular arithmetic side just consists of -1 and all primes up to B. The factorbase on the algebraic side is the topic of the next section.

11.5 Prime Ideals and the Algebraic Factorbase

This last example worked well, but we've been lucky again! Here is an example where the same method as the last example appears to work, but then something goes wrong.

Consider the following example:

Example 11.9 Let $N = 9019 = 95^2 - 6$. There is a ring homomorphism from $\mathbb{Z}[\sqrt{6}]$ to $\mathbb{Z}/9019\mathbb{Z}$, which sends $a + b\sqrt{6}$ to $a + 95b \pmod{9019}$. Consider the following two rows of the matrix (just including the relevant columns):

(a, b)	$a^2 - 6b^2$	-1	5	$a + 11b$	2
$(-17, 7)$	-5	1	1	$2^3 \cdot 3^4$	1
$(-7, 23)$	-5^5	1	1	$2 \cdot 3^2 \cdot 11^2$	1

The product $(-17 + 7\sqrt{6})(-7 + 23\sqrt{6}) = 1085 - 440\sqrt{6}$ is not a square, even though the product of the two norms is a square.

We can use the factorisation of ideals into prime ideals to see what is going wrong in the above example. In $\mathbb{Z}[\sqrt{6}]$, ideals factor in the following manner:

$$\langle 2 \rangle = \langle 2 + \sqrt{6} \rangle^2$$
$$\langle 3 \rangle = \langle 3 + \sqrt{6} \rangle^2$$
$$\langle 5 \rangle = \langle \sqrt{6} + 1 \rangle \langle \sqrt{6} - 1 \rangle$$
$$\langle 7 \rangle = \langle 7 \rangle$$

Let's write $\mathfrak{p}_5 = \langle \sqrt{6} + 1 \rangle$ and $\mathfrak{p}'_5 = \langle \sqrt{6} - 1 \rangle$. Then, in terms of the generators of these ideals, we can factor the numbers appearing in the last example as

$$-17 + 7\sqrt{6} = (\sqrt{6} - 1)(5 - 2\sqrt{6})$$
$$-7 + 23\sqrt{6} = (\sqrt{6} + 1)^5(5 - 2\sqrt{6}),$$

where $5 - 2\sqrt{6}$ is a unit in $\mathbb{Z}[\sqrt{6}]$. So the product is

$$(\sqrt{6}-1)(\sqrt{6}+1)^5(5-2\sqrt{6})^2 = 5(\sqrt{6}+1)^4(5-2\sqrt{6})^2,$$

5 times a square. The problem in terms of ideals is that

$$\langle -17+7\sqrt{6}\rangle = \mathfrak{p}_5', \qquad \langle -7+23\sqrt{6}\rangle = \mathfrak{p}_5^5,$$

and so the product ideal is $\mathfrak{p}_5^5 \mathfrak{p}_5'$, not the square of an ideal.

Here, there are two ideals, $\mathfrak{p}_5$ and $\mathfrak{p}_5'$, with the same norm, and their product therefore has square norm. However, the ideal itself is not square, since it is the product of an odd power of two different ideals.

The left-hand side of the matrix should therefore consist of the powers of the prime ideals involved in the factorisation of $\langle a + b\alpha\rangle$. The factorbase on the algebraic side should therefore consist of prime ideals of K lying above rational primes which are less than or equal to B. The matrix should really read:

(a, b)	$\langle a+b\sqrt{6}\rangle$	$\mathfrak{p}_5$	$\mathfrak{p}_5'$	$a + 11b$	2
$(-17, 7)$	$\langle -17+7\sqrt{6}\rangle$	0	1	$2^3 \cdot 3^4$	1
$(-7, 23)$	$\langle -7+23\sqrt{6}\rangle$	1	0	$2 \cdot 3^2 \cdot 11^2$	1

and we can read off that the product ideal is *not* the square of an ideal; this means in particular that the product of the elements cannot be a square.

This certainly suggests that considering the norm is not sufficient on the algebraic side (although it may be a useful first check), and that we should instead be considering the factorisations of the elements in the number field itself.

But there is another more serious reason why we should consider factorisation into prime ideals on the algebraic side, namely the possible failure of unique factorisation. As it happens, $\mathbb{Z}[\sqrt{6}]$ has unique factorisation, but in general, factorisation of elements in number fields is of course not unique.

There is a danger that this non-uniqueness might make us miss relations. To take one of our examples from Chap. 4, we might find a relation in $\mathbb{Z}[\sqrt{10}]$, say, involving 2 and 3 and another involving $4 + \sqrt{10}$ and $4 - \sqrt{10}$ without realising that if we combine them, the equality $2 \times 3 = (4 + \sqrt{10})(4 - \sqrt{10})$ would give us a relation.

The only way we can be really sure that we get all relations is by using the uniqueness of factorisation of *ideals* into prime ideals in the number field, and forming the algebraic factorbase by taking some collection of prime *ideals*.

Let us now consider the prime ideals which do appear in the algebraic factorbase. We will consider prime ideals whose norm is of the form p^k with p a prime which is at most the smoothness bound B.

Recall also that we are restricting attention to the special case where the ring is $\mathbb{Z}[\sqrt{d}]$ for some d with d a squarefree integer (as in all our examples at the moment), although this is never the case in practice!

Then the norm of an element $a + b\sqrt{d}$ is therefore $a^2 - db^2$, and this norm is divisible by p if $a^2 - db^2$ is divisible by p; it is easy to see that this can happen if and only if d is a square modulo p.

Lemma 11.10 *Suppose that a and b are given, with $(a, b) = 1$. Then $p | a^2 - db^2$ if and only if $ab^{-1} \equiv r \pmod{p}$, where r is a square root of d modulo p.*

Proof Indeed, if $a^2 - db^2 \equiv 0 \pmod{p}$, then $a^2 \equiv db^2 \pmod{p}$. If $b \equiv 0 \pmod{p}$, then we would need $a^2 \equiv 0 \pmod{p}$, and so $a \equiv 0 \pmod{p}$; we see that $p | b$ and $p | a$, and this contradicts the coprimality of a and b.

So b is invertible modulo p, with inverse b^{-1}, say. We then derive that $(ab^{-1})^2 \equiv d \pmod{p}$, and so d will be a square modulo p. Furthermore, ab^{-1} must be a root of $x^2 - d$ modulo p.

If $\mathfrak{p}$ is a nonzero prime ideal in $\mathbb{Z}_K$, then $\mathbb{Z}_K / \mathfrak{p}$ must be a finite field (as nonzero prime ideals are maximal, and using Lemma 5.10). The ideal $\mathfrak{p}$ must have norm p or p^2, depending on the factorisation of $\langle p \rangle$ in $\mathbb{Z}_K$. We want to consider those primes $\mathfrak{p}$ which can occur in the factorisation of $\langle a + b\sqrt{d} \rangle$, and claim that the primes which can occur must have norm p.

Proposition 11.11 *If $\mathfrak{p}$ is a nonzero prime ideal in $\mathbb{Z}_K$ appearing in the factorisation of $\langle a + b\sqrt{d} \rangle$ for some coprime pair (a, b), then $\mathfrak{p}$ has norm p, a prime. Furthermore, the prime ideal $\mathfrak{p}$ can be written $\langle p, \sqrt{d} - r \rangle$ where $ab^{-1} \equiv r \pmod{p}$ is a square root of d modulo p.*

Proof The map $\phi : \mathbb{Z}_K \longrightarrow \mathbb{Z}_K / \mathfrak{p}$ is a surjection from $\mathbb{Z}_K$ to the finite field $\mathbb{Z}_K / \mathfrak{p}$ and this must contain the finite field $\mathbb{F}_p = \mathbb{Z} / p\mathbb{Z}$. We'll claim that the image is just $\mathbb{F}_p$. Indeed, certainly $\mathbb{Z} \subset \mathbb{Z}_K$ maps to $\mathbb{F}_p$, so we just need to check that $\sqrt{d}$ maps to something in $\mathbb{F}_p$ also.

But $a + b\sqrt{d}$ lies in $\mathfrak{p}$, and so maps to 0 in the quotient. Thus $\phi(a) + \phi(b)\phi(\sqrt{d}) = 0$, and so $\phi(\sqrt{d}) = -\phi(ab^{-1}) = -ab^{-1} \pmod{p}$, and we have already noted that this lies in $\mathbb{F}_p$. So the image of ϕ is $\mathbb{F}_p$—but ϕ was a surjection onto $\mathbb{Z}_K / \mathfrak{p}$. It follows that $\mathfrak{p}$ is an ideal of $\mathbb{Z}_K$ of norm p.

We also know that $\phi(a + b\sqrt{d}) = 0$; as $a + br \equiv 0 \pmod{p}$, we see also that $\phi(a + br) = 0$. So $\phi(\sqrt{d}) = \phi(r)$ as ϕ is a homomorphism, showing that $\phi(\sqrt{d} - r) = 0$, or in other words that $\sqrt{d} - r \in \mathfrak{p}$.

So $\mathfrak{p}$ is a prime ideal containing p and also $\sqrt{d} - r$. It is easy to see that

$$\mathfrak{p} = \langle p, \sqrt{d} - r \rangle,$$

as the right-hand side is in the left-hand side, and both sides have norm p.   □

We have shown the following:

Theorem 11.12 *Prime ideals occurring in the factorisation of any $\langle a + b\sqrt{d} \rangle$ with $(a, b) = 1$ are of the form $\langle p, \sqrt{d} - r \rangle$, where*

- *p is a prime number p with $(\frac{d}{p}) = 1$;*
- *r is a square root of d modulo p.*

Write $\mathfrak{p}_{[p,r]}$ for the prime ideal in $\mathbb{Z}[\sqrt{d}]$ generated by p and $\sqrt{d} - r$.

The argument above shows that $\mathfrak{p}_{[p,r]}$ should divide $\langle a + b\sqrt{d}\rangle$ if and only if $ab^{-1} \equiv r \pmod{p}$. Furthermore, given that $ab^{-1} \pmod{p}$ is unique, only one prime ideal of norm p can divide $\langle a + b\sqrt{d}\rangle$, and as it has norm p, its exponent can simply be read off as the exponent of p in $N_{K/\mathbb{Q}}(\langle a+b\sqrt{d}\rangle) = N_{K/\mathbb{Q}}(a+b\sqrt{d}) = a^2 - db^2$.

Then the factorbase on the algebraic side is

$$\{\mathfrak{p}_{[p,r]} = \langle p, \sqrt{d} - r\rangle \mid r^2 \equiv d \pmod{p}, \ p \leq B \text{ is prime}\}.$$

Exercise 11.5 Suppose we wish to factor $N = 9019$ as above with the Number Field Sieve by writing it as $95^2 - 6$. Put $B = 20$. What is the factorbase?

11.6 Further Obstructions

At this stage, the strategy is to try lots of pairs (a, b) to find those which are smooth on the algebraic side, in the sense that $\langle a + b\alpha\rangle$ factors only with prime ideals in the algebraic factorbase, and also smooth on the modular arithmetic side, in the sense that $a + bm \pmod{N}$ is a B-smooth integer.

Suppose that $\mathcal{B}$ denotes the complete factorbase, made up of both the algebraic factorbase and the modular arithmetic factorbase. If we can find more than $|\mathcal{B}|$ pairs (a, b) which are smooth on both sides, then mod 2 linear algebra on the exponent vectors will give us some combination of pairs $S = \{(a, b)\}$ such that $\prod\langle a + b\alpha\rangle$ is a square ideal, and $\prod a + bm \pmod{N}$ is a square number.

However, there is still quite a lot that can go wrong on the algebraic side. What we really need is that $\prod a + b\alpha$ is a square in $\mathbb{Z}[\alpha]$ so that we can apply the map ϕ.

Example 11.13 Consider $N = 33499 = 183^2 + 10$, and use the Number Field Sieve with polynomial $f(X) = X^2 + 10$ and $m = 183$. There is a homomorphism

$$\phi : \mathbb{Z}[\sqrt{-10}] \longrightarrow \mathbb{Z}/33499\mathbb{Z}.$$

$$a + b\sqrt{-10} \mapsto a + 183b \pmod{33499}$$

Consider just the pairs $(a, b) = (9, 2)$ and $(a, b) = (3, 4)$.

The pair $(a, b) = (9, 2)$ corresponds to the element $9 + 2\sqrt{-10}$ of norm 121, and the ideal $\langle 9 + 2\sqrt{-10}\rangle$ factors as the square of an ideal of norm 11. In fact, $\langle 9 + 2\sqrt{-10}\rangle = \mathfrak{p}_{[11,10]}^2$ (notation as above). It maps to $9 + 183 \times 2 = 375 = 3 \times 5^3$.

The pair $(a, b) = (3, 4)$ corresponds to the element $3 + 4\sqrt{-10}$ of norm 169, and the ideal $\langle 3 + 4\sqrt{-10}\rangle$ factors as the square of an ideal of norm 13: $\langle 3 + 4\sqrt{-10}\rangle = \mathfrak{p}_{[13,4]}^2$. It maps to $3 + 183 \times 4 = 735 = 3 \times 5 \times 7^2$.

Let's consider the matrix of factorisations with irrelevant rows and columns removed. Since both pairs generate square ideals, there will be no nontrivial columns on the algebraic side, and the only nontrivial columns will be on the modular arithmetic side, where 3 and 5 are the only primes to divide the images to an odd power.

(a, b)	$\langle a + b\sqrt{-10}\rangle$	$a + 183b$	3	5
$(9, 2)$	$\mathfrak{p}^2_{[11,10]}$	3×5^3	1	1
$(3, 4)$	$\mathfrak{p}^2_{[13,4]}$	$3 \times 5 \times 7^2$	1	1

It is clear that combining these two pairs will give a square ideal that maps to a square. Further

$$\langle 9 + 2\sqrt{-10}\rangle\langle 3 + 4\sqrt{-10}\rangle = (\mathfrak{p}_{[11,10]}\mathfrak{p}_{[13,4]})^2.$$

Suppose that $(9 + 2\sqrt{-10})(3 + 4\sqrt{-10}) = \gamma^2$ for some element $\gamma \in \mathbb{Z}[\sqrt{-10}]$. Then the product ideal would be of the form $\langle\gamma^2\rangle = \langle\gamma\rangle^2$, and by unique factorisation, $\langle\gamma\rangle = \mathfrak{p}_{[11,10]}\mathfrak{p}_{[13,4]}$.

In fact, $\mathfrak{p}_{[11,10]}$ is principal and $\mathfrak{p}_{[13,4]}$ is not principal. ($\mathbb{Q}(\sqrt{-10})$ does not have unique factorisation, and the class group has order 2.) This means that there can be no element γ as above.

One way to see that $\mathfrak{p}_{[11,10]}$ is principal is to consider another factorisation:

$$\langle -23 + \sqrt{-10}\rangle = \mathfrak{p}^2_{[7,5]}\mathfrak{p}_{[11,10]};$$

as the class group has order 2, the ideal $\mathfrak{p}^2_{[7,5]}$ is principal; clearly the left-hand side is principal, and so $\mathfrak{p}_{[11,10]}$ must also be principal.

Similarly, the factorisation

$$\langle -5 + 2\sqrt{-10}\rangle = \mathfrak{p}_{[5,0]}\mathfrak{p}_{[13,4]}$$

can be used to see that $\mathfrak{p}_{[13,4]}$ is not principal, since it is easy to verify that $\mathfrak{p}_{[5,0]} = \langle 5, \sqrt{-10}\rangle$ is not principal.

Thus the class group gives an obstruction; the ideal $I = \prod\langle a + b\alpha\rangle$ may be the square of an ideal J, but that J may not be principal, and so we cannot write $J = \langle\gamma\rangle$.

However, the class group is a finite group, and if we gather enough such examples with squares of nonprincipal ideals, it will be possible to find a product of some of them which gives something principal.

But even if $J = \langle\gamma\rangle$, further issues may arise.

Example 11.14 Now let $N = 3019 = 55^2 - 6$, and try to use the Number Field Sieve with $f(X) = X^2 - 6$ and $m = 55$. There is a homomorphism

$$\phi : \mathbb{Z}[\sqrt{6}] \longrightarrow \mathbb{Z}/3019\mathbb{Z}.$$

$$a + b\sqrt{6} \mapsto a + 55b \pmod{3019}$$

Fix a smoothness bound of $B = 20$. The ring $\mathbb{Z}[\sqrt{6}]$ is the ring of integers of $\mathbb{Q}(\sqrt{6})$, and has unique factorisation. Then the class number is 1, and so every ideal is principal. This shows that the problem preventing Example 11.13 from working is not relevant to this example.

We find that the algebraic factorbase consists of 6 prime ideals:

$$\mathfrak{p}_{[2,0]}, \quad \mathfrak{p}_{[3,0]}, \quad \mathfrak{p}_{[5,1]}, \quad \mathfrak{p}_{[5,4]}, \quad \mathfrak{p}_{[19,5]}, \quad \mathfrak{p}_{[19,14]},$$

and the modular arithmetic factorbase contains $\{2, 3, 5, 7, 11, 13, 17, 19\}$.

Consider just the pairs $(a, b) = (-5, 1)$ and $(a, b) = (13, 5)$.

The pair $(a, b) = (-5, 1)$ corresponds to the element $-5 + \sqrt{6}$ in $\mathbb{Z}[\sqrt{6}]$, and maps to $-5 + 55 = 50 = 2 \times 5^2$. The element $-5 + \sqrt{6}$ has norm 19, so the same is true for the ideal $\langle -5 + \sqrt{6} \rangle$. It therefore just has a prime ideal of norm 19 in its factorisation; using Theorem 11.2, it is $\mathfrak{p}_{[19,14]}$, as $(-5).1^{-1} \equiv 14 \pmod{19}$. (As already remarked, $\mathfrak{p}_{[19,14]}$ is principal, and one can show that both 19 and $\sqrt{6} - 14$ are multiples of $5 - \sqrt{6}$, and deduce that $\mathfrak{p}_{[19,14]} = \langle 5 - \sqrt{6} \rangle$.)

Similarly, $(a, b) = (13, 5)$ corresponds to $13 + 5\sqrt{6}$ in $\mathbb{Z}[\sqrt{6}]$, and maps to $13 + 55.5 = 288 = 2^5 3^2$. The element $13 + 5\sqrt{6}$ has norm 19, so $\langle 13 + 5\sqrt{6} \rangle$ just has a prime ideal of norm 19 in its factorisation; it is again $\mathfrak{p}_{[19,14]}$, as $13.5^{-1} \equiv 14 \pmod{19}$.

The matrix (with irrelevant rows and columns removed) is simply:

(a, b)	$\langle a + b\sqrt{6} \rangle$	$\mathfrak{p}_{[19,14]}$	$a + 55b$	2
$(-5, 1)$	$\langle -5 + \sqrt{6} \rangle$	1	2×5^2	1
$(13, 5)$	$\langle 13 + 5\sqrt{6} \rangle$	1	$2^5 \times 3^4$	1

However, $(-5 + \sqrt{6})(13 + 5\sqrt{6}) = -35 - 12\sqrt{6}$, which is not the square of an element. (If it were the square of an element, it would have to be positive.)

So even though the ideal I generated by the product of the elements is a square ideal, the product itself is not a square. In fact,

$$-35 - 12\sqrt{6} = (-5 - 2\sqrt{6})(5 - \sqrt{6})^2,$$

so

$$\langle -5 + \sqrt{6} \rangle . \langle 13 + 5\sqrt{6} \rangle = \langle 5 - \sqrt{6} \rangle^2$$

is the square of a principal ideal.

But $-5 - 2\sqrt{6}$ is a unit which is not a square, and does not show up when working with the corresponding ideals.

It is therefore not enough that the product ideal should be the square of a principal ideal to see that the product of the corresponding elements is the square of an element in $\mathbb{Z}[\alpha]$, as units may be present.

In this example, $\mathbb{Z}[\sqrt{6}]$ has units of the form

$$\{\pm(5+2\sqrt{6})^n \mid n \in \mathbb{Z}\}$$

as $\eta = 5+2\sqrt{6}$ is a fundamental unit. The square units are

$$\{+(5+2\sqrt{6})^{2n} \mid n \in \mathbb{Z}\},$$

and these have index 4 in the full unit group. It follows that if I is principal, given by $\langle \beta \rangle$, say, then β can be written in one of the forms γ^2, $-\gamma^2$, $\eta\gamma^2$ or $-\eta\gamma^2$. If we were to gather sufficiently many examples where nonsquare units appeared, it would be possible to multiply them together to get a square. (In fact, it is easy to see that three examples suffices in this case.)

In general, Dirichlet's Unit Theorem gives the structure of the full unit group, and for any $\mathbb{Z}[\alpha]$, there will be a bound on the number of examples required to guarantee that a combination of some subset will produce a square.

Exercise 11.6 Suppose that $f(X)$ is actually a cubic. Use Dirichlet's Unit Theorem to conclude that any set of 4 units in $\mathbb{Q}(\alpha)$ contains a subset whose product is a square.

There is a technique ("Adleman columns") for adding extra columns to the matrix which can deal with both the class group obstruction and the unit obstruction. We will say a little more about this in the next section, but the reader is referred to [4] for more details.

Example 11.15 Let $N = 6893 = 83^2 + 4$. As usual, this means that $m = 83$ is a root of the polynomial $f(x) = x^2 + 4$ modulo 6893, and this gives a homomorphism $\mathbb{Z}[2i] \longrightarrow \mathbb{Z}/6893\mathbb{Z}$ given by $a + b \cdot 2i \mapsto a + 83b \pmod{6893}$. (Of course, we are choosing $2i$ as our root of $x^2 + 4$.)

Now let $a = 3$ and $b = 2$, and observe that

$$3 + 2 \cdot 2i = 3 + 4i \mapsto 3 + 2 \cdot 83 = 169 = 13^2.$$

However, although $3 + 4i$ is a square in $\mathbb{Z}[i]$, since $3 + 4i = (2 + i)^2$, it is not a square in the ring $\mathbb{Z}[2i]$.

There is a simple remedy in this case; we simply multiply our square root by a suitable factor so that it is in our ring—we use $2(2 + i) = 4 + 2i$ instead of $2 + i$. This corresponds to scaling our (a, b) by 4, so we use $a = 12$, $b = 8$:

$$12 + 8 \cdot 2i = 12 + 16i = (4 + 2i)^2,$$

and $12 + 8 \cdot 2i \mapsto 12 + 8 \cdot 83 = 26^2$. Also, $4 + 2i \mapsto 4 + 1 \cdot 83 = 87$, so we get a congruence $26^2 \equiv 87^2 \pmod{6893}$, giving the factorisation $6893 = 61 \times 113$, as required.

Having to use a ring which is a proper subring of the ring of integers of the number field arising can cause problems. These subrings may not share good arithmetic properties of $\mathbb{Z}_K$; we have not proven unique factorisation of ideals into prime ideals for these subrings, for example.

In general, the situation is even worse than this; we have only been working with monic polynomials for simplicity, so that the root α is an algebraic integer. In this case, if K denotes the field of fractions of $\mathbb{Z}[\alpha]$, then $\mathbb{Z}[\alpha] \subseteq \mathbb{Z}_K$, but in general when f need not be monic, we need have no such inclusion.

These are, essentially, all the possible obstructions. However, one problem remains; even if we know that $\prod a + b\alpha = \gamma^2$ for some element $\gamma \in \mathbb{Z}[\alpha]$, it is not generally a trivial matter to work out what γ is, expressed as a polynomial in α.

11.7 The General Case

The reader who has followed the discussion so far should now realise how the general case will proceed. We will give a fairly complete description of the algorithm, but there are numerous refinements which would be made in actual implementations.

Suppose that a number N is given to be factorised. The aim is to find integers x and y such that $x^2 \equiv y^2 \pmod{N}$.

Step 1: Find a Polynomial $f(X) \in \mathbb{Z}[X]$ and an Integer m

We start by trying to find a polynomial $f(X) \in \mathbb{Z}[X]$ and an integer m such that $f(m) \equiv 0 \pmod{N}$. Then if α denotes a root of $f(X)$ in $\mathbb{C}$, the algebraic side of the Number Field Sieve will take place in $\mathbb{Z}[\alpha]$.

Remark 11.16 One never needs to compute an actual value for α; one just has to know the polynomial $f(X)$, and deduce properties of $\mathbb{Z}[\alpha]$ and $\mathbb{Q}(\alpha)$.

The norm on the algebraic side is closely related to the values taken by the polynomial, and therefore to the size of its coefficients. Thus we are more likely to find smooth values on the algebraic side if the size of the coefficients can be made as small as possible.

Let's justify this comment now.

We are going to consider only elements of $\mathbb{Z}[\alpha]$ of the form $a + b\alpha$. One reason is that it will be easy to see whether or not $a + b\alpha$ is B-smooth. Indeed, if $\alpha_1 = \alpha$,

$\alpha_2, \ldots \alpha_d$ are the roots of f, which we will again suppose monic for simplicity, then we have

$$f(X) = (X - \alpha)(X - \alpha_2) \cdots (X - \alpha_d).$$

The norm of $a + b\alpha$ is the product of the conjugates of $a + b\alpha$, and these are $a + b\alpha$, $a + b\alpha_2, \ldots, a + b\alpha_d$. Then

$$N(a + b\alpha) = (a + b\alpha)(a + b\alpha_2) \cdots (a + b\alpha_d)$$
$$= (-b)^d \left(-\frac{a}{b} - \alpha\right) \left(-\frac{a}{b} - \alpha_2\right) \cdots \left(-\frac{a}{b} - \alpha_d\right)$$
$$= (-b)^d f \left(-\frac{a}{b}\right),$$

so it is very easy to compute the norm of $a + b\alpha$. Indeed, if we write $f(X) = X^d + c_{d-1}X^{d-1} + \cdots + c_0$, let $F(X, Y) = X^d + c_{d-1}X^{d-1}Y + \cdots + c_0 Y^d$ be the corresponding homogeneous degree d polynomial; then $F(X, Y) = Y^d f(\frac{X}{Y})$ and the formula above shows that $N(a + b\alpha) = F(a, -b)$.

Remark 11.17 It may seem as if we are losing possible useful information by considering only numbers of the form $a + b\alpha \in \mathbb{Z}[\alpha]$, but it is less easy to compute norms of other numbers in this ring, and in practice, there seems to be little to be gained by using other elements of $\mathbb{Z}[\alpha]$.

Here's one way to find a polynomial with fairly small coefficients. We begin by selecting the degree d of the polynomial (we will comment on the choice of d in the final section). Next, we choose m to be $\lfloor N^{\frac{1}{d}} \rfloor$, and write N in base m, as

$$N = m^d + c_{d-1}m^{d-1} + \cdots + c_0,$$

where each c_i is between 0 and $m - 1$. We then simply choose f to be

$$f(X) = X^d + c_{d-1}X^{d-1} + \cdots + c_0.$$

For example, if $N = 5731$, and we chose d to be 3, our formula gives $m = 17$. Then we write N in base 17:

$$5731 = 17^3 + 2 \cdot 17^2 + 14 \cdot 17 + 2,$$

and let $f(x) = x^3 + 2x^2 + 14x + 2$. Then $f(17) = 5731 \equiv 0 \pmod{5731}$, and so 17 is a root of $f(x)$ modulo 5731, as required.

Since the size of m is about the dth root of N, and the coefficients are all less than m, we can see that the coefficient size is of the order $N^{\frac{1}{d}}$.

Remark 11.18 While this is one of the easiest ways to produce polynomials with coefficients of the right sort of size, there are better techniques which are actually used in practice.

Step 2: Form the Homomorphism ϕ

We claimed above that there is a ring homomorphism $\phi : \mathbb{Z}[\alpha] \longrightarrow \mathbb{Z}/N\mathbb{Z}$. We will suppose that f is a monic polynomial for simplicity here—as it is irreducible, this means that f will be the minimal polynomial of α. (If α is not an algebraic integer, the argument below will work as long as the coefficients of f are integers with no common factor, even though it may not be monic.)

Lemma 11.19 *There is a ring homomorphism* $\phi : \mathbb{Z}[\alpha] \longrightarrow \mathbb{Z}/N\mathbb{Z}$ *taking* α *to m.*

Proof We need to check that ϕ is well-defined, and is a homomorphism. Once we show that it is well-defined, it will be easy to see that it is a homomorphism, so we will just show that it is well-defined.

Suppose $g(\alpha)$ and $h(\alpha)$ are two elements of $\mathbb{Z}[\alpha]$, that is, they are polynomial expressions in α with integer coefficients. We need to check that if $g(\alpha)$ and $h(\alpha)$ have the same value, then $\phi(g(\alpha)) = \phi(h(\alpha))$.

But if $g(\alpha) = h(\alpha)$, we see that α is a root of $g - h$. Since f is the minimal polynomial of α, we have $f | g - h$. This divisibility is in $\mathbb{Z}[x]$, not just $\mathbb{Q}[x]$, by Gauss's Lemma (Remark 2.23). So $g - h = fq$ for some polynomial $q(x) \in \mathbb{Z}[x]$. In particular,

$$g(m) - h(m) = f(m)q(m);$$

as $f(m) \equiv 0 \pmod{N}$, the right-hand side is therefore divisible by N. So $g(m) \equiv h(m) \pmod{N}$. This shows that ϕ is well-defined. It is clear that it is a homomorphism from the definition. $\square$

Step 3: Compile a Factorbase

Select a smoothness bound B. (In practice, one might want different smoothness bound on the algebraic side and the modular arithmetic side.) The choice of B will be discussed in the final section.

The proof earlier that the algebraic factorbase only consists of prime ideals of norm p, a prime, continues to be valid, as does the description of the prime ideals involved. So we form the algebraic factorbase from the prime ideals in the ring of integers of $\mathbb{Q}(\alpha)$ of the form $\mathfrak{p}_{[p,r]} = \langle p, \alpha - r \rangle$, where $p \le B$ and $f(r) \equiv 0 \pmod{p}$.

The modular arithmetic factorbase will consist of all the prime numbers up to B, together with -1.

Then the complete factorbase consists of the union of these two factorbases.

Step 4: Sieve to Find Relations

Now we select a region consisting of many pairs (a, b). In order to perform sieving to find relations, it is helpful if the region has a convenient shape, perhaps a rectangular region of the form

$$\{(x, y) \mid -L \le x \le L, \ 0 < y \le W\}$$

for some length L and width W.

For every pair (a, b) in the region with $(a, b) = 1$, compute the norm $N(a + b\alpha) = F(a, -b)$ as above (a sieving process), and hence factor the ideal $\langle a + b\alpha \rangle$, if possible, in the algebraic factorbase. A prime p divides $N(a + b\alpha)$ if and only if a prime $\mathfrak{p}_{[p,r]}$ divides $\langle a + b\alpha \rangle$ for some r. Since a and b are coprime, either a or b is not divisible by p. It is easy to see (as above) that $F(a, -b) \equiv 0 \pmod{p}$ means that $a + br \equiv 0 \pmod{p}$ for some r which is a root of $f(x)$ modulo p. Since r is uniquely determined by a and b, there is a unique prime above p dividing $\langle a + b\alpha \rangle$, and since it has norm p, its exponent must be the same as the exponent of p in $N(a + b\alpha) = F(a, -b)$.

Also factor the integer $a + bm \pmod{N}$, if possible, in the modular arithmetic factorbase.

Say that a pair (a, b) is *smooth* if $\langle a + b\alpha \rangle$ and $a + bm \pmod{N}$ both factor in the factorbase.

Every smooth pair (a, b) gives a row of a matrix whose columns consist of elements of the factorbase (and also Adleman columns, discussed as part of the next step).

Remark 11.20 One might want to use a large prime variant, as in Remark 11.5 for the quadratic sieve.

Step 5: Linear Algebra

The aim of the linear algebra modulo 2 is to produce from all the smooth pairs a subset S such that $I = \prod_{(a,b)\in S}\langle a + b\alpha \rangle$ is a square ideal in $\mathbb{Z}[\alpha]$, and also $\prod_{(a,b)\in S}(a + bm) \pmod{N}$ is a square integer. Write $\beta = \prod_{(a,b)\in S}(a + b\alpha)$, so that $I = \langle \beta \rangle$.

As already seen in the quadratic examples above, quite a lot can still go wrong!

1. I may be the square of an ideal in $\mathbb{Z}[\alpha]$, but this differs in general from the ring of integers $\mathbb{Z}_K$ of $K = \mathbb{Q}(\alpha)$.
2. Even if I is the square of an ideal J in $\mathbb{Z}_K$, the ideal J may not be principal.
3. Even if J is principal, so that $\langle \beta \rangle = \langle \gamma \rangle^2$, it may not be true that $\beta = \gamma^2$.
4. Next, if $\beta = \gamma^2$ for some γ, we only know that $\gamma \in \mathbb{Z}_K$, and it may not be true that $\gamma \in \mathbb{Z}[\alpha]$.

As already mentioned in Example 11.15, the final difficulty is dealt with by scaling, and is easy to overcome.

In practice, the first three obstructions are ignored, but extra tests are built in to the linear algebra stage to try to guarantee that $\beta = \gamma^2$ for some γ. The method used is to add extra columns ("Adleman columns") to the linear algebra matrix. We are seeking a set S so that $\beta = \prod_{(a,b)\in S}(a + b\alpha)$ is a square.

Suppose that q is a prime, and s is chosen so that $f(s) \equiv 0 \pmod{q}$. Just as for the map ϕ, there will be a homomorphism $\phi_q : \mathbb{Z}[\alpha] \longrightarrow \mathbb{Z}/q\mathbb{Z}$ sending α to s.

If β is a square, then $\phi_q(\beta) = \prod_{(a,b)\in S}(a + bs) \pmod{q}$ will be a square modulo q, and this can be checked with Legendre symbols: if β is a square, then $\prod_{(a,b)\in S}\left(\frac{a+bs}{q}\right) = 1$.

We add columns to the matrix corresponding to several pairs (q, s) as above. For each smooth pair (a, b), as well as storing its factorisations with respect to the factorbase, we also store the values of $\left(\frac{a+bs}{q}\right)$ for each pair (q, s). (More precisely, put '0' in the matrix under the column (q, s) if $\left(\frac{a+bs}{q}\right) = 1$ and '1' if $\left(\frac{a+bs}{q}\right) = -1$.)

In the linear algebra stage, we find a set S such that $\langle\beta\rangle = \prod_{(a,b)\in S}\langle a + b\alpha\rangle$ is a square ideal, $\prod_{(a,b)\in S}(a + bm) \pmod{N}$ is a square and also the Adleman columns satisfy $\prod_{(a,b)\in S}\left(\frac{a+bs}{q}\right) = 1$ for each (q, s) pair.

If β is a square, it would be automatic that $\prod_{(a,b)\in S}\left(\frac{a+bs}{q}\right) = 1$ for each (q, s) pair, but if not, there should be no reason why this should be $+1$ or -1 for any (q, s) column. Thus if we choose enough (q, s) pairs, and all of the results $\prod_{(a,b)\in S}\left(\frac{a+bs}{q}\right) = 1$, it is very likely that β is a square.

Step 6: Square Roots

With the notation of the previous step, if $\beta = \gamma^2$ for some $\gamma \in \mathbb{Z}[\alpha]$, it is not clear how to go about finding γ in general (in our quadratic cases, this is easy, but in higher degree number fields, it is not so clear). This is a difficult problem, but not insurmountable. One approach sketched in [4] is to choose primes p such that $f(X)$ is irreducible modulo p. Then we can reduce $\beta \in \mathbb{Z}[\alpha]$ to $\overline{\beta} \in \mathbb{F}_p[\alpha]$, and try to solve $\overline{\gamma}^2 = \overline{\beta}$ as a power series in α with coefficients mod p. With enough primes, it is possible to reconstruct $\gamma \in \mathbb{Z}[\alpha]$ using the Chinese Remainder Theorem. The authors remark, however, that in practice, various refinements of this technique are used.

Step 7: Factorise N—or Not!

Once we have $\gamma \in \mathbb{Z}[\alpha]$, we should be able to find integers x and y, and a congruence $x^2 \equiv y^2$ (mod N). At this stage, we know $\beta = \gamma^2$ in $\mathbb{Z}[\alpha]$, and $\beta = \prod_{(a,b) \in S}(a + b\alpha)$ maps to a square $y^2 = \prod_{(a,b) \in S}(a + bm)$ (mod N). Write $x = \phi(\gamma)$; as $\phi(\beta) = \phi(\gamma)^2$, we get $y^2 = x^2$ in $\mathbb{Z}/N\mathbb{Z}$, which is the desired congruence $x^2 \equiv y^2$ (mod N).

Now $N | x^2 - y^2 = (x + y)(x - y)$, and if we are lucky, N may split into two parts, one of which divides $x + y$, and the other dividing $x - y$. These may be computed using Euclid's algorithm on N and $x \pm y$, and if both are smaller than N, we get a factorisation of N. (However, it may be that $N | x + y$ or $N | x - y$, and then one has to find another set S of pairs with the same properties.)

Exercise 11.7 Make up lots of examples of your own; the best way to understand the algorithm is to try to work through it and see what can go wrong, and how to put it right. All of the features of the algorithm should be apparent in the quadratic case, and I recommend treating only the quadratic case until you are confident about all the issues.

Readers with a suitable background may wish to try to program parts of the algorithm.

11.8 Closing Comments

The Number Field Sieve looks a very complicated algorithm, and one may ask why we use it in preference to, say, the quadratic sieve.

The answer is that it allows us to factor larger numbers. When considering the two algorithms, one can see that both depend on having a significant supply of smooth numbers. The approach of the number field sieve produces more numbers of small norm; so we should get more numbers on the algebraic side which stand a chance of being smooth. Indeed, for the quadratic sieve, we need to find x near $\sqrt{N}$ and consider whether $x^2 - N$ is B-smooth. But if $x \approx \sqrt{N}$, then $x^2 - N$ is of magnitude about $\sqrt{N}$. On the other hand, the number field sieve requires both $N(a + b\alpha)$ and $a + bm$ (mod N) to be smooth; the first is equivalent to $F(a, -b)$ being smooth, but $F(a, -b)$ has coefficients of order $N^{\frac{1}{d}}$, and, with suitable choices of m, and the region of (a, b) pairs, we need two numbers to be smooth which are both of order about $N^{\frac{1}{d}}$. Informally at least, this gives more chances of finding smooth relations, at least if d is sufficiently large ($d \geq 5$ should suffice).

For more details of the running time analysis, the reader should consult either of the two books [4] or [8] on the Number Field Sieve mentioned at the end of the book.

Given a number N, there are some choices to be made. We need to find a polynomial of some degree d, and then choose a smoothness bound B. The approximate running time of the algorithm can be computed in terms of d and B, and this can be optimised by making appropriate choices for these parameters. Various approximations are made in the course of this analysis, and if X and Y are two expressions depending on N, we will use the notation $X \sim Y$ to mean that $X/Y \to 1$ as $N \to \infty$. We may say that X tends to Y *asymptotically* in this case.

It turns out that the best value for d is:

$$d \sim \left(\frac{3 \ln N}{\ln \ln N}\right)^{1/3}.$$

The value of B should be chosen to be about

$$B \sim \exp\left((\tfrac{8}{9})^{\frac{1}{3}} (\ln N)^{\frac{1}{3}} (\ln \ln N)^{\frac{2}{3}}\right),$$

and these choices give a running time of (asymptotically) about

$$\exp\left((\tfrac{64}{9})^{\frac{1}{3}} (\ln N)^{\frac{1}{3}} (\ln \ln N)^{\frac{2}{3}}\right).$$

(Actually, this is not quite right; it is really

$$\exp\left(\left(\tfrac{64}{9} + o(1)\right)^{\frac{1}{3}} (\ln N)^{\frac{1}{3}} (\ln \ln N)^{\frac{2}{3}}\right),$$

meaning that it is better than $\exp\left(c(\ln N)^{\frac{1}{3}} (\ln \ln N)^{\frac{2}{3}}\right)$ for any $c > \sqrt[3]{\tfrac{64}{9}}$.)

In comparison, the corresponding figures for the quadratic sieve (here there is no parameter d, just a smoothness bound B):

$$B \sim \exp(\tfrac{1}{2}\sqrt{\ln N \ln \ln N}),$$

and the expected running time tends asymptotically to $\exp(\sqrt{\ln N \ln \ln N})$.

One can readily see that as N gets bigger and bigger, the Number Field Sieve begins to outshine the quadratic sieve; in practice, this happens when one expects to choose $d \geq 5$ in the Number Field Sieve, meaning that N should have around 100 digits. The quadratic sieve begins to become impractical for numbers with too many more digits; the Number Field Sieve has factored numbers with over 200 digits.

These running time estimates arise from an estimate of the number of pairs (a, b) such that $a + b\alpha$ is smooth; by the earlier result, this means that $F(a, -b)$ should be smooth, and the estimates above suppose that the coefficients of F can be chosen to be about $N^{1/d}$. Sometimes one wants to use these methods to factor numbers of a very particular form, for which one can choose the coefficients of f to be very small. For example, one of the first applications of the method was to factor the ninth Fermat number, $N = 2^{2^9} + 1$; the polynomial $f(x) = x^5 + 8$ was used, with $m = 2^{103}$, and $f(m) = 2^{515} + 8 = 8(2^{512} + 1) \equiv 0 \pmod{N}$. When we can choose our polynomials to have smaller coefficients than we expect, we can improve our running time—if the coefficients of f are of size $N^{e/d}$, then the running time becomes

$$\exp\left(\left(\tfrac{32(1+e)}{9} + o(1)\right)^{\frac{1}{3}} (\ln N)^{\frac{1}{3}} (\ln \ln N)^{\frac{2}{3}}\right),$$

so the best we can do is when $e = 0$, with a running time of

$$\exp\left(\left(\tfrac{32}{9} + o(1)\right)^{\frac{1}{3}} (\ln N)^{\frac{1}{3}} (\ln \ln N)^{\frac{2}{3}}\right).$$

When we are dealing with numbers where we can choose e near 0 (like the Fermat number mentioned above), the Number Field Sieve is known as the Special Number Field Sieve. In general, however, we expect the coefficients to be about $N^{1/d}$, with running time given above, and this is the General Number Field Sieve.

Appendix A
Solutions and Hints to Exercises

Chapter 1

1.1 1. Suppose that there are only finitely many prime $p \equiv 3 \pmod 4$, and label
them $p_1, \ldots, p_n$.
Then consider $N = 4p_1 \ldots p_n - 1$. This number is clearly odd, and has
remainder 3 after dividing by 4. So its prime divisors are all odd, and they
can't all be 1 (mod 4), as the product of numbers which are 1 (mod 4) is
again 1 (mod 4). So N has a prime divisor $p \equiv 3 \pmod 4$. But p must be
different from all the p_i, since $p|N$ but $p_i \nmid N$ for all $i = 1, \ldots, n$. So there
must be infinitely many primes $p \equiv 3 \pmod 4$.

2. Suppose that there are only finitely many prime $p \equiv 5 \pmod 6$, and label
them $p_1, \ldots, p_n$.
Then consider $N = 6p_1 \ldots p_n - 1$. This number is clearly odd, and has
remainder 5 after dividing by 6. So $(6, N) = 1$, and the prime divisors of
N can't all be 1 (mod 6), as the product of numbers which are 1 (mod 6) is
again 1 (mod 6). So N has a prime divisor $p \equiv 5 \pmod 6$. But p must be
different from all the p_i, since $p|N$ but $p_i \nmid N$ for all $i = 1, \ldots, n$. So there
must be infinitely many primes $p \equiv 5 \pmod 6$.

1.2 1. Suppose that $p|x^2 + 1$. If x is even, clearly p is odd. Then $x^2 \equiv -1 \pmod p$
(and as $p \neq 2$, we have $x^2 \not\equiv 1 \pmod p$), and so $x^4 \equiv 1 \pmod p$. Thus x
has order 4 modulo p. So the order of the multiplicative group is a multiple
of 4, which is the order of x. But the multiplicative group has order $p - 1$.
Thus $4|p - 1$, and so $p \equiv 1 \pmod 4$.

2. Suppose that there are only finitely many prime $p \equiv 1 \pmod 4$, and label
them $p_1, \ldots, p_n$.
Then consider $N = (2p_1 \ldots p_n)^2 + 1$. This number is clearly odd. Let p
be a prime divisor of N. By (1), $p \equiv 1 \pmod 4$. But p must be different
from all the p_i, since $p|N$ but $p_i \nmid N$ for all $i = 1, \ldots, n$. So there must be
infinitely many primes $p \equiv 1 \pmod 4$.

1.3 Suppose that $x \not\equiv 1 \pmod 3$. Then $x^2 + x + 1 \equiv 1 \pmod 6$. Let $p|x^2 + x + 1$.
Then $(x^2 + x + 1)(x - 1) = x^3 - 1 \equiv 0 \pmod p$. Thus $x^3 \equiv 1 \pmod p$. As

F. Jarvis, *Algebraic Number Theory*, Springer Undergraduate
Mathematics Series, DOI: 10.1007/978-3-319-07545-7,
© Springer International Publishing Switzerland 2014

above, $3 | p - 1$, and so $p \equiv 1 \pmod 3$.

Now suppose that there are only finitely many prime $p \equiv 1 \pmod 6$, and label them $p_1, \ldots, p_n$.

Then consider $N = (6p_1 \ldots p_n)^2 + (6p_1 \ldots p_n) + 1$. Let p be a prime divisor of N. By the above, $p \equiv 1 \pmod 3$. Also, N is odd, so p must be odd too, and so $p \equiv 1 \pmod 6$. But p must be different from all the p_i, since $p | N$ but $p_i \nmid N$ for all $i = 1, \ldots, n$. So there must be infinitely many primes $p \equiv 1 \pmod 6$.

1.4 1. If $r_{n+1} < r_n/2$, then certainly $r_{n+2} < r_{n+1}$. If $r_{n+1} = r_n/2$, in the next step, $r_{n+2} = 0$. Finally, if $r_{n+1} > r_n/2$, in the next step, $r_{n+2} = r_n - r_{n+1} < r_n/2$. So in all cases, $r_{n+2} < r_n/2$.

2. For every two steps of the algorithm, the remainder halves (at least).

3. If $b < 2^n$, this means that the remainder after $2n$ steps is less than 1, so must be 0, and the algorithm has terminated. Thus numbers which are at most 2^n require at most $2n$ applications of the division algorithm.

1.5 1. Allowing negative remainders, we have

$$
\begin{array}{rcrcrcr}
630 & = & 5 & \times & 132 & - & 30 \\
132 & = & 4 & \times & 30 & + & 12 \\
30 & = & 2 & \times & 12 & + & 6 \\
12 & = & 2 & \times & 6 & + & 0
\end{array}
$$

(so only one step is saved in this case).

2. Similar to the last exercise.

1.6 The existence of the division algorithm just follows from long division of polynomials, and the Euclidean algorithm is again simply a repetition of the division algorithm. For the example, the first step is

$$x^5 + 4x^4 + 10x^3 + 15x^2 + 14x + 6 = x(x^4 + 4x^3 + 9x^2 + 12x + 9) + x^3 + 3x^2 + 5x + 6$$

and then repeating in the same way as the Euclidean algorithm gives

$$x^4 + 4x^3 + 9x^2 + 12x + 9 = (x + 1)(x^3 + 3x^2 + 5x + 6) + x^2 + x + 3$$

and finally

$$x^3 + 3x^2 + 5x + 6 = (x + 2)(x^2 + x + 3) + 0,$$

so the highest common factor is $x^2 + x + 3$.

1.7 We'll use the modified algorithm of Exercise 1.5 as an example.

$$
\begin{array}{rcrcrcr}
999 & = & 1 & \times & 700 & + & 299 \\
700 & = & 2 & \times & 299 & + & 102 \\
299 & = & 3 & \times & 102 & - & 7 \\
102 & = & 15 & \times & 7 & - & 3 \\
7 & = & 2 & \times & 3 & + & 1
\end{array}
$$

Then

$$1 = 7 - 2 \times 3 = 7 - 2 \times (15 \times 7 - 102) = 2 \times 102 - 29 \times 7$$
$$= 2 \times 102 - 29 \times (3 \times 102 - 299) = 29 \times 299 - 85 \times 102$$
$$= 29 \times 299 - 85 \times (700 - 2 \times 299) = 199 \times 299 - 85 \times 700$$
$$= 199 \times (999 - 700) - 85 \times 700 = 199 \times 999 - 284 \times 700$$

1.8 As in Lemma 1.20.

1.9 As in Theorem 1.19, using the result of the last exercise to derive unique factorisation in $\mathbb{Z}[\sqrt{-2}]$.

1.10 Given α and β in $\mathbb{Z}[\sqrt{2}]$, write $\alpha/\beta = x + y\sqrt{2}$ with $x, y \in \mathbb{Q}$, and let m and n be the closest integers to x and y respectively. Put $\kappa = m + n\sqrt{2}$, and then $N(\alpha/\beta - \kappa) = |(x - m)^2 - 2(y - n)^2| \leq \frac{1}{2}$. The result follows as in Lemma 1.20.

1.11 We can take $\alpha = \sqrt{-7}$ and $\beta = 2$. Then the imaginary part of α/β is $\frac{\sqrt{7}}{2}$, and this is at least $\frac{\sqrt{7}}{2} > 1$ from any point of $\mathbb{Z}[\sqrt{-7}]$, since points in $\mathbb{Z}[\sqrt{-7}]$ have imaginary parts which are integer multiples of $\sqrt{7}$. We will see later that the problem is that we are working with the wrong ring!

1.12 We find $(\frac{2}{1009}) = (\frac{3}{1009}) = (\frac{5}{1009}) = (\frac{7}{1009}) = 1$, but $(\frac{11}{1009}) = -1$ (either by using known properties of Legendre symbols or using Euler's formula $a^{(p-1)/2} \equiv (\frac{a}{p}) \pmod{p}$). Compute $11^{(1009-1)/4} = 11^{252} \equiv 469 \pmod{1009}$. Then $x = 469$ is a solution (and so is $x = -469 \equiv 540 \pmod{1009}$).

1.13 1. The elements except for those where $f(s) = s$ may be partitioned into pairs $\{s, f(s)\}$. So S is the union of $\mathrm{Fix}_S(f)$ with all the pairs $\{s, f(s)\}$, and so $|S|$ is the sum of $|\mathrm{Fix}_S(f)|$ and $2r$, where r is the number of pairs.

   2. This is rather fiddly! First take (x, y, z) with $x < y - z$. Then $f(x, y, z) = (X, Y, Z)$, where $X = x + 2z$, $Y = z$ and $Z = y - x - z$. Then $f(f(x, y, z)) = f(X, Y, Z)$. But clearly $X = x + 2z > 2z = 2Y$, and as $X > 2Y$, $f(X, Y, Z) = (X - 2Y, X - Y + Z, Y) = (x, y, z)$. The other two cases are similar.

   3. Suppose that $f(x, y, z) = (x, y, z)$. It is easy to see that (x, y, z) must satisfy $y - z < x < 2y$, as the other two possibilities get mapped to triples with different conditions. Then $f(x, y, z) = (2y - x, y, x - y + z) = (x, y, z)$, so $x = y$. Then $(x, x, z) \in S$ means that $x^2 + 4xz = p$, so $p = x(x+4z)$. But p is prime, so $x = 1$ and then $z = k$ (where $p = 4k+1$). So the unique fixed point is $(1, 1, k)$. By the first part, $|S|$ is odd.

   4. It is clear that f' is another involution on S. As $|S|$ is odd (by the last part), f' has a fixed point (by the first part).

   5. If (x, y, z) is a fixed point of f', then $(x, z, y) = (x, y, z)$, so $y = z$. So there is a point of S of the form (x, y, y), and so $x^2 + 4y^2 = p$, and thus p is the sum of two squares.

1.14 For Step 1, we have already found $x = 469$ in Exercise 1.12. The Euclidean algorithm for 1009 and 469 is:

$$
\begin{aligned}
1009 &= 2 \times 469 + 71 \\
469 &= 6 \times 71 + 43 \\
71 &= 1 \times 43 + 28 \\
43 &= 1 \times 28 + 15 \\
28 &= 1 \times 15 + 13 \\
15 &= 1 \times 13 + 2 \\
13 &= 6 \times 2 + 1 \\
2 &= 2 \times 1 + 0
\end{aligned}
$$

and the algorithm gives $1009 = 28^2 + 15^2$.

1.15 Write $x^3 = (y + \sqrt{-2})(y - \sqrt{-2})$. If x were even, we would have $y^2 + 2 \equiv 0 \pmod 8$, which is not possible, so x is odd, and then y will also be odd.

If α is a common divisor of $y + \sqrt{-2}$ and $y - \sqrt{-2}$, then $\alpha | 2\sqrt{-2}$, and so $N(\alpha) | N(2\sqrt{-2}) = 8$. So α is a unit, or $\sqrt{-2} | \alpha$. But if $\sqrt{-2} | \alpha$, and $\alpha | y + \sqrt{-2}$, then $\sqrt{-2} | y$, and then y would be even. So this is not possible, and α is a unit.

Then we have $y + \sqrt{-2} = u\beta^3$, $y - \sqrt{-2} = v\gamma^3$ for units u and v. The only units in $\mathbb{Z}[\sqrt{-2}]$ are ± 1, so we need

$$
y + \sqrt{-2} = \pm(a + b\sqrt{-2})^3.
$$

Expanding, and taking the imaginary parts gives $1 = \pm(3a^2 b - 2b^3)$, or $b(3a^2 - 2b^2) = \pm 1$. This is solved with $b = \pm 1$, $a = \pm 1$, so $y + \sqrt{-2} = \pm(1 \pm \sqrt{-2})^3$, and we can read off $y = \pm 5$, and then it is easy to recover $x = 3$.

1.16 This is very similar to the example in the chapter.

Chapter 2

2.1 These numbers are all transcendental for the same reasons as for Liouville's number (and you will be able to see that many other numbers are transcendental for the same reason). Suppose there are countably many such numbers, $\alpha_1, \alpha_2, \ldots$, where $\alpha_i = \sum_{k=1}^{\infty} s_k^{(i)} 10^{-k!}$. Then let $t_k = 1$ if $s_k^{(k)} = -1$ and vice versa. The number $\beta = \sum_{k=1}^{\infty} t_k 10^{-k!}$ differs from any given α_i because it has a different coefficient of $10^{-i!}$, and so is a new number in the set. Thus any list indexed by the natural numbers must be incomplete, and so the set must be uncountable.

2.2 Let $\alpha = \sqrt{2} + \sqrt{3}$. Then $\alpha^2 = 5 + 2\sqrt{6}$, and so $(\alpha^2 - 5) = 2\sqrt{6}$. Squaring gives $\alpha^4 - 10\alpha^2 + 25 = 24$, and so α is a root of $X^4 - 10X^2 + 1 = 0$. This is the minimal polynomial over $\mathbb{Q}$; this is not completely obvious, but one way to see this is to observe that the other roots are $\pm\sqrt{2} \pm \sqrt{3}$, and to try polynomials of smaller degree, such as $(X - (\sqrt{2} + \sqrt{3}))(X - (\sqrt{2} - \sqrt{3}))$ and see that none of them have rational coefficients. (There are better ways available given some knowledge of Galois

Theory.)

Over $\mathbb{Q}(\sqrt{2})$, we simply need to eliminate the $\sqrt{3}$. Write $\alpha - \sqrt{2} = \sqrt{3}$ and square to get $\alpha^2 - 2\sqrt{2}\alpha + 2 = 3$; so α is a root of $X^2 - 2\sqrt{2}X - 1 = 0$, which is the minimal polynomial over $\mathbb{Q}(\sqrt{2})$. (Since $\alpha \notin \mathbb{Q}(\sqrt{2})$, the minimal polynomial must be at least quadratic, so this quadratic must be the minimal polynomial over $\mathbb{Q}(\sqrt{2})$.)

2.3 $\sqrt{2}\alpha = 1 + i$. Squaring gives $2\alpha^2 = 2i$, and the minimal polynomial over $\mathbb{Q}(i)$ is therefore $X^2 - i$. Alternatively, $\sqrt{2}\alpha - 1 = i$, and squaring gives $2\alpha^2 - 2\sqrt{2}\alpha + 1 = -1$. Thus α is a root of $X^2 - \sqrt{2}X + 1 = 0$, and this is the minimal polynomial for α over $\mathbb{Q}(\sqrt{2})$. Now multiply the top and bottom of the fraction by i to see that $\alpha = \frac{i-1}{\sqrt{-2}}$. Then $\sqrt{-2}\alpha + 1 = i$, and squaring gives $-2\alpha^2 + 2\sqrt{-2}\alpha + 1 = -1$, so α is a root of $X^2 - \sqrt{-2}X - 1 = 0$, and this is the minimal polynomial over $\mathbb{Q}(\sqrt{-2})$. (In these cases, it is easy to see that these really are the minimal polynomials; clearly α is not in any of these quadratic fields, and so the polynomial of smallest degree with α as a root must be at least quadratic.)

2.4 Suppose that $a + b\sqrt{2} + c\sqrt{3} + d\sqrt{6} = 0$. Rearrange this as

$$a + b\sqrt{2} = -c\sqrt{3} - d\sqrt{6},$$

so that

$$\sqrt{3} = -\frac{a + b\sqrt{2}}{c + d\sqrt{2}},$$

and so $\sqrt{3}$ would lie in $\mathbb{Q}(\sqrt{2})$, whose elements are expressible in the form $p + q\sqrt{2}$ for $p, q \in \mathbb{Q}$. Let's show that $\sqrt{3} \notin \mathbb{Q}(\sqrt{2})$. If $\sqrt{3} = p + q\sqrt{2}$, we can square to get $p^2 + 2q^2 + 2pq\sqrt{2} = 3$, and the irrationality of $\sqrt{2}$ implies that $pq = 0$. This leads to the conclusion that either $\sqrt{3} \in \mathbb{Q}$, or is a rational multiple of $\sqrt{2}$; the first is well-known to be false, and the second (after scaling by $\sqrt{2}$) implies that $\sqrt{6}$ would be rational, which is again known to be false.

2.5 We can use $\alpha = \sqrt{2} + \sqrt{3}$, and note that $\sqrt{2} - \sqrt{3}$ is not of this form. ($\alpha = \sqrt[3]{2}$ is another natural choice.)

2.6 As $\alpha^3 - 2 = 0$, we use the Euclidean algorithm on $X^3 - 2$ and $X + 2$. We find that $X^3 - 2 = (X^2 - 2X + 4)(X + 2) - 10$, and substituting α gives $(\alpha^2 - 2\alpha + 4)(\alpha + 2) = 10$. Then $\frac{1}{\alpha+2} = \frac{\alpha^2 - 2\alpha + 4}{10}$, and so

$$\frac{\alpha^2 - 1}{\alpha + 2} = \frac{(\alpha^2 - 1)(\alpha^2 - 2\alpha + 4)}{10} = \frac{3\alpha^2 + 4\alpha - 8}{10},$$

using the fact that $\alpha^3 = 2$.

2.7 Similar to 2.6. When you do the Euclidean algorithm, you will notice that fractions appear in the calculations. You should find that

$$\frac{1}{\alpha^2 - 2\alpha + 2} = \frac{2\alpha^3 + 3\alpha^2 + 2\alpha + 2}{5},$$

and then the answer to the exercise is $(2\alpha^4 + 5\alpha^3 + 5\alpha^2 + 4\alpha + 2)/5$, and using $\alpha^4 = -2\alpha - 1$, this simplifies to $\alpha^3 + \alpha^2$.

2.8 The degree of $\mathbb{Q}(\sqrt[3]{2}, \sqrt{2})$ over $\mathbb{Q}$ is 6.

2.9 If $\alpha = \sqrt{3} + \sqrt{5}$, then $\alpha^3 = 18\sqrt{3} + 14\sqrt{5}$. Then $\sqrt{3} = (\alpha^3 - 14\alpha)/4$, $\sqrt{5} = (18\alpha - \alpha^3)/4$, and the rest of the argument follows Example 2.19.

2.10 If $\alpha = \frac{1+\sqrt{5}}{2}$, then $\alpha^2 - \alpha - 1 = 0$.

2.11 If $\alpha = \frac{1+\sqrt{3}}{\sqrt{2}}$, then $\alpha\sqrt{2} = 1 + \sqrt{3}$, so $2\alpha^2 = 4 + 2\sqrt{3}$, and $(\alpha^2 - 2)^2 = 3$, giving $\alpha^4 - 4\alpha^2 + 1 = 0$.

2.12 $(\alpha - 1/3)^3 = \alpha^3 - \alpha^2 + \alpha/3 - 1/27$. But $\alpha - 1/3 = \frac{a^{1/3} + a^{2/3}}{3}$, so

$$(\alpha - 1/3)^3 = \frac{a^2 + 3a^{5/3} + 3a^{4/3} + a}{27} = \frac{a^2 - 2a}{27} + \frac{3a + 3a^{4/3} + 3a^{5/3}}{27}.$$

Equating these gives

$$\alpha^3 - \alpha^2 + \frac{\alpha}{3} - \frac{1}{27} = \frac{a^2 - 2a}{27} + \frac{a\alpha}{3},$$

which simplifies to the given equation. Clearly, if $a \equiv 1 \pmod 9$, then this equation has integral coefficients, so that α is an algebraic integer.

2.13 The calculation of the characteristic polynomial is standard. If $\theta = (\sqrt{5} + \sqrt{-3})/2$, then $\theta^2 = \frac{1+\sqrt{-15}}{2}$, and so $(\theta^2 - \frac{1}{2})^2 = -\frac{15}{4}$. This expands to $\theta^4 - \theta^2 + \frac{1}{4} = -\frac{15}{4}$, which gives the solution.

2.14 Write $\alpha = \sqrt{2}$ and $\beta = \sqrt[3]{2}$. Let $\mathbf{v} = (1 \ \alpha \ \beta \ \beta\alpha \ \beta^2 \ \beta^2\alpha)^T$, and note that if

$$A = \begin{pmatrix} 0 & 1 & 0 & 0 & 0 & 0 \\ 2 & 0 & 0 & 0 & 0 & 0 \\ 0 & 0 & 0 & 1 & 0 & 0 \\ 0 & 0 & 2 & 0 & 0 & 0 \\ 0 & 0 & 0 & 0 & 0 & 1 \\ 0 & 0 & 0 & 0 & 2 & 0 \end{pmatrix}, \quad B = \begin{pmatrix} 0 & 0 & 1 & 0 & 0 & 0 \\ 0 & 0 & 0 & 1 & 0 & 0 \\ 0 & 0 & 0 & 0 & 1 & 0 \\ 0 & 0 & 0 & 0 & 0 & 1 \\ 2 & 0 & 0 & 0 & 0 & 0 \\ 0 & 2 & 0 & 0 & 0 & 0 \end{pmatrix},$$

that $A\mathbf{v} = \alpha\mathbf{v}$ and $B\mathbf{v} = \beta\mathbf{v}$. Then $(A+B)\mathbf{v} = (\alpha+\beta)\mathbf{v}$, so $\alpha+\beta$ is a root of the characteristic polynomial of $A+B$, which is $X^6 - 6X^4 - 4X^3 + 12X^2 - 24X - 4$.

2.15 If $\alpha = a + b(\frac{1+\sqrt{-3}}{2})$, then $\alpha = \frac{2a+b}{2} + \frac{b\sqrt{-3}}{2}$, and so its modulus is $\left(\frac{2a+b}{2}\right)^2 + \frac{3b^2}{4} = a^2 + ab + b^2$ as required.

We need to solve $a^2 + ab + b^2 = 19$. Let's complete the square, and write this as $(2a+b)^2 + 3b^2 = 76$. If $b \geq 6$, the term $3b^2$ is too big. Trying $b = 0, \ldots, 5$ gives solutions for (a, b): $(3, 2), (-5, 2), (2, 3), (-5, 3), (-2, 5), (-3, 5)$ and if (a, b) is a solution, so is $(-a, -b)$. These correspond to elements $\pm 4 \pm \sqrt{-3}$, $\frac{\pm 7 \pm 3\sqrt{-3}}{2}$ and $\frac{\pm 1 \pm 5\sqrt{-3}}{2}$.

An element $a + b\sqrt{-2} \in \mathbb{Z}[\sqrt{-2}]$ has modulus $a^2 + 2b^2$, and a similar but easier calculation gives the elements as $\pm 1 \pm 3\sqrt{-2}$.

2.16 We just need to check that $\alpha = \frac{1+\sqrt{3}}{\sqrt{2}}$ is not in $\mathbb{Z}[\sqrt{2}, \sqrt{3}]$. But $\alpha = \frac{1}{2}\sqrt{2} + \frac{1}{2}\sqrt{6}$, whereas elements of $\mathbb{Z}[\sqrt{2}, \sqrt{3}]$ should be of the form $a + b\sqrt{2} + c\sqrt{3} + d\sqrt{6}$ with $a, b, c, d \in \mathbb{Z}$.

Chapter 3

3.1 Given $\alpha = a + bi$, it is a root of $X^2 - 2aX + (a^2 + b^2)$; the other root of this equation is $\overline{\alpha} = a - bi$.

3.2 The minimal polynomial of $\sqrt{2}$ is $X^2 - 2$, and the roots are $\pm\sqrt{2}$.

3.3 The minimal polynomial of $\sqrt{2} + \sqrt{3}$ was worked out above (in Exercise 2.2); it is $X^4 - 10X^2 + 1$. Looking at the construction of the polynomial, it is clear that $\pm\sqrt{2} \pm \sqrt{3}$ are also roots, and so these are the conjugates.

3.4 The minimal polynomial is $X^3 - 2$ (this is irreducible by Eisenstein's criterion with $p = 2$). The roots are $\sqrt[3]{2}$, $\omega \sqrt[3]{2}$ and $\omega^2 \sqrt[3]{2}$, where $\omega = e^{2\pi i/3} = \frac{-1+\sqrt{-3}}{2}$, and so these are the conjugates. Note that in this example, the conjugates of $\sqrt[3]{2}$ do *not* lie in $\mathbb{Q}(\sqrt[3]{2})$.

3.5 The embeddings are those where we map $\sqrt{2} \in \mathbb{Q}(\sqrt{2}, \sqrt{3})$ to the positive square root $+1.4142\ldots \in \mathbb{R}$ or to the negative square root $-1.4142\ldots \in \mathbb{R}$ and similarly for $\sqrt{3}$. They are given by

$$\sigma_1 : a + b\sqrt{2} + c\sqrt{3} + d\sqrt{6} \in \mathbb{Q}(\sqrt{2}, \sqrt{3}) \mapsto a + b\sqrt{2} + c\sqrt{3} + d\sqrt{6} \in \mathbb{C}$$
$$\sigma_2 : a + b\sqrt{2} + c\sqrt{3} + d\sqrt{6} \in \mathbb{Q}(\sqrt{2}, \sqrt{3}) \mapsto a + b\sqrt{2} - c\sqrt{3} - d\sqrt{6} \in \mathbb{C}$$
$$\sigma_3 : a + b\sqrt{2} + c\sqrt{3} + d\sqrt{6} \in \mathbb{Q}(\sqrt{2}, \sqrt{3}) \mapsto a - b\sqrt{2} + c\sqrt{3} - d\sqrt{6} \in \mathbb{C}$$
$$\sigma_4 : a + b\sqrt{2} + c\sqrt{3} + d\sqrt{6} \in \mathbb{Q}(\sqrt{2}, \sqrt{3}) \mapsto a - b\sqrt{2} - c\sqrt{3} + d\sqrt{6} \in \mathbb{C}.$$

1. We can pick any $\alpha \in \mathbb{Q}$, such as $\alpha = 1$, say.
2. Clearly $\beta = \sqrt{2} + \sqrt{3}$ is one such element.
3. With the order above, we can choose $\gamma = 1 + \sqrt{2}$, say.

Many other choices are possible in each case.

3.6 The minimal polynomial of $\alpha = 1$ is just $X - 1$, so $d_\alpha = 1$ and $r_\alpha = 4$. Its image under all four embeddings is 1, so $\prod_{i=1}^{4}(X - \sigma_k(1)) = (X - 1)^4$.

The minimal polynomial of $\beta = \sqrt{2} + \sqrt{3}$ is $X^4 - 10X^2 + 1$, so $d_\beta = 4$ and $r_\beta = 1$. It can also be written as $\prod_{i=1}^{4}(X - \sigma_k(\sqrt{2} + \sqrt{3}))$, since this is the same as

$$(X - (\sqrt{2} + \sqrt{3}))(X - (\sqrt{2} - \sqrt{3}))(X - (-\sqrt{2} + \sqrt{3}))(X - (-\sqrt{2} - \sqrt{3})).$$

Finally, if $\gamma = 1+\sqrt{2}$, its minimal polynomial is $X^2 - 2X - 1$, so $d_\gamma = r_\gamma = 2$. Under two embeddings, γ is mapped to $1 + \sqrt{2}$, and under the other two, it is mapped to $1 - \sqrt{2}$. Then

$$\prod_{i=1}^{4}(X - \sigma_k(\gamma)) = (X - (1+\sqrt{2}))^2 (X - (1 - \sqrt{2}))^2 = (X^2 - 2X - 1)^2.$$

3.7 We can choose a basis $\{1, i\}$ for $\mathbb{Q}(i)$ as a $\mathbb{Q}$-vector space. Then $(a + bi)1 = a.1 + b.i$ and $(a + bi)i = -b.1 + ai$, so that the matrix of the multiplication by $a + bi$ is given by $\begin{pmatrix} a & b \\ -b & a \end{pmatrix}$, which has trace $2a$ and determinant $a^2 + b^2$.
With the reformulation of Proposition 3.16, we compute the conjugates, namely $a + bi$ and $a - bi$; the norm is their product $(a + bi)(a - bi) = a^2 + b^2$, and the trace is their sum $(a + bi) + (a - bi) = 2a$.

3.8 Choose a basis $\{1, \sqrt{2}, \sqrt{3}, \sqrt{6}\}$ for $\mathbb{Q}(\sqrt{2}, \sqrt{3})$ over $\mathbb{Q}$. Then

$$\begin{aligned}
(\sqrt{2}+\sqrt{3})1 &= 0.1 + 1.\sqrt{2} + 1.\sqrt{3} + 0.\sqrt{6} \\
(\sqrt{2}+\sqrt{3})\sqrt{2} &= 2.1 + 0.\sqrt{2} + 0.\sqrt{3} + 1.\sqrt{6} \\
(\sqrt{2}+\sqrt{3})\sqrt{3} &= 3.1 + 0.\sqrt{2} + 0.\sqrt{3} + 1.\sqrt{6} \\
(\sqrt{2}+\sqrt{3})\sqrt{6} &= 0.1 + 3.\sqrt{2} + 2.\sqrt{3} + 0.\sqrt{6}
\end{aligned}$$

so the multiplication is represented by the matrix $\begin{pmatrix} 0 & 1 & 1 & 0 \\ 2 & 0 & 0 & 1 \\ 3 & 0 & 0 & 1 \\ 0 & 3 & 2 & 0 \end{pmatrix}$, which has determinant 1 and trace 0.

Alternatively, the conjugates are $\pm\sqrt{2} \pm \sqrt{3}$, which sum to 0, and

$$(\sqrt{2}+\sqrt{3})(-\sqrt{2}+\sqrt{3})(\sqrt{2}-\sqrt{3})(-\sqrt{2}-\sqrt{3}) = 1$$

as follows easily from the formula $x^2 - y^2 = (x + y)(x - y)$.

3.9 The matrix M is given by $M = \begin{pmatrix} 1 & \sqrt{2} & \sqrt{3} & \sqrt{6} \\ 1 & \sqrt{2} & -\sqrt{3} & -\sqrt{6} \\ 1 & -\sqrt{2} & \sqrt{3} & -\sqrt{6} \\ 1 & -\sqrt{2} & -\sqrt{3} & \sqrt{6} \end{pmatrix}$, of determinant -96, so the discriminant is 9216.

With the reformulation as in Lemma 3.19, we have $T_{K/\mathbb{Q}}(\omega_1\omega_1) = 4$, $T_{K/\mathbb{Q}}(\omega_1\omega_2) = 0$, $T_{K/\mathbb{Q}}(\omega_1\omega_3) = 0$, $T_{K/\mathbb{Q}}(\omega_1\omega_4) = 0$, $T_{K/\mathbb{Q}}(\omega_2\omega_1) = 0$, $T_{K/\mathbb{Q}}(\omega_2\omega_2) = 8$, $T_{K/\mathbb{Q}}(\omega_2\omega_3) = 0$, $T_{K/\mathbb{Q}}(\omega_2\omega_4) = 0$, $T_{K/\mathbb{Q}}(\omega_3\omega_1) = 0$, $T_{K/\mathbb{Q}}(\omega_3\omega_2) = 0$, $T_{K/\mathbb{Q}}(\omega_3\omega_3) = 12$, $T_{K/\mathbb{Q}}(\omega_3\omega_4) = 0$, $T_{K/\mathbb{Q}}(\omega_4\omega_1) = 0$,

$T_{K/\mathbb{Q}}(\omega_4\omega_2) = 0, T_{K/\mathbb{Q}}(\omega_4\omega_3) = 0, T_{K/\mathbb{Q}}(\omega_4\omega_4) = 24$, and so the discrimi-

nant is the determinant of $\begin{pmatrix} 4 & 0 & 0 & 0 \\ 0 & 8 & 0 & 0 \\ 0 & 0 & 12 & 0 \\ 0 & 0 & 0 & 24 \end{pmatrix}$, which is again 9216.

3.10 The conjugates are $\gamma_1 = \sqrt{2} + \sqrt{3}, \gamma_2 = \sqrt{2} - \sqrt{3}, \gamma_3 = -\sqrt{2} + \sqrt{3}$ and $\gamma_4 = -\sqrt{2} - \sqrt{3}$. Then $\gamma_1 - \gamma_2 = 2\sqrt{2}, \gamma_1 - \gamma_3 = 2\sqrt{3}, \gamma_1 - \gamma_4 = 2(\sqrt{2} + \sqrt{3})$, $\gamma_2 - \gamma_3 = 2(\sqrt{2} - \sqrt{3})$, $\gamma_2 - \gamma_4 = 2\sqrt{2}$, $\gamma_3 - \gamma_4 = 2\sqrt{3}$, and so $\prod_{i<j}(\gamma_i - \gamma_j) = -384$, and $\prod_{i<j}(\gamma_i - \gamma_j)^2 = 147456$.

We also have $\gamma = 0.1 + 1.\sqrt{2} + 1.\sqrt{3} + 0.\sqrt{6}, \gamma^2 = 5.1 + 0\sqrt{2} + 0.\sqrt{3} + 2.\sqrt{6}$ and $\gamma^3 = 0.1 + 11\sqrt{2} + 9\sqrt{3} + 0.\sqrt{6}$. So the change of

basis matrix is $\begin{pmatrix} 1 & 0 & 0 & 0 \\ 0 & 1 & 1 & 0 \\ 5 & 0 & 0 & 2 \\ 0 & 11 & 9 & 0 \end{pmatrix}$, of determinant 4. Then we expect that

$\Delta\{1, \gamma, \gamma^2, \gamma^3\} = 4^2 \Delta\{1, \sqrt{2}, \sqrt{3}, \sqrt{6}\}$; we computed $\Delta\{1, \sqrt{2}, \sqrt{3}, \sqrt{6}\} = 9216$ in Exercise 3.9, and $4^2 \times 9216 = 147456$, as expected.

3.11 We simply need to work out

$$\begin{vmatrix} \sigma_1(1) & \sigma_1(\sqrt{d}) \\ \sigma_2(1) & \sigma_2(\sqrt{d}) \end{vmatrix}^2 = \begin{vmatrix} 1 & \sqrt{d} \\ 1 & -\sqrt{d} \end{vmatrix}^2 = (-2\sqrt{d})^2 = 4d.$$

3.12 The minimal polynomial is $f(X) = X^3 - 2$. Then $f'(X) = 3X^2$, and $f'(\sqrt[3]{2}) = 3\sqrt[3]{4}$, whose conjugates are $3\sqrt[3]{4}, 3\omega\sqrt[3]{4}$ and $3\omega^2\sqrt[3]{4}$, where $\omega = e^{2\pi i/3}$. The norm is the product of the conjugates, which is 108. So $D_K = (-1)^{3(3-1)/2} 108 = -108$.

3.13 If $K = \mathbb{Q}(\sqrt{2}, \sqrt{5})$, put $K_1 = \mathbb{Q}(\sqrt{2})$ and $K_2 = \mathbb{Q}(\sqrt{5})$. These fields satisfy the conditions of Proposition 3.34; an integral basis is therefore $\{1, \sqrt{2}, \frac{1+\sqrt{5}}{2}, \frac{\sqrt{2}+\sqrt{10}}{2}\}$. As the discriminant of K_1 is 8, and the discriminant of K_2 is 5, the Proposition gives $D_K = (8^2 5^2) = 1600$. We can also work it out from the given integral basis: it will be

$$\begin{vmatrix} 1 & \sqrt{2} & \frac{1+\sqrt{5}}{2} & \frac{\sqrt{2}+\sqrt{10}}{2} \\ 1 & \sqrt{2} & \frac{1-\sqrt{5}}{2} & \frac{\sqrt{2}-\sqrt{10}}{2} \\ 1 & -\sqrt{2} & \frac{1+\sqrt{5}}{2} & \frac{-\sqrt{2}-\sqrt{10}}{2} \\ 1 & -\sqrt{2} & \frac{1-\sqrt{5}}{2} & \frac{-\sqrt{2}+\sqrt{10}}{2} \end{vmatrix}^2 = (-40)^2 = 1600.$$

3.14 Let's compute D_K from the definition: it is

$$\begin{vmatrix} 1 & \sqrt{2} & \sqrt{3} & \frac{\sqrt{2}+\sqrt{6}}{2} \\ 1 & \sqrt{2} & -\sqrt{3} & \frac{\sqrt{2}-\sqrt{6}}{2} \\ 1 & -\sqrt{2} & \sqrt{3} & \frac{-\sqrt{2}-\sqrt{6}}{2} \\ 1 & -\sqrt{2} & -\sqrt{3} & \frac{-\sqrt{2}+\sqrt{6}}{2} \end{vmatrix}^2 = (-48)^2 = 2304.$$

With the reformulation of Lemma 3.19, we compute

$$D_K = \begin{vmatrix} T_{K/\mathbb{Q}}(1) & T_{K/\mathbb{Q}}(\sqrt{2}) & T_{K/\mathbb{Q}}(\sqrt{3}) & T_{K/\mathbb{Q}}(\frac{\sqrt{2}+\sqrt{6}}{2}) \\ T_{K/\mathbb{Q}}(\sqrt{2}) & T_{K/\mathbb{Q}}(2) & T_{K/\mathbb{Q}}(\sqrt{6}) & T_{K/\mathbb{Q}}(1+\sqrt{3}) \\ T_{K/\mathbb{Q}}(\sqrt{3}) & T_{K/\mathbb{Q}}(\sqrt{6}) & T_{K/\mathbb{Q}}(3) & T_{K/\mathbb{Q}}(\frac{\sqrt{6}+3\sqrt{2}}{2}) \\ T_{K/\mathbb{Q}}(\frac{\sqrt{2}+\sqrt{6}}{2}) & T_{K/\mathbb{Q}}(1+\sqrt{3}) & T_{K/\mathbb{Q}}(\frac{\sqrt{3}+3\sqrt{2}}{2}) & T_{K/\mathbb{Q}}(2+\sqrt{3}) \end{vmatrix}$$

$$= \begin{vmatrix} 4 & 0 & 0 & 0 \\ 0 & 8 & 0 & 4 \\ 0 & 0 & 12 & 0 \\ 0 & 4 & 0 & 8 \end{vmatrix}$$

$$= 2304.$$

Now write $\gamma = \gamma_1 = \frac{\sqrt{2}+\sqrt{6}}{2}$. Its conjugates are $\gamma_2 = \frac{\sqrt{2}-\sqrt{6}}{2}$, $\gamma_3 = \frac{-\sqrt{2}-\sqrt{6}}{2}$, and $\gamma_4 = \frac{-\sqrt{2}+\sqrt{6}}{2}$, so Example 3.21 gives

$$D_K = \prod_{i<j}(\gamma_i - \gamma_j)^2$$
$$= (\sqrt{2})^2(\sqrt{6})^2(\sqrt{2}+\sqrt{6})^2(\sqrt{2})^2(\sqrt{2}-\sqrt{6})^2(-\sqrt{6})^2$$
$$= 2304.$$

Finally, the minimal polynomial of γ is easily computed to be $f(X) = X^4 - 4X^2 + 1$ (see Exercise 2.11), so we conclude that $D_K = \pm N_{K/\mathbb{Q}}(f'(\gamma)) = (-1)^{4(4-1)/2}N_{K/\mathbb{Q}}(4\gamma^3 - 8\gamma) = 2304$.

3.15 A basis for $K = \mathbb{Q}(\sqrt{-2}, \sqrt{-5})$ over $\mathbb{Q}$ is clearly $\{1, \sqrt{-2}, \sqrt{-5}, \sqrt{10}\}$; write $\alpha = a + b\sqrt{-2} + c\sqrt{-5} + d\sqrt{10} \in K$. If $\alpha \in \mathbb{Z}_K$, so is $\alpha + \alpha_2$, where $\alpha_2 = a + b\sqrt{-2} - c\sqrt{-5} - d\sqrt{10}$ (since α_2 is a conjugate of α, so they satisfy the same minimal polynomial). Then $2a + 2b\sqrt{-5}$ is an algebraic integer, so $A = 2a$ and $B = 2b \in \mathbb{Z}$. Similar arguments show that $C = 2c$ and $D = 2d \in \mathbb{Z}$.

Next, $\alpha\alpha_2 = \left(\frac{A^2 + 2B^2 - 5C^2 - 10D^2}{4}\right) + \left(\frac{AC - 2BD}{2}\right)\sqrt{-5}$ is also an algebraic integer. So $2|AC - 2BD$, and so $2|AC$ and at least one of A and C is even. As $2|A^2 + 2B^2 - 5C^2 - 10D^2$, both A and C must be even. Then in order that $4|A^2 + 2B^2 - 5C^2 - 10D^2$, we need $2|B^2 - 5D^2$, so both B and D are even, or both are odd. Such elements are then $\mathbb{Z}$-linear combinations of the given basis $\{1, \sqrt{-2}, \sqrt{-5}, \frac{\sqrt{-2}+\sqrt{10}}{2}\}$; all these elements are integers, and so this is an integral basis.

3.16 The condition that $3|g(\gamma)$ means that $g(\gamma) = 3a(\gamma)$ for some $a(X) \in \mathbb{Z}[X]$. Thus γ is a root of $g - 3a$, and so $f|(g - 3a)$. So this shows that $g(X) = 3a(X) + f(X)b(X)$ for some $b(X) =\in \mathbb{Z}[X]$. Reducing modulo 3, we see that $\overline{f}|\overline{g}$. All these steps are easily seen to be reversible, and the result follows. (Readers with knowledge of ring theory might rephrase this argument, using the First Isomorphism Theorem to deduce an isomorphism $\mathbb{Z}[X]/\langle f \rangle \cong \mathbb{Z}[\gamma]$, and using this to see that $3|g(\gamma)$ would imply that $g \in \langle f, 3 \rangle$.)

3.17 Put $\theta_1 = \theta = \dfrac{a + b\alpha + c\alpha'}{3}$, so that $\theta_2 = \dfrac{a + b\alpha\omega + c\alpha'\omega^2}{3}$ and $\theta_3 = \dfrac{a + b\alpha\omega^2 + c\alpha'\omega}{3}$ are the conjugates. Then

$$\theta_1\theta_2 + \theta_2\theta_3 + \theta_3\theta_1 = \frac{a^2 - 35bc}{3},$$

so that $3|a^2 - 35bc$, and

$$\theta_1\theta_2\theta_3 = \frac{a^3 + 175b^3 + 245c^3 - 105abc}{27},$$

so that $27|a^3 + 175b^3 + 245c^3 - 105abc$. If $3|a$, it follows easily that $3|b$ and $3|c$. Suppose $3 \nmid a$. Then $a^2 \equiv 1 \pmod 3$, so that $bc \equiv 2 \pmod 3$. The only solutions that also make $9|a^3 + 175b^3 + 245c^3 - 105abc$ are $a \equiv b \equiv 1 \pmod 3$ and $c \equiv 2 \pmod 3$, and $a \equiv b \equiv 2 \pmod 3$ and $c \equiv 1 \pmod 3$. For the first case, put $a = 1 + 3l$, $b = 1 + 3m$, $c = 2 + 3n$, and then $a^3 + 175b^3 + 245c^3 - 105abc \equiv 9 \pmod{27}$; for the second, put $a = 2 + 3l$, $b = 2+3m$, $c = 1+3n$, and then $a^3+175b^3+245c^3-105abc \equiv 18 \pmod{27}$, so that we never find any integers. Thus it follows that $3|a$, $3|b$ and $3|c$.

Chapter 4

4.1 $441 = 21 \times 21 = 9 \times 49$.

4.2 The irreducible elements are $a + ib$, where $a^2 + b^2 = p$ for some prime $p \equiv 1 \pmod 4$ or $p = 2$, and $\pm p$ and $\pm ip$, where $p \equiv 3 \pmod 4$.

4.3 1. Clearly the set $\{a + b\omega \mid a, b \in \mathbb{Z}\}$ is closed under addition. Further

$$(a + b\omega)(c + d\omega) = (ac - bd) + (ad + bc - bd)\omega,$$

so it is also closed under multiplication. The remaining axioms follow as it is a subset of $\mathbb{C}$.

2. Consider $\left(\frac{1+\sqrt{3}}{2}\right)^2 = 2 + \frac{\sqrt{3}}{2}$; this does not belong to the given set.

3. Neither quotient $(1 + \sqrt{-3})/2$ or $2/(1 + \sqrt{-3}) = (1 - \sqrt{-3})/2$ lies in $\mathbb{Z}[\sqrt{-3}]$, so they are not associates, and therefore give different factorisations.

4.4 Note that
$$\frac{4 + \sqrt{5}}{2\sqrt{5} - 3} = \frac{(4 + \sqrt{5})(2\sqrt{5} + 3)}{11} = 2 + \sqrt{5},$$

a unit, and so $4 + \sqrt{5}$ is a unit multiplied by $2\sqrt{5} - 3$. Similarly, $4 - \sqrt{5} = (\sqrt{5} - 2)(2\sqrt{5} + 3)$.

4.5 Using the multiplicativity of the norm, if $\alpha = \beta\gamma$, and $N(\alpha) = p$, then $N(\beta) = 1$ or $N(\gamma) = 1$, and so either β or γ must be prime. If $\alpha = 3$, then $N(\alpha) = 9$, but there are no elements of norm 3, so 3 has no factorisation into elements with norm bigger than 1.

In $\mathbb{Z}[\sqrt{-2}]$, a similar conclusion can be drawn for elements of prime norm (for the same reasons). If $\alpha = a + b\sqrt{-2}$ had norm 5, then $a^2 + 2b^2 = 5$, and this has no solutions. So there are no elements of norm 5, and so 5 is irreducible, although its norm is 25.

4.6 The norms in these factorisations are $10 \times 10 = 4 \times 25$, so the elements on the two sides are not associate; further, there are no elements in $\mathbb{Z}[\sqrt{10}]$ of norm 2 (consider $a^2 - 10b^2 = \pm 2$ modulo 10) so $\sqrt{10}$ and 2 are irreducible. (In fact, 5 is also irreducible; one can see that $a^2 - 10b^2 = \pm 5$ has no solution by working modulo 5 to see that $5|a$, and then modulo 25 to see that there is no solution for b.)

4.7 $6 = 2 \times 3 = (-\sqrt{-6})\sqrt{-6}$. The norms are $4 \times 9 = 6 \times 6$, and there are no elements of norm 2, since there is no solution to $a^2 + 6b^2 = 2$, and no elements of norm 3 (use $a^2 + 6b^2 = 3$).

4.8 $14 = 2 \times 7 = (1 + \sqrt{-13})(1 - \sqrt{-13})$. Again the norms are different, and there are no elements of norm 2 or 7.

4.9 The equalities are clear. The norms are $9 \times 9 \times 9 = 27 \times 27$. However, there are no elements of norm 3, so each factor is irreducible.

4.10 $2(a + b\sqrt{10}) + (4 + \sqrt{10})(c + d\sqrt{10}) = (2a + 4c + 10d) + \sqrt{10}(2b + c + 4d)$. We claim that given any $A + B\sqrt{10} \in \mathbb{Z}[\sqrt{10}]$ we can write it in this form precisely when $2|A$. This condition is clearly necessary. Conversely, we can choose c and d so that $c + 4d = B$, then pick $b = 0$, and finally choose a so that $2a = A - (4c - 10d)$. (There are other ways to do this.)

4.11 If $\alpha|2$ and $\alpha|4 + \sqrt{10}$, then taking norms gives $N(\alpha)|4$ and $N(\alpha)|6$. Thus α has norm 1 or norm 2. But α cannot have norm 2, since $\mathbb{Z}[\sqrt{10}]$ has no elements of norm 2. So α is a unit. But $\mathfrak{a}_1$ cannot be a unit; it generates the ideal of Exercise 4.10, and this is a proper ideal, as $1 \notin \mathfrak{a}_1$. So $\mathfrak{a}_1$ is not an element of $\mathbb{Z}[\sqrt{10}]$.

4.12 Given $a \in \mathbb{Z}$, we need to write a as $6p + 10q + 15r$. But $1 = 6 + 10 - 15$, so $a = 6a + 10a + 15(-a)$. However, $\{6, 10\}$ is not enough to generate any integer, as all linear combinations are even. Similarly, every linear combination of $\{6, 15\}$ is divisible by 3, and every linear combination of $\{10, 15\}$ is divisible by 5, so $\{6, 10, 15\}$ is minimal.

4.13 $\langle 12 \rangle \cap \langle 20 \rangle = \langle 60 \rangle$; $\langle 12 \rangle \langle 20 \rangle = \langle 240 \rangle$; $\langle 12 \rangle + \langle 20 \rangle = \langle 4 \rangle$, and $\langle 12, 20 \rangle = \langle 4 \rangle$.

4.14　1. An element of $I + J$ is the sum of something in I and something in J; an element of I is a linear combination of $a_1, \ldots, a_m$, and an element of J is a linear combination of $b_1, \ldots, b_n$; clearly their sum is a linear combination of $a_1, \ldots, a_m, b_1, \ldots, b_n$. Conversely, given such a linear combination, we split into a sum of those multiples of $a_1, \ldots, a_m$ and those of $b_1, \ldots, b_n$; this sum lies in $I + J$. (Hopefully you got the same answers to the last two parts of the previous exercise; this explains why.)

　　　2. An element of IJ is the sum of terms of the form ij with i a linear combination of $a_1, \ldots, a_m$ and j a linear combination of $b_1, \ldots, b_n$; clearly ij is then a linear combination of the given set. Conversely, each generator $a_r b_s$ lies in IJ, and so any linear combination of them does.

4.15　1. $2(a + b\sqrt{-5}) + (1 + \sqrt{-5})(c + d\sqrt{-5}) = (2a + c - 5d) + (2b + c + d)\sqrt{-5}$, and so is of the form $A + B\sqrt{-5}$ with $A \equiv B \pmod 2$ as claimed. Conversely, given $A + B\sqrt{-5}$ with $A \equiv B \pmod 2$, choose any c and d with $c + d = B$, and let $b = 0$; then we need to find a so that $2a + c - 5d = A$; this means solving $2a = (A - B) + 6d$, which can be solved. Thus $\mathfrak{a} = \{A + B\sqrt{-5} \mid A \equiv B \pmod 2\}$. In particular, $1 = 1 + 0\sqrt{-5} \notin \mathfrak{a}$.

　　　2. Suppose that $\mathfrak{a} = \langle \alpha \rangle$. Then $N(\mathfrak{a}) = N(\alpha)$; as $N(\mathfrak{a}) \mid N(2) = 4$ and $N(\mathfrak{a}) \mid N(1 + \sqrt{-5}) = 6$, we see that $N(\alpha) \mid 2$. But $\mathfrak{a} \neq R$, since $1 \notin \mathfrak{a}$; we conclude that $N(\alpha) > 1$, and so $N(\alpha) = 2$. But there are no elements of norm 2.

　　　3. Let $\mathfrak{a}' = \langle 2, 1 - \sqrt{5} \rangle$. As $2 \in \mathfrak{a}$, and $1 - \sqrt{5} = 2 + (-1)(1 - \sqrt{5}) \in \mathfrak{a}$, both generators of $\mathfrak{a}'$ lie in $\mathfrak{a}$, so that $\mathfrak{a}' \subseteq \mathfrak{a}$. The other inclusion is similar.

4.16　1. $3(a + b\sqrt{-5}) + (1 - \sqrt{-5})(c + d\sqrt{-5}) = (3a + c + 5d) + (3b - c + d)\sqrt{-5}$, and the remaining claims follow as in Exercise 4.15(1).

　　　2. Again, this follows as any generator would have to have norm dividing $N(3) = 9$ and $N(1 \pm \sqrt{-5}) = 6$. But there are no elements of norm 3, and the generator could not be a unit, since the ideals are not the whole ring (by the characterisation in the previous part, 1 does not belong to either ideal).

4.17　1. There are no solutions to $a^2 + 6b^2 = 2$ or 5. The norms on both sides are different, and if any factor were reducible, the factors would have to have norms 2 or 5, which is impossible.

　　　2. If $\langle 2, \sqrt{-6} \rangle$ were principal, any generator would need to divide both $N(2) = 4$ and $N(\sqrt{-6}) = 6$, so the norm of any generator would be 1 or 2; it cannot have norm 1, otherwise it would be a unit, and the ideal it generated would be the whole ring (but 1 is not in the ideal), and there are no elements of norm 2.

　　　3. $\mathfrak{a}^2 = \langle 2^2, 2\sqrt{-6}, (\sqrt{-6})^2 \rangle = \langle 4, -6, \sqrt{-6} \rangle$, and each factor is divisible by 2. Thus $\mathfrak{a}^2 \subseteq \langle 2 \rangle$. Conversely, $2 \in \mathfrak{a}^2$, as 4 and -6 lie in $\mathfrak{a}^2$, and $2 = (-1)4 + (-1)(-6)$.

4. $\mathfrak{b}_1\mathfrak{b}_2 = \langle 25, 5(2 + \sqrt{-6}), 5(2 - \sqrt{-6}), 10\rangle$, so $\mathfrak{b}_1\mathfrak{b}_2 \subseteq \langle 5\rangle$; conversely, as $5 = 25 - 2 \times 10$, we see $5 \in \mathfrak{b}_1\mathfrak{b}_2$. So $\mathfrak{b}_1\mathfrak{b}_2 = \langle 5\rangle$. Then

$$\langle 10\rangle = \mathfrak{a}^2\mathfrak{b}_1\mathfrak{b}_2.$$

5. The two factorisations are

$$\langle 10\rangle = (\mathfrak{a}^2)(\mathfrak{b}_1\mathfrak{b}_2) = (\mathfrak{a}\mathfrak{b}_1)(\mathfrak{a}\mathfrak{b}_2);$$

the equalities $\langle 2 + \sqrt{-6}\rangle = \mathfrak{a}\mathfrak{b}_1$ and $\langle 2 - \sqrt{-6}\rangle = \mathfrak{a}\mathfrak{b}_2$ are proven as above.

4.18 We have already seen that $\mathbb{Z}[\sqrt{2}]$ has a Euclidean algorithm, and that this implies that it has unique factorisation and is a PID.

Chapter 5

5.1 Suppose that $r \in I$. Let's see that $r + I = I$. Given any $i \in I$, we can write $i = r + (i - r) \in r + I$, so $I \subseteq r + I$. Conversely, if $i \in I$, then the element $r + i$ of $r + I$ must be in I by the definition of ideal. So $r + I \subseteq I$, and $r + I = I$, as required. Conversely, if $r + I = I$, then every element of $r + I$ lies in I, and in particular $r + 0 = r$ is in I.

5.2 Suppose that R is a PID, and that p is an irreducible element. We claim that $\langle p\rangle$ is maximal. If not, there is an ideal $\langle p\rangle \subset I \subset R$, with both inclusions proper. As R is a PID, $I = \langle a\rangle$ for some a. As $\langle p\rangle \subset \langle a\rangle$, we see $p \in \langle a\rangle$, so that $p = ab$ for some $b \in R$. As $\langle p\rangle \subset \langle a\rangle$ is proper, we cannot have that b is a unit. As $\langle a\rangle \subset R$ is proper, clearly a is not a unit either. So $p = ab$ for two non-units a and b, so p was not irreducible, a contradiction.

5.3 This argument is the same as that of Example 5.11 (the existence of a Euclidean algorithm for $\mathbb{Z}[i]$ means that it is a PID, and the analogue of Lemma 4.13 holds).

5.4 Similarly, the prime ideals in $\mathbb{Z}[i]$ are $\langle \alpha\rangle$ where α is an element of prime norm $p \equiv 1 \pmod 4$ and also $\langle p\rangle$ where $p \equiv 3 \pmod 4$ is a prime, as well as the zero ideal $\langle 0\rangle$.

5.5 Almost all the field axioms hold because we are working in a quotient of an integral domain; we just need to check that every nonzero element is invertible. This follows because:

$$1 \times 1 = 1$$
$$2 \times 2 = 1$$
$$\sqrt{2} \times 2\sqrt{2} = 1$$
$$(1 + \sqrt{2}) \times (2 + \sqrt{2}) = 1$$
$$(1 + 2\sqrt{2}) \times (2 + 2\sqrt{2}) = 1,$$

It follows that every element is invertible. (You might like to check that the multiplication table is that of a cyclic group; it is a fact that the multiplicative group of any finite field is cyclic.) Thus $\mathbb{Z}_K/\mathfrak{a}$ is a finite field, and so $\mathfrak{a}$ is prime.

5.6 Now $\mathbb{Z}_K/\mathfrak{a}$ is not a field, since $\mathfrak{a} \subset \langle\sqrt{2}\rangle$ is a proper inclusion, so that $\mathfrak{a}$ is not maximal. (Alternatively, $\mathbb{Z}_K/\mathfrak{a}$ has zero-divisors: $\sqrt{2} \times \sqrt{2} = 0$ in the quotient, so $\mathbb{Z}_K/\mathfrak{a}$ is not even an integral domain, so $\mathfrak{a}$ is not prime.)

For $\mathfrak{b} = \langle 7 \rangle$, one can argue similarly, using the factorisation $7 = (3+\sqrt{2})(3-\sqrt{2})$ to see that either $\mathfrak{b} \subset \langle 3 \pm \sqrt{2}\rangle$ is an inclusion of $\mathfrak{b}$ into a proper ideal, or that $\mathbb{Z}_K/\mathfrak{b}$ has zero-divisors.

5.7 From the previous exercise, $\mathfrak{a} = \langle 2 \rangle = \mathfrak{p}_2^2$, where $\mathfrak{p}_2 = \langle\sqrt{2}\rangle$, and $\mathfrak{b} = \langle 7 \rangle = \mathfrak{p}_7\mathfrak{p}_7'$, where $\mathfrak{p}_7 = \langle 3+\sqrt{2}\rangle$ and $\mathfrak{p}_7' = \langle 3-\sqrt{2}\rangle$.

5.8 We need to check two things; firstly, given $\mathfrak{a}$ and $\mathfrak{b}$ as in the exercise, that the highest common factor is $\prod_{\mathfrak{p}} \mathfrak{p}^{\min(a_{\mathfrak{p}}, b_{\mathfrak{p}})}$, and secondly, that $\mathfrak{a}+\mathfrak{b}$ also satisfies the defining property to be the highest common factor. (Then this shows that $\mathfrak{a}+\mathfrak{b}$ is also equal to the given expression.)

Write $\mathfrak{h}$ for $\prod_{\mathfrak{p}} \mathfrak{p}^{\min(a_{\mathfrak{p}}, b_{\mathfrak{p}})}$, and we will show that it satisfies the conditions required to be the highest common factor. It is clear that $\mathfrak{h}|\mathfrak{a}$ and that $\mathfrak{h}|\mathfrak{b}$, since the exponents of any given prime ideal in $\mathfrak{h}$ is at most the corresponding exponents in $\mathfrak{a}$ and $\mathfrak{b}$. If $\mathfrak{c} = \prod_{\mathfrak{p}} \mathfrak{p}^{c_{\mathfrak{p}}}$, and $\mathfrak{c}|\mathfrak{a}$ and $\mathfrak{c}|\mathfrak{b}$, then $c_{\mathfrak{p}} \le a_{\mathfrak{p}}$ and $c_{\mathfrak{p}} \le b_{\mathfrak{p}}$, and so $c_{\mathfrak{p}} \le \min(a_{\mathfrak{p}}, b_{\mathfrak{p}})$, so that $\mathfrak{c}|\mathfrak{h}$ as required. (This is an analogue of the well-known result for elements of $\mathbb{Z}$.)

Also, notice that $\mathfrak{a}+\mathfrak{b}|\mathfrak{a}$ and $\mathfrak{a}+\mathfrak{b}|\mathfrak{b}$ and that if $\mathfrak{c}|\mathfrak{a}$ and $\mathfrak{c}|\mathfrak{b}$, then $\mathfrak{c}|\mathfrak{a}+\mathfrak{b}$, so that $\mathfrak{a}+\mathfrak{b}$ satisfies the two requirements to be the highest common factor.

5.9 If m is coprime to h_K, then there are integers s and t such that $sm + th_K = 1$. Then $\mathfrak{a} = \mathfrak{a}^{sm}\mathfrak{a}^{th_K}$. As $\mathfrak{a}^{h_K}$ is principal, and $\mathfrak{a}^m$ is principal, we conclude that $\mathfrak{a}$ is principal.

5.10 We saw in Chap. 3 that the ring of integers is given by $\mathbb{Z}[\sqrt[3]{2}]$.

By Proposition 5.42, it suffices to see how the polynomial $X^3 - 2$ factorises modulo 5, 7 and 11. But

$$X^3 - 2 \equiv (X-3)(X^2 - 2X - 1) \pmod 5$$
$$X^3 - 2 \equiv X^3 - 2 \pmod 7$$
$$X^3 - 2 \equiv (X-4)(X-7)(X-20) \pmod{31},$$

where each factor is irreducible. So

$$\langle 5 \rangle = \mathfrak{p}_5\mathfrak{p}_5',$$

where $\mathfrak{p}_5$ is a prime ideal of norm 5, and $\mathfrak{p}'_5$ is a prime ideal of norm 25. Further, we can read off from Remark 5.43 that $\mathfrak{p}_5 = \langle p, \sqrt[3]{2} - 3 \rangle$ and that $\mathfrak{p}'_5 = \langle p, \sqrt[3]{2}^2 - 2\sqrt[3]{2} - 1 \rangle$.

Similarly, $\langle 7 \rangle$ is already a prime ideal of $\mathbb{Q}(\sqrt[3]{2})$, and

$$\langle 31 \rangle = \mathfrak{p}_{31}\mathfrak{p}'_{31}\mathfrak{p}''_{31},$$

where $\mathfrak{p}_{31} = \langle 31, \sqrt[3]{2} - 4 \rangle$, $\mathfrak{p}'_{31} = \langle 31, \sqrt[3]{2} - 7 \rangle$ and $\mathfrak{p}''_{31} = \langle 31, \sqrt[3]{2} - 20 \rangle$.

5.11 The discriminant of the polynomial (or the number field) is -8, so the only ramified prime is 2, and $\langle 2 \rangle = \langle \sqrt{-2} \rangle^2$.

We have to consider how $X^2 + 2$ factorises modulo p for all primes p. Now $X^2 + 2$ is irreducible if -2 is not a square modulo p, i.e., if $\left(\frac{-2}{p}\right) = -1$. The expressions for $\left(\frac{-1}{p}\right)$ and $\left(\frac{2}{p}\right)$ are well-known; combining these shows that $X^2 + 2$ is irreducible modulo p if $p \equiv 5 \pmod 8$ or $p \equiv 7 \pmod 8$. By Proposition 5.42, these are the inert primes for K. The primes $p \equiv 1 \pmod 8$ and $p \equiv 3 \pmod 8$ are split. (Notice that the splitting behaviour is given by a congruence condition modulo D_K; this is a general phenomenon.)

5.12 Suppose $\mathbb{Z}_K = \mathbb{Z}[\gamma]$. If $\langle 2 \rangle$ factors as the product of three distinct prime ideals of norm 2, then Proposition 5.42 shows that the minimal polynomial of γ (a cubic polynomial) would have three distinct linear factors modulo 2. But the only linear polynomials modulo 2 are X and $X - 1$, so this is a contradiction.

5.13 Using Proposition 5.42 again, we need to see how the minimal polynomial of γ factorises, where γ is such that $\mathbb{Z}_K = \mathbb{Z}[\gamma]$. In this case, we can take $\gamma = \frac{1+\sqrt{5}}{2}$, and its minimal polynomial is $X^2 - X - 1$. The discriminant of this field is 5, so the only ramified prime is 5.

For $p = 2$, $X^2 - X - 1$ is irreducible, so $\langle 2 \rangle$ is inert.

For all other primes, we have

$$X^2 - X - 1 = (X - \tfrac{1}{2})^2 - \tfrac{5}{4},$$

and this is soluble if $\left(\frac{5}{p}\right) = 1$. This happens when $\left(\frac{p}{5}\right) = 1$, by Quadratic Reciprocity, which is when $p \equiv 1 \pmod 5$ or $p \equiv 4 \pmod 5$. So these primes split, whereas the primes $p \equiv 2 \pmod 5$ and $p \equiv 3 \pmod 5$ are inert.

Chapter 6

6.1 The line joining $(0, 0)$ to $(\frac{1}{2}, \frac{\sqrt{|d|}}{2})$ has gradient $\sqrt{|d|}$, so its perpendicular bisector will have gradient $-1/\sqrt{|d|}$. Its equation is therefore $Y = -X/\sqrt{|d|} + c$, and it passes through the midpoint $(\frac{1}{4}, \frac{\sqrt{|d|}}{4})$ of the two given points. We can then read off $c = \frac{|d|+1}{4\sqrt{|d|}}$. By symmetry, the line meets the perpendicular bisector of the line joining $(0, 0)$ to $(-\frac{1}{2}, \frac{\sqrt{|d|}}{2})$ when $X = 0$, which is the point $Y = c$, i.e., at $\left(0, \frac{|d|+1}{4\sqrt{|d|}}\right)$. The bisector of the line joining $(0, 0)$ to $(1, 0)$ is just $X = \frac{1}{2}$, and then we can read off $Y = \frac{|d|-1}{4\sqrt{|d|}}$.

6.2 We can take $\alpha = \sqrt{-19}$ and $\beta = 4$. Then $\alpha/\beta = \sqrt{-19}/4$ has imaginary part $\sqrt{-19}/4 = (1.0897\ldots)i$. However, every integer of $\mathbb{Q}(\sqrt{-19})$ has imaginary part which is an integer multiple of $\sqrt{-19}.2 = (2.1794\ldots)i$; it follows that α/β differs from every $\kappa \in \mathbb{Z}_K$ by at least 1. So for every $\kappa \in \mathbb{Z}_K$, we have $N_{K/\mathbb{Q}}\left(\frac{\alpha}{\beta} - \kappa\right) > 1$. Multiplying up, $N_{K/\mathbb{Q}}(\alpha - \kappa\beta) > N_{K/\mathbb{Q}}(\beta)$ for every $\kappa \in \mathbb{Z}_K$.

6.3 When $\mathbb{Z}_K = \mathbb{Z}[\sqrt{d}]$ and $d < -2$, Theorem 6.1 gives that the units are $\{1, -1\}$. Suppose that a universal side-divisor exists as in Theorem 6.2. Now take $\alpha = 2$, and apply the theorem; either $u|\alpha$ or $u|\alpha \pm 1$. So $u|1$ (impossible as u is not a unit) or $u|2$ or $u|3$. But 2 and 3 are irreducible, as any factor would have to have norm 2 or 3. Notice that $d \leq -5$ (as $d = -3$ is covered by the case $d \equiv 1 \pmod 4$), and the norm of $a + b\sqrt{d}$ is $a^2 + (-d)b^2$, and this can never equal 2 or 3 when $d \leq -5$. So $u = \pm 2$ or ± 3. But if $\alpha = \sqrt{d}$, it is easy to see that none of α or $\alpha \pm 1$ is divisible by ± 2 or ± 3, so u cannot exist, and we conclude from Corollary 6.3 that $\mathbb{Z}_K$ is not Euclidean.

6.4 Suppose that f and g are equivalent. Then $g(px + qy, rx + sy) = f(x, y)$ for some p, q, r and s with $ps - qr = \pm 1$. Suppose that $g(x, y) = ax^2 + bxy + cy^2$, with discriminant $D_g = b^2 - 4ac$. Then

$$f(x, y) = a(px + qy)^2 + b(px + qy)(rx + sy) + c(rx + sy)^2,$$

which expands to

$$(ap^2 + bpr + cr^2)x^2 + (2apq + bps + bqr + 2crs)xy + (aq^2 + bqs + cs^2)y^2,$$

and this has discriminant

$$D_f = (2apq + bps + bqr + 2crs)^2 - 4(ap^2 + bpr + cr^2)(aq^2 + bqs + cs^2);$$

it is easy to expand this and simplify, to get $D_f = (b^2 - 4ac)(ps - qr)^2$, so that $D_f = D_g$.

6.5 A tedious calculation shows that if $f(x, y) = g(px + qy, rx + sy)$ for some matrix $M_1 = \begin{pmatrix} p & q \\ r & s \end{pmatrix}$, and $g(x, y) = h(tx + uy, vx + wy)$ for some matrix $M_2 = \begin{pmatrix} t & u \\ v & w \end{pmatrix}$, then $f(x, y) = h(kx + ly, mx + ny)$, where

$$\begin{pmatrix} k & l \\ m & n \end{pmatrix} = M_2 M_1 = \begin{pmatrix} pt + ru & qt + su \\ pv + rw & qv + sw \end{pmatrix}.$$

Both relations are clearly reflexive (use $p = s = 1, q = r = 0$). They are also symmetric: if $f(x, y) = g(px + qy, rx + sy)$ for some matrix $M = \begin{pmatrix} p & q \\ r & s \end{pmatrix}$, then the calculation shows that $g(x, y) = f(p'x + q'y, r'x + s'y)$, where $M^{-1} = \begin{pmatrix} p' & q' \\ r' & s' \end{pmatrix}$ (so we can take $p' = s, q' = -q, r' = -r$ and $s' = p$) so that $g(x, y) = f(sx - qy, -rx + py)$. The calculation above (together with the multiplicativity of the determinant) shows immediately that both relations are also transitive.

6.6 From the calculations in the previous exercise, $g(x, y) = f(sx - qy, -rx + py)$. Then $f(x, y) = n$ if and only if $g(u, v) = n$, where $\begin{pmatrix} u \\ v \end{pmatrix} = \begin{pmatrix} p & q \\ r & s \end{pmatrix} \begin{pmatrix} x \\ y \end{pmatrix}$, and that this is equivalent to $\begin{pmatrix} x \\ y \end{pmatrix} = \begin{pmatrix} s & -q \\ -r & p \end{pmatrix} \begin{pmatrix} u \\ v \end{pmatrix}$.

If $X = \{(x, y) \mid f(x, y) = n\}$ and $U = \{(u, v) \mid g(u, v) = n\}$, we see that $(x, y) \in X$ if and only if $(u, v) \in U$, where $\begin{pmatrix} u \\ v \end{pmatrix} = \begin{pmatrix} p & q \\ r & s \end{pmatrix} \begin{pmatrix} x \\ y \end{pmatrix}$, and the inverse matrix gives a map in the other direction, so these are inverse bijections.

6.7 1. $(6, -7, 8) \mapsto (6, 5, 7)$ which is reduced. We just needed one application of the rule corresponding to the matrix $\begin{pmatrix} 1 & 1 \\ 0 & 1 \end{pmatrix}$.

2. $(13, 12, 11) \mapsto (11, -12, 13) \mapsto (11, 10, 12)$, which is reduced. We needed $\begin{pmatrix} 0 & 1 \\ -1 & 0 \end{pmatrix}$ followed by $\begin{pmatrix} 1 & 1 \\ 0 & 1 \end{pmatrix}$, so the matrix is the product, $\begin{pmatrix} 0 & 1 \\ -1 & -1 \end{pmatrix}$.

3. $(43, 71, 67) \mapsto (43, -15, 39) \mapsto (39, 15, 43)$, which is reduced. The same steps are used as in the previous example, so the matrix is again $\begin{pmatrix} 0 & 1 \\ -1 & -1 \end{pmatrix}$.

6.8 We had $m = 469$, and $469^2 = 218 \times 1009 - 1$. So we consider the quadratic form $1009x^2 + 938xy + 218y^2$, of discriminant -4. Let us reduce this form:

$(1009, 938, 218) \mapsto (218, -938, 1009) \mapsto (218, -502, 289) \mapsto (218, -66, 5)$
$\mapsto (5, 66, 218) \mapsto (5, 56, 157) \mapsto (5, 46, 106) \mapsto (5, 36, 65)$
$\mapsto (5, 26, 34) \mapsto (5, 16, 13) \mapsto (5, 6, 2) \mapsto (2, -6, 5)$
$\mapsto (2, -2, 1) \mapsto (1, 2, 2) \mapsto (1, 0, 1).$

A calculation shows that the matrix used to do the reduction is $\begin{pmatrix} 28 & 13 \\ 15 & 7 \end{pmatrix}$, so that

$$1009x^2 + 938xy + 218y^2 = (28x + 13y)^2 + (15x + 7y)^2.$$

Now put $x = 1, y = 0$, and we see that $1009 = 28^2 + 15^2$.

6.9 $\mathfrak{p}_2$ has generators 2 and $\sqrt{-6}$ as a $\mathbb{Z}$-module, and has norm 2, so the associated form is

$$N_{\mathbb{Q}(\sqrt{-6})/\mathbb{Q}}(2x + \sqrt{-6}y)/2 = (4x^2 + 6y^2)/2 = 2x^2 + 3y^2.$$

$\mathfrak{p}_3$ has generators 3 and $\sqrt{-6}$ as a $\mathbb{Z}$-module, and has norm 3, so the associated form is

$$N_{\mathbb{Q}(\sqrt{-6})/\mathbb{Q}}(3x + \sqrt{-6}y)/3 = (9x^2 + 6y^2)/3 = 3x^2 + 2y^2.$$

This form is not reduced, but $(3, 0, 2)$ reduces to $(2, 0, 3)$ via $(a, b, c) \mapsto (c, -b, a)$.

$\langle 2 \rangle$ has generators 2 and $2\sqrt{-6}$ as a $\mathbb{Z}$-module, and has norm 4, so the associated form is

$$N_{\mathbb{Q}(\sqrt{-6})/\mathbb{Q}}(2x + 2\sqrt{-6}y)/4 = (4x^2 + 24y^2)/4 = x^2 + 6y^2.$$

$\langle 3 \rangle$ has generators 3 and $3\sqrt{-6}$ as a $\mathbb{Z}$-module, and has norm 9, so the associated form is

$$N_{\mathbb{Q}(\sqrt{-6})/\mathbb{Q}}(3x + 3\sqrt{-6}y)/9 = (9x^2 + 54y^2)/9 = x^2 + 6y^2.$$

$\langle \sqrt{-6} \rangle$ has generators -6 and $\sqrt{-6}$ as a $\mathbb{Z}$-module, and has norm 6, so the associated form is

$$N_{\mathbb{Q}(\sqrt{-6})/\mathbb{Q}}(-6x + \sqrt{-6}y)/6 = (36x^2 + 6y^2)/6 = 6x^2 + y^2.$$

This form is not reduced, but $(6, 0, 1)$ reduces to $(1, 0, 6)$ via $(a, b, c) \mapsto (c, -b, a)$.

6.10 We need to compute the number of distinct reduced quadratic forms of discriminant -56. If $b^2 - 4ac = -56$, then $3ac \le 56 \le 4ac$, so $14 \le ac \le 18$.

When $ac = 14$, we need $b^2 = 0$, and we get 2 reduced forms, $(1, 0, 14)$ and $(2, 0, 7)$.

When $ac = 15$, we need $b^2 = 4$, and we get 2 reduced forms, $(3, 2, 5)$ and $(3, -2, 5)$. (Note that $(1, \pm2, 15)$ are not reduced.)

When $ac = 16$, we need $b^2 = 8$, which has no solutions. When $ac = 17$, we need $b^2 = 12$, which has no solutions. When $ac = 18$, we need $b^2 = 16$, but if $ac = 18$, and $a \le c$, we need $a \le 3$, and if $b = \pm4$, there are no reduced forms.

We get 4 distinct reduced forms, and conclude that the class number of $\mathbb{Q}(\sqrt{-14})$ is 4.

6.11 The discriminant of $\mathbb{Q}(\sqrt{-43})$ is -43, so we are looking for reduced quadratic forms of discriminant -43. We need to solve $3ac \le 43 \le 4ac$, so $11 \le ac \le 14$. If $ac = 11$, $b^2 = 1$, and we get the form $(1, \pm 1, 11)$; note that $(1, -1, 11)$ is equivalent to $(1, 1, 11)$ via $(a, b, c) \sim (a, b+2a, c+b+a)$, so we get 1 reduced form. If $ac = 12$, we need $b^2 = 5$, which has no solutions; if $ac = 13$, we need $b^2 = 9$, but we also need $a = 1$, so there are no reduced forms; if $ac = 14$, we need $b^2 = 13$, which has no solutions.

The arguments for $\mathbb{Q}(\sqrt{-67})$ and $\mathbb{Q}(\sqrt{-163})$ are similar.

The discriminant of $X^2 + X + 41$ is -163. If we consider the values of ac when $b^2 - 4ac = -163$, we see that if $b = 1$, we get $ac = 41$; if $b = 3$, then $ac = 43$; if $b = 5$, then $ac = 47$; if $b = 7$, then $ac = 53$, etc. The values of ac obtained are $41, 43, 47, 53, \ldots$, exactly the values taken on by $X^2 + X + 41$. Since they are all prime, the only factorisations have $a = 1$, and it follows that the resulting forms cannot be reduced. So the only reduced form can arise when $b = 1$ and $a = 1$, namely $(1, 1, 41)$.

The polynomial $X^2 + X + 11$ is prime for $X = 0, \ldots, 9$, and the polynomial $X^2 + X + 17$ is prime for $X = 0, \ldots, 15$. Again it follows that $\mathbb{Q}(\sqrt{-43})$ and $\mathbb{Q}(\sqrt{-67})$ both have class number 1.

These are also related to the curious observations that $e^{\pi\sqrt{43}}$, $e^{\pi\sqrt{67}}$ and $e^{\pi\sqrt{163}}$ are extremely close to integers. Indeed, $e^{\pi\sqrt{163}}$ differs from an integer by less than 10^{-12}. The explanation lies beyond the scope of this book! However, the reader might like to compute the values of $e^{\pi\sqrt{n}}$ for many values of n; the results are striking.

6.12 The forms are $f_1(x, y) = x^2 + 14y^2$, $f_2(x, y) = 2x^2 + 7y^2$, $f_3(x, y) = 3x^2 + 2xy + 5y^2$ and $f_4(x, y) = 3x^2 - 2xy + 5y^2$. Notice that

$$(a + b\sqrt{-14})(c + d\sqrt{-14}) = (ac - 14bd) + (ad + bc)\sqrt{-14}$$

shows that

$$(a^2 + 14b^2)(c^2 + 14d^2) = (ac - 14bd)^2 + 14(ad + bc)^2,$$

so that $f_1(a, b)f_1(c, d) = f_1(ac - 14bd, ad + bc)$.

As $f_2(x, y) = 2x^2 + 7y^2$, we can think of $f_2(x, y) = (4x^2 + 14y^2)/2 = f_1(2x, y)/2$. Then

$$\begin{aligned}
f_2(a, b)f_2(c, d) &= f_1(2a, b)f_1(2c, d)/4 \\
&= f_1(4ac - 14bd, 2ad + 2bc)/4 \\
&= f_1(2ac - 7bd, ad + bc).
\end{aligned}$$

Thus f_2 has order 2 in the class group.

Similarly, $f_3(x, y) = 3x^2+2xy+5y^2 = (9x^2+6xy+15y^2)/3 = ((3x+y)^2+14y^2)/3 = f_1(3x + y, \pm y)/3$. Use this to write $f_3(a, b) = f_1(3a + b, b)/3$ and $f_3(c, d) = f_1(3c + d, -d)/3$ (note the minus sign). Then

$$
\begin{aligned}
f_3(a, b) f_3(c, d) &= f_1(3a + b, b) f_1(3c + d, -d)/9 \\
&= f_1((3a + b)(3c + d) + 14bd, -(3a + b)d + b(3c + d))/9 \\
&= f_1(9ac + 3bc + 3ad + 15bd, -3ad + 3bc)/9 \\
&= f_1(3ac + bc + ad + 5bd, -ad + bc)
\end{aligned}
$$

So f_3 also has order 2 in the class group. At this point, we can deduce that the class group looks like the Klein 4-group; if it were cyclic, it would only have 1 nontrivial element of order 2.

For completeness, we can work out $f_4(a, b) f_4(c, d)$ similarly: $f_4(x, y) = f_1(3x - y, \pm y)/3$, so that $f_4(a, b) = f_1(3a - b, b)/3$ and $f_4(c, d) = f_1(3c - d, -d)/3$, and

$$
f_4(a, b) f_4(c, d) = f_1(3ac - bc - ad + 5bd, -ad + bc).
$$

Chapter 7

7.1 The convexity should not be a surprise, if you try to visualise the shape in $\mathbb{R}^n$. However, a formal proof is a little tedious! Let $x, y \in X_t$. We need to see that if $\lambda\mu \geq 0$ and $\lambda + \mu = 1$, then $\lambda x + \mu y \in X_t$. At real embeddings $(i = 1, \ldots, r_1)$, it is certainly true that

$$
|\lambda x_i + \mu y_i| \leq \lambda |x_i| + \mu |y_i|
$$

by the triangle inequality. The same inequality holds for complex embeddings $(i = r_1 + 1, \ldots, r_1 + r_2)$; again, the triangle inequality holds for complex numbers. Now

$$
\sum_{i=1}^{r_1+r_2} |\lambda x_i + \mu y_i| \leq \sum_{i=1}^{r_1+r_2} \lambda |x_i| + \mu |y_i| = \lambda \sum_{i=1}^{r_1+r_2} |x_i| + \mu \sum_{i=1}^{r_1+r_2} |y_i| < \lambda t + \mu t = t,
$$

using the facts that x and y belong to X_t, and that $\lambda + \mu = 1$. Thus $\lambda x + \mu y \in X_t$, and so X_t is convex.

7.2 Make a change of variable in $I_{r,s}(t) = \int_{Y_t} R_1 \cdots R_{r_2} \, dx_1 \ldots dx_{r_1} dR_1 \ldots dR_{r_2}$ so that $x_i = tx_i'$ and $R_i = tR_i'$. Then $(x_1, \ldots, x_{r_1}, R_1, \ldots, R_{r_2}) \in Y_t$ if and only if $(x_1', \ldots, x_{r_1}', R_1', \ldots, R_{r_2}') \in Y_1$. Also, $dx_i = t \, dx_i'$ and $R_i \, dR_i = t^2 R_i' \, dR_i'$. The first equality follows.

Note that

$$I_{r,s}(1) = \int_0^1 I_{r-1,s}(1 - x_r)\, dx_r.$$

From the first relation, $I_{r-1,s}(1 - x_r) = (1 - x_r)^{r-1+2s} I_{r-1,s}(1)$, so

$$I_{r,s}(1) = \int_0^1 (1 - x_r)^{r-1+2s} I_{r-1,s}(1)\, dx_r = I_{r-1,s}(1) \int_0^1 (1 - x_r)^{r-1+2s}\, dx_r.$$

A change of variable $x = 1 - x_r$ now gives

$$I_{r,s}(1) = I_{r-1,s}(1) \int_0^1 x^{r-1+2s}\, dx = \frac{I_{r-1,s}(1)}{r + 2s}.$$

Finally, in the same way,

$$I_{0,s}(1) = \int_0^1 I_{0,s-1}(1 - R_s) R_s\, dR_s = \int_0^1 (1 - R_s)^{2(s-1)} I_{0,s-1}(1) R_s\, dR_s.$$

Again, write $R = 1 - R_s$:

$$I_{0,s}(1) = I_{0,s-1}(1) \int_0^1 R^{2s-2}(1 - R)\, dR = I_{0,s-1}(1) \left[\frac{R^{2s-1}}{2s - 1} - \frac{R^{2s}}{2s} \right]_0^1,$$

and the result follows.

7.3 This is just like Corollary 7.22. The condition on k implies that x is odd (when $k \equiv 2 \pmod 4$, if y is even, then $y^2 + k \equiv 2 \pmod 4$, so cannot be the cube of an even number). Again x and y must be coprime, using the hypothesis that k is squarefree, and the next part of the argument goes through to see that $y + \sqrt{-k} = u\alpha^3$ for some unit $u \in \mathbb{Z}[\sqrt{-k}]$. All the fields under consideration have ± 1 as the only units, and again we conclude that $y + \sqrt{-k} = (a + b\sqrt{-k})^3$ for some integers a and b. Comparing imaginary parts gives $1 = b(3a^2 - kb^2)$, so $b = \pm 1$, and then the condition on k implies that there are no solutions for a.

7.4 All the fields have $n = 2$ and $r_2 = 1$. The Minkowski bound for $\mathbb{Q}(\sqrt{-1})$ is

$$\frac{2!}{2^2} \left(\frac{4}{\pi} \right) |4|^{1/2} = 1.273\ldots.$$

For $\mathbb{Q}(\sqrt{-2})$, $\mathbb{Q}(\sqrt{-3})$ and $\mathbb{Q}(\sqrt{-7})$ the corresponding values are $1.800\ldots$, $1,102\ldots$, $1.684\ldots$, the only differences in the calculations coming from replacing $|4|^{1/2}$ with $|D_K|^{1/2}$. The answer follows as in Example 7.19.

7.5 All the fields have $n = 2$ and $r_2 = 0$. The Minkowski bound is then

$$\frac{2!}{2^2}|D_K|^{1/2};$$

for $\mathbb{Q}(\sqrt{2})$ we get a Minkowski bound of $\sqrt{8}/2 = 1.414\ldots$; for $\mathbb{Q}(\sqrt{3})$ we get $\sqrt{12}/2 = 1.732\ldots$; for $\mathbb{Q}(\sqrt{13})$ the bound is $\sqrt{13}/2 = 1.802\ldots$.

7.6 The Minkowski bound for $\mathbb{Q}(\sqrt{6})$ is $\sqrt{24}/2 = 2.449\ldots$, and the Minkowski bound for $\mathbb{Q}(\sqrt{-6})$ is $3.118\ldots$. For $\mathbb{Q}(\sqrt{6})$, this shows that every ideal is equivalent to an ideal of norm 1 or 2. As any ideal of norm 2 would divide $\langle 2 \rangle$, and $\langle 2 \rangle = \langle \sqrt{6}+2 \rangle^2$, the only ideal of norm 2 is principal. Thus, as every ideal of norm 1 or 2 is principal, $\mathbb{Q}(\sqrt{6})$ has class number 1. For $\mathbb{Q}(\sqrt{-6})$, we need to consider ideals of norm at most 3. As $\langle 2 \rangle = \langle 2, \sqrt{-6} \rangle^2$ and $\langle 3 \rangle = \langle 3, \sqrt{-6} \rangle^2$, we see that not every ideal is principal. We can see that ideals of norm at most 3 all have order dividing 2 in the class group, and as there are only two distinct ideals which might represent different elements in the class group, the class group has at most 3 elements. We conclude that the class group is cyclic with 2 elements, and thus the class number is 2. (Alternatively, we could say that $\langle 2, \sqrt{-6} \rangle \langle 3, \sqrt{-6} \rangle = \langle \sqrt{-6} \rangle$ is principal, so that $\langle 2, \sqrt{-6} \rangle$ and $\langle 3, \sqrt{-6} \rangle$ are inverse in the class group – as both have order 2, they must therefore be the same element, and the class number is 2.)

7.7 In this case, we have $n = 3$, $r_2 = 1$ and $|D_K| = 243$. This gives the Minkowski bound of $4.410\ldots$, so every ideal is equivalent to one of norm at most 4. Write $\alpha = \sqrt[3]{3}$.

We show that every such ideal is principal. Clearly it suffices to treat the prime ideals below this bound, as every ideal is a product of prime ideals. Any prime ideal of norm 2 or 4 will divide $\langle 2 \rangle$. Proposition 5.42 shows that the prime factorisation of $\langle 2 \rangle$ is the product of one prime of norm 2, and one of norm 4. But we also see that $2 = \alpha^3 - 1 = (\alpha - 1)(\alpha^2 + \alpha + 1)$, so $\langle 2 \rangle = \langle \alpha - 1 \rangle \langle \alpha^2 + \alpha + 1 \rangle$, and so these are the desired primes, and both are principal. Similarly, every prime ideal of norm 3 divides $\langle 3 \rangle$, but obviously $3 = \alpha^3$, and so $\langle 3 \rangle = \langle \alpha \rangle^3$, and so there is just one prime ideal of norm 3, and it is also principal. Thus the class number is 1.

7.8 Do some!

7.9 We know that $r_1 + 2r_2 = n$. If n is odd, it follows that r_1 is odd, and so K has a real embedding. So there is a map $K \hookrightarrow \mathbb{R}$; as the only roots of unity in $\mathbb{R}$ are ± 1, these are the only roots of unity lying in K, and so $\mu(K) = \{\pm 1\}$ (of course, every field has ± 1).

7.10 The unit group has infinitely many elements if $r = r_1 + r_2 - 1 > 0$. We can only have $r_1 + r_2 - 1 = 0$ if $r_1 = 1$, $r_2 = 0$, when we have a field of degree $r_1 + 2r_2 = 1$ which must be $\mathbb{Q}$, or if $r_1 = 0$, $r_2 = 1$, when we have a field K

of degree $r_1 + 2r_2 = 2$, so must be quadratic – and then as there are no real embeddings, K must be imaginary quadratic.

Chapter 8

8.1 Use the Euclidean algorithm: $999 = 1 \times 700 + 299$, $700 = 2 \times 299 + 102$, $299 = 2 \times 102 + 95$, $102 = 1 \times 95 + 7$, $95 = 13 \times 7 + 4$, $7 = 1 \times 4 + 3$, $4 = 1 \times 3 + 1$ and $3 = 3 \times 1 + 0$. We can compute the convergents using the table:

		1	2	2	1	13	1	1	3
0	1	1	3	7	10	137	147	284	999
1	0	1	2	5	7	96	103	199	700

and the convergents are $\frac{1}{1}, \frac{3}{2}, \frac{7}{5}, \frac{10}{7}, \frac{137}{96}, \frac{147}{103}$ and $\frac{284}{199}$.

8.2 Easy from the previous exercise.

8.3 $\sqrt{21} = [4, \overline{1, 1, 2, 1, 1, 8}]$ and $\sqrt{71} = [8, \overline{2, 2, 1, 7, 1, 2, 2, 16}]$.

8.4 $\sqrt{31} = [5, \overline{1, 1, 3, 5, 3, 1, 1, 10}]$, so the table of convergents begins

	a_n	5	1	1	3	5	3	1	1	10
0	1	5	6	11	39	206	657	863	1520	16063
1	0	1	1	2	7	37	118	155	273	2885
$p_n^2 - dq_n^2$		-6	5	-3	2	-3	5	-6	1	-6

8.5 $\sqrt{11} = [3, \overline{3, 6}]$, and the first solution to $x^2 - 11y^2 = 1$ is $10^2 - 11 \times 3^2 = 1$. All solutions are given by $x_n + y_n\sqrt{11} = \pm(10 + 3\sqrt{11})^n$. Similarly, $\sqrt{31} = [5, \overline{1, 1, 3, 5, 3, 1, 1, 10}]$ as in Exercise 8.4; from the table of N_n, we find the first solution $1520 + 273\sqrt{31}$; all solutions are given by $x_n + y_n\sqrt{31} = \pm(1520 + 273\sqrt{31})^n$.

8.6 $\sqrt{7} = [2, \overline{1, 1, 1, 4}]$, and it is then easy to find the unit $8 + 3\sqrt{7}$ from the convergents to $\sqrt{7}$. Then every number of the form $\pm(8 + 3\sqrt{7})^n$ also is a unit.

8.7 We have already seen (Exercise 8.6) that the fundamental unit for $\mathbb{Q}(\sqrt{11})$ is $10 + 3\sqrt{11}$. Similarly, $\sqrt{51} = [7, \overline{7, 14}]$, and one can find the fundamental unit as $50 + 7\sqrt{51}$; $\sqrt{58} = [7, \overline{1, 1, 1, 1, 1, 1, 14}]$, and the fundamental unit is $99 + 13\sqrt{58}$ (note that the corresponding N_n here is -1).

8.8 $\sqrt{29} = [5, \overline{2, 1, 1, 2, 10}]$ and we find quickly that $5^2 - 29 \times 1^2 = -4$, so that $\frac{5 + \sqrt{29}}{2}$ is a unit with norm -1. $\sqrt{33} = [5, \overline{1, 2, 1, 10}]$, and the first unit we find is $23 + 4\sqrt{33}$.

8.9 The fundamental unit is $2143295 + 221064\sqrt{94}$.

8.10 Here we compute the discriminants as in Chap. 3. For simplicity, we just do one case, (2) say. Then $\omega_1 = 1$, $\omega_2 = \frac{1 + \sqrt{m}}{2}$, $\omega_3 = \sqrt{n}$ and $\omega_4 = \frac{\sqrt{n} + \sqrt{k}}{2}$, where $m \equiv 1 \pmod 4$. With the usual embeddings, the discriminant is the square of the determinant

$$\begin{vmatrix} 1 & \frac{1+\sqrt{m}}{2} & \sqrt{n} & \frac{\sqrt{n}+\sqrt{k}}{2} \\ 1 & \frac{1-\sqrt{m}}{2} & \sqrt{n} & \frac{\sqrt{n}-\sqrt{k}}{2} \\ 1 & \frac{1+\sqrt{m}}{2} & -\sqrt{n} & \frac{-\sqrt{n}-\sqrt{k}}{2} \\ 1 & \frac{1-\sqrt{m}}{2} & -\sqrt{n} & \frac{-\sqrt{n}+\sqrt{k}}{2} \end{vmatrix}.$$

Subtract row 1 from row 2, and row 3 from row 4 to get

$$\begin{vmatrix} 1 & \frac{1+\sqrt{m}}{2} & \sqrt{n} & \frac{\sqrt{n}+\sqrt{k}}{2} \\ 0 & -\sqrt{m} & 0 & -\sqrt{k} \\ 1 & \frac{1+\sqrt{m}}{2} & -\sqrt{n} & \frac{-\sqrt{n}-\sqrt{k}}{2} \\ 0 & -\sqrt{m} & 0 & \sqrt{k} \end{vmatrix}.$$

Subtract row 1 from row 3, and row 2 from row 4 to get

$$\begin{vmatrix} 1 & \frac{1+\sqrt{m}}{2} & \sqrt{n} & \frac{\sqrt{n}+\sqrt{k}}{2} \\ 0 & -\sqrt{m} & 0 & -\sqrt{k} \\ 0 & 0 & -2\sqrt{n} & -\sqrt{k} \\ 0 & 0 & 0 & 2\sqrt{k} \end{vmatrix}$$

which is upper triangular, so the determinant is the product of the diagonal entries, namely $4\sqrt{kmn}$; thus the discriminant is $16kmn$, as required.

Chapter 9

9.1 If ζ and ζ' are any two primitive nth roots of unity, then $\zeta' = \zeta^r$ for some r, and conversely $\zeta = \zeta'^s$, where $rs \equiv 1 \pmod{n}$. Then $\mathbb{Q}(\zeta') \subseteq \mathbb{Q}(\zeta)$ as ζ' is a power of ζ, and the other inclusion follows as ζ is a power of ζ'.

9.2 We claim that ζ is a primitive nth root of 1 if and only if $-\zeta$ is a primitive $2n$th root of 1.

If ζ is a primitive nth root of 1, then $(-\zeta)^n = (-1)^n\zeta^n = -1$ as n is odd. Then $(-\zeta)^{2n} = 1$. So $-\zeta$ is a $2n$th root of 1. Let m be the order of $-\zeta$, so that $m|2n$. As $(-\zeta)^m = 1$, squaring shows that $\zeta^{2m} = 1$. But ζ has order n, so $n|2m$. Also, n is odd, so we conclude that $n|m$. So $m = n$ or $m = 2n$. But $m \neq n$, as $(-\zeta)^n = -1$. So the order of $-\zeta$ is $2n$.

Conversely, if $-\zeta$ is a primitive $2n$th root of 1, then $(-\zeta^n)^2 = 1$, so $(-\zeta)^n = \pm 1$. But $-\zeta$ is a *primitive* $2n$th root of unity, so $(-\zeta)^n \neq 1$. Thus $(-\zeta)^n = -1$, and thus, as n is odd, $\zeta^n = 1$. In a similar way to the above, ζ must have order exactly n (if its order is m, then squaring shows that $(-\zeta)^{2m} = \zeta^{2m} = 1$, but $-\zeta$ has order $2n$, so $2n|2m$, so $n|m$).

Finally,

$$\lambda_{2n}(x) = \prod_{\text{primitive } 2n\text{th roots}} (x - \xi)$$

$$= \prod_{\text{primitive } n\text{th roots}} (x + \zeta)$$

(by the above remarks)

$$= (-1)^{\phi(n)} \prod_{\text{primitive } n\text{th roots}} (-x - \zeta)$$

$$= (-1)^{\phi(n)} \lambda_n(-x)$$

If n is odd, $\phi(n)$ is even, unless $n = 1$. So, for n odd and greater than 1, $\lambda_{2n}(x) = \lambda_n(-x)$. (Note that for $n = 1$, the statement is false—instead, $\lambda_2(x) = -\lambda_1(x)$.)

9.3   1. The degree of $\lambda_{mn}(X)$ is $\phi(mn)$; the degree of $\lambda_n(X)$ is $\phi(n)$. Recall that $\phi(n) = n \prod_{p|n} \left(1 - \frac{1}{p}\right)$; as the primes dividing mn are exactly the same as those dividing n, it easily follows that $\phi(mn) = m\phi(n)$. As $\lambda_n(X)$ has degree $\phi(n)$, then $\lambda_n(X^m)$ has degree $m\phi(n) = \phi(mn)$ as required.

   2. We claim that if ξ is any mth root of ζ, then ζ is a primitive nth root of unity if and only if ξ is a primitive mnth root of unity. Perhaps the easiest way to do this (although there are others) is to argue in the following way.

ζ is a primitive nth root of unity if and only if $\zeta = e^{\frac{2\pi i k}{n}}$ for some k prime to n. Then the possible values of ξ are $e^{\frac{2\pi i (k+tn)}{mn}}$ for $t = 0, \ldots, m - 1$. But k is coprime to n, and therefore so is $k + tn$; as $m|n$, it follows that $k + tn$ is coprime to mn. Thus ξ is a primitive mnth root of unity.

Conversely, if ξ is a primitive mnth root of unity, then $\xi = e^{\frac{2\pi i k}{mn}}$ for some k coprime to mn. Then $\xi^m = e^{\frac{2\pi i k}{n}}$, which is a primitive nth root of unity as k is coprime to n. Then

$$\lambda_{mn}(x) = \prod_{\text{primitive } mn\text{th roots}} (x - \xi)$$

$$= \prod_{\text{primitive th roots} \zeta} \prod_{\xi^m = \zeta} (x - \xi)$$

(by the above remarks)

$$= \prod_{\text{primitive } n\text{th roots}} (x^m - \zeta)$$

$$= \lambda_n(x^m)$$

   3. Put $m = p$, $n = p$ to deduce that $\lambda_{p^2}(X) = \lambda_p(X^p)$. Next use $m = p^2$, $n = p$ to get $\lambda_{p^3}(X) = \lambda_{p^2}(X^p)$, and we know that this is $\lambda_p(X^{p^2})$. Repeat (or write a proof using induction).

9.4 $X^q - 1 = (X - 1)\lambda_q(X)$, so $X^{pq} - 1 = (X^p - 1)\lambda_q(X^p)$. We also know that $X^{pq} - 1 = \lambda_{pq}(X)\lambda_p(X)\lambda_q(X)\lambda_1(X)$, and that $X^p - 1 = \lambda_p(X)\lambda_1(X)$, so $X^{pq} - 1 = (X^p - 1)\lambda_q(X)\lambda_{pq}(X)$, and the result follows by comparing with the earlier expression for $X^{pq} - 1$.

9.5 1. Simply because $p | \lambda_n(x)$ and $\lambda_n(x) | x^n - 1$.

2. The roots of $X^n - 1$ are all the nth roots of unity; the roots of $X^k - 1$ are all the kth roots of unity. So the roots of the quotient are all the nth roots of unity which are not kth roots of unity. Since these include all primitive nth roots of unity (as $k < n$), we deduce the result.

3. As $x^k \equiv 1 \pmod p$, we see that

$$x^{k(l-1)} + x^{k(l-2)} + \cdots + 1 \equiv l \pmod p$$

as each of the l terms is congruent to 1 (mod p). As $x^n \equiv 1 \pmod p$, and $n | x$, we see that $p \nmid n$. Also, $l | n$, so $p \nmid l$. Therefore $p \nmid \frac{x^n - 1}{x^k - 1}$. However, $\lambda_n(x) | \frac{x^n - 1}{x^k - 1}$, and p was chosen as a factor of $\lambda_n(x)$. This is a contradiction.

4. Let x be as above, and choose p to be a divisor of $\lambda_n(x)$. Then $x^n \equiv 1 \pmod p$. Let k be the order of x modulo p; this shows that $k | n$. If $k < n$, the previous part gives a contradiction, so $k = n$. Thus x has order n in the group of integers modulo p of order $p - 1$, and so $n | p - 1$. This shows that $p \equiv 1 \pmod n$. As $x^n \equiv 1 \pmod p$, we see that $p \nmid x$, so that p is not any of $p_1, \ldots, p_r$.

9.6 This is almost identical to Lemma 9.14; of course, the difference in sign is due to the properties of the quadratic residue symbol.

9.7 Recall that $\mathfrak{p} = \langle 2, \frac{1+\sqrt{-23}}{2} \rangle$. Write $\rho = \frac{1+\sqrt{-23}}{2}$. Then $\mathfrak{p} = \langle 8, 4\rho, 2\rho^2, \rho^3 \rangle$. But $4\rho = 2 + 2\sqrt{-23}$, $2\rho^2 = -11 + \sqrt{-23}$, and $\rho^3 = -\frac{17+5\sqrt{-23}}{2}$. It is easy to check that

$$8 = \left(\frac{3 + \sqrt{-23}}{2} \right) \left(\frac{3 - \sqrt{-23}}{2} \right)$$

$$2 + 2\sqrt{-23} = \left(\frac{-5 + \sqrt{-23}}{2} \right) \left(\frac{3 - \sqrt{-23}}{2} \right)$$

$$-11 + \sqrt{-23} = \left(\frac{-7 - \sqrt{-23}}{2} \right) \left(\frac{3 - \sqrt{-23}}{2} \right)$$

$$-\frac{17 + 5\sqrt{-23}}{2} = (2 - \sqrt{-23}) \left(\frac{3 - \sqrt{-23}}{2} \right)$$

Thus $\mathfrak{p}^3 \subseteq \langle \frac{3-\sqrt{-23}}{2} \rangle$. The norm of $\mathfrak{p}$ is 2, so the norm of $\mathfrak{p}^3$ is 8. Also, the norm of $\frac{3-\sqrt{-23}}{2}$ is just its modulus as a complex number, and this is easily seen to be 8 also. It follows that $\mathfrak{p}^3 = \langle \frac{3-\sqrt{-23}}{2} \rangle$.

9.8 Simply multiply out the brackets. You should get 49 terms, and should find that each power ζ^a occurs an odd number of times in the product. However, $1 + \zeta + \zeta^2 + \cdots + \zeta^{22} = 0$, since ζ is a root of the 23rd cyclotomic polynomial λ_{23}. After substituting this into the expression for the product, every term now occurs an even number of times, and so there is a factor of 2.

9.9 $B_2 = \frac{1}{6}$, $B_4 = -\frac{1}{30}$, $B_6 = \frac{1}{42}$, $B_8 = -\frac{1}{30}$, $B_{10} = \frac{5}{66}$, $B_{12} = -\frac{691}{2730}$, $B_{14} = \frac{7}{6}$, $B_{16} = -\frac{3617}{510}$. All $B_{2k+1} = 0$, except for $B_1 = -\frac{1}{2}$. By the proposition, both 691 and 3617 are irregular primes.

Chapter 10

10.1 Write $\epsilon_1, \ldots, \epsilon_r$ for the fundamental units, and $\tau_1, \ldots, \tau_{r_1+r_2}$ for the embeddings. As ϵ_i is a unit, we have $\text{tr}(\lambda(\epsilon_j)) = 0$ for each i. In particular, $\sum_{i=1}^{r_1+r_2} \lambda_i(\epsilon_j) = 0$. This implies that the absolute value of the determinant of the submatrix got by deleting one column is independent of the column.

10.2 We have 2 real embeddings, and from the previous exercise can choose either in calculating the regulator. Let us make the natural choice of embedding so that $\sqrt{d}$ maps to the positive real square root of d in $\mathbb{R}$. With this embedding, we can choose the fundamental unit $\eta > 1$, and the regulator formula shows that R_K is $|\log \eta|$. We have chosen $\eta > 1$, so that $\log \eta > 0$, and so $R_K = \log \eta$.

For $\mathbb{Q}(\sqrt{2})$, we know that the fundamental unit is $1 + \sqrt{2}$, and the regulator is therefore $\log(1 + \sqrt{2})$.

10.3 If $K = \mathbb{Q}(\sqrt{3})$, the fundamental unit is $2 + \sqrt{3}$, and so the regulator is $\log(2 + \sqrt{3})$. The discriminant is 12, and the corresponding Dirichlet character is $\chi(1) = \chi(11) = 1$, $\chi(5) = \chi(7) = -1$. Then

$$L(s, \chi) = 1 - \frac{1}{5^s} - \frac{1}{7^s} + \frac{1}{11^s} + \frac{1}{13^s} - \frac{1}{17^s} - \frac{1}{19^s} + \frac{1}{23^s} + \cdots ;$$

we compute

$$h_K = \frac{\sqrt{12}}{2 \log(2 + \sqrt{3})} L(1, \chi) \approx 1.315 L(1, \chi),$$

and

$$L(1, \chi) = 1 - \left(\frac{1}{5} + \frac{1}{7} - \frac{1}{11} - \frac{1}{13} \right) - \left(\frac{1}{17} + \frac{1}{19} - \frac{1}{23} - \frac{1}{25} \right) - \cdots < 1,$$

so $h_K = 1$.

10.4 The fundamental unit for $K = \mathbb{Q}(\sqrt{17})$ is $4 + \sqrt{17} = 8.123 \ldots$. Its discriminant is 17, so we need to take the product over $0 < a < 17/2$. We compute $\left(\frac{1}{17} \right) = \left(\frac{2}{17} \right) = \left(\frac{4}{17} \right) = \left(\frac{8}{17} \right) = 1$, and $\left(\frac{3}{17} \right) = \left(\frac{5}{17} \right) = \left(\frac{6}{17} \right) = \left(\frac{7}{17} \right) = -1$. Then we compute

$$\frac{\sin\left(\frac{\pi}{17}\right)\sin\left(\frac{2\pi}{17}\right)\sin\left(\frac{4\pi}{17}\right)\sin\left(\frac{8\pi}{17}\right)}{\sin\left(\frac{3\pi}{17}\right)\sin\left(\frac{5\pi}{17}\right)\sin\left(\frac{6\pi}{17}\right)\sin\left(\frac{7\pi}{17}\right)} = 0.1231\ldots,$$

and we see that this is ϵ^{-1}, so that $h_K = 1$.

For $K = \mathbb{Q}(\sqrt{-23})$, we have $D_K = -23$, and can compute the values $\chi(a) = \left(\frac{a}{23}\right)$. We find that $\chi(2) = 1$, and that $\sum_{a=1}^{11} \chi(a) = 3$, so that $h_K = 3$.

10.5 By multiplicativity, $|1| \cdot |1| = |1|$, so $|1| = 0$ or $|1| = 1$. However, $|x| = 0$ only for $x = 0$, so $|1| = 1$.

Similarly, $|-1|.|-1| = |1| = 1$, so $|-1| = \pm 1$. However, norms can only take non-negative values, so $|-1| = 1$.

10.6 Every nonzero element of K can be written $\gamma = \alpha/\beta$ for α and β nonzero elements of $\mathbb{Z}_K$. Then $\alpha = \beta\gamma$, and so $|\alpha| = |\beta| \cdot |\gamma|$. If $|x| = 1$ for all elements of $\mathbb{Z}_K$, this gives $|\alpha| = |\beta| = 1$, and then $|\gamma| = 1$ also, so that every nonzero element of K has norm 1, and the norm is trivial.

10.7 It is clear that $|1|_p = 1$ and that $|ab|_p = |a|_p|b|_p$. It remains to check the triangle inequality. In fact, we will prove the stronger inequality

$$|a + b|_p \le \max\{|a|_p, |b|_p\},$$

known as the *ultrametric inequality*. Suppose $a = p^s \frac{m}{n}$ and $b = p^t \frac{m'}{n'}$ are two rational numbers, and that $p \nmid mn$ and $p \nmid m'n'$. Then $|a|_p = p^{-s}$ and $|b|_p = p^{-t}$. Clearly $\max\{|a|_p, |b|_p\} = p^{-\min\{s,t\}}$. On the other hand, it is clear that if $a + b = p^u \frac{q}{r}$ where $p \nmid qr$, that $u \ge \min\{s, t\}$ (if $s \ne t$, then this is an equality, but if $s = t$, we might have a strict inequality). Then $|a + b|_p = p^{-u} \le p^{-\min\{s,t\}} = \max\{|a|_p, |b|_p\}$, as required.

10.8 If $x = -\frac{360}{91} = \frac{2^3 \times 3^2 \times 5}{7 \times 13}$, then $|x|_2 = 2^{-3}$, $|x|_3 = 3^{-2}$, $|x|_5 = 5^{-1}$, $|x|_7 = 7$ and $|x|_{13} = 13$. For all other primes p, we have $|x|_p = 1$. Since $|x|_\infty = \frac{360}{91}$, we have

$$|x|_\infty \times \prod_p |x|_p = \frac{360}{91}.2^{-3}3^{-2}5^{-1}7 \times 13 = 1,$$

as required.

Chapter 11

11.1 $\sqrt{12707} = 112.725\ldots$, so we start with 113^2. We note that $113^2 - 12707 = 62$, not a square, but that $114^2 - 12707 = 289 = 17^2$. Then we find the factorisation $12707 = 97 \times 131$.

11.2 $132^2 - 21311 = -3887 = -13^2 \times 23$ and $144^2 - 21311 = -575 = -5^2 \times 23$, so

$$(132 \times 144)^2 \equiv (5 \times 13 \times 23)^2 \pmod{21311},$$

or $19008^2 \equiv 1495^2 \pmod{21311}$. Then $(21311, 19008 + 1495) = 101$ and $(21311, 19008 - 1495) = 211$, giving $21311 = 101 \times 211$.

11.3 There is a homomorphism $\mathbb{Z}[\sqrt{2}] \longrightarrow \mathbb{Z}/527\mathbb{Z}$ given by $a + b\sqrt{2} \mapsto a + 23b \pmod{527}$, and $3 + 2\sqrt{2}$ maps to $49 = 7^2$. Also, $3 + 2\sqrt{2} = (1 + \sqrt{2})^2$, and $1 + \sqrt{2} \mapsto 24$. So $7^2 \equiv 24^2 \pmod{527}$, and we get that $527 = 31 \times 17$.

11.4 There is a homomorphism $\mathbb{Z}[\sqrt{2}] \longrightarrow \mathbb{Z}/1679\mathbb{Z}$ given by $a + b\sqrt{2} \mapsto a + 41b \pmod{1679}$, and $-1 + 2\sqrt{2}$ maps to $81 = 3^4$, while $5 + 4\sqrt{2}$ maps to $169 = 13^2$. Neither $-1 + 2\sqrt{2}$ nor $5 + 4\sqrt{2}$ is a square in $\mathbb{Z}[\sqrt{2}]$, but their product is $11 + 6\sqrt{2} = (3 + \sqrt{2})^2$. We also have $3 + \sqrt{2} \mapsto 44$, and we conclude that

$$3^4 \times 13^2 \equiv 44^2 \pmod{1679},$$

or that $117^2 \equiv 44^2 \pmod{1679}$, leading to $1679 = 73 \times 23$.

11.5 On the modular arithmetic side, the factorbase is $-1, 2, 3, 5, 7, 11, 13, 17, 19$. Harder is to compute the factorbase on the algebraic side. We factor the prime numbers at most B.

$\langle 2 \rangle = \mathfrak{p}_{[2,0]}^2$, where $\mathfrak{p}_{[2,0]} = \langle 2, \sqrt{6} \rangle = \langle \sqrt{6} + 2 \rangle$.

$\langle 3 \rangle = \mathfrak{p}_{[3,0]}^2$, where $\mathfrak{p}_{[3,0]} = \langle 3, \sqrt{6} \rangle = \langle \sqrt{6} + 3 \rangle$.

$\langle 5 \rangle = \mathfrak{p}_{[5,1]}\mathfrak{p}_{[5,4]}$, where $\mathfrak{p}_{[5,1]} = \langle 5, \sqrt{6} - 1 \rangle$, $\mathfrak{p}_{[5,4]} = \langle 5, \sqrt{6} - 4 \rangle$. (Note that $\mathbb{Z}[\sqrt{6}]$ has unique factorisation, so these ideals will be principal.)

$\langle 7 \rangle$, $\langle 11 \rangle$, $\langle 13 \rangle$ and $\langle 17 \rangle$ are all inert, so have norm p^2, and are thus not part of the factorbase.

$\langle 19 \rangle = \mathfrak{p}_{[19,5]}\mathfrak{p}_{[19,14]}$, where $\mathfrak{p}_{[19,5]} = \langle 19, \sqrt{6} - 5 \rangle$, $\mathfrak{p}_{[19,14]} = \langle 19, \sqrt{6} - 14 \rangle$.

Thus the algebraic factorbase consists of $\mathfrak{p}_{[2,0]}$, $\mathfrak{p}_{[3,0]}$, $\mathfrak{p}_{[5,1]}$, $\mathfrak{p}_{[5,4]}$, $\mathfrak{p}_{[19,5]}$, $\mathfrak{p}_{[19,14]}\}$.

11.6 There are two cases: $f(X)$ has 3 real roots, or $f(X)$ has 1 real root and 2 complex roots. In both cases, there is a real embedding, so that the only roots of unity in the field $\mathbb{Q}(\alpha)$ are ± 1. In the first case, every unit is of the form $\pm \epsilon_1^{a_1} \epsilon_2^{a_2}$, where $\{\epsilon_1, \epsilon_2\}$ are fundamental units, and these are square if the sign is $+$, and both a_1 and a_2 are even. It is easy to see that any set of 4 such units contain a subset whose product is a square. Indeed, we can identify units with 3-dimensional vectors modulo 2, where the first co-ordinate is 1 if the sign is $-$ and 0 if it is $+$, and the second and third co-ordinates are the values of a_1 and a_2 modulo 2. Linear algebra tells us that any 4 vectors in a 3-dimensional vector space are linearly dependent, so given 4 units, we can find some subset whose product is a square.

In the second case, $r_1 = r_2 = 1$, and every unit is of the form $\pm \epsilon^a$, where ϵ is a fundamental unit. As above, any set of 3 units will contain a subset whose product is a square. (You should now see how to do the general case.)

11.7 Do some examples!

References

1. Borevich, Z., Shafarevich, I.: Number Theory. Academic Press, New York (1966)
2. Cassels, J.W.S., Fröhlich, A.: Algebraic Number Theory. London Mathematical Society, London (2010)
3. Cohen, H.: A Course in Computational Algebraic Number Theory. Graduate Texts in Mathematics, vol. 138. Springer, Berlin (1993)
4. Crandall, R., Pomerance, C.: Prime Numbers: A Computational Perspective. Springer, New York (2005)
5. Fröhlich, A., Taylor, M.: Algebraic Number Theory. Cambridge University Press, Cambridge (1991)
6. Howie, J.: Fields and Galois Theory. Undergraduate Mathematics Series. Springer, London (2006)
7. Jones, G., Jones, M.: Elementary Number Theory. Undergraduate Mathematics Series. Springer, Berlin (1998)
8. Lenstra, A., Lenstra, H. (eds.): The Development of the Number Field Sieve. Lecture Notes in Mathematics, vol. 1554. Springer, Berlin (1993)
9. Marcus, D.: Number Fields. Springer, New York (1977)
10. Neukirch, J.: Algebraic Number Theory. Springer, Berlin (1999)
11. Niven, I.: The transcendence of π. Am. Math. Monthly **46**, 469–471 (1939)
12. Niven, I., Zuckerman, H., Montgomery, H.: An Introduction to the Theory of Numbers, 5th edn. Wiley, New York (1991)
13. Stewart, I., Tall, D.: Algebraic Number Theory and Fermat's Last Theorem, 3rd edn. Peters, Natick (2002)
14. Taylor, R., Wiles, A.: Ring-theoretic properties of certain Hecke algebras. Ann. Math. **141**, 553–572 (1995)
15. Wagon, S.: The Euclidean algorithm strikes again. Am. Math. Monthly **97**, 125–129 (1990)
16. Washington, L.: Introduction to Cyclotomic Fields. Graduate Texts in Mathematics, vol. 83. Springer, New York (1982)
17. Wiles, A.: Modular elliptic curves and Fermat's last theorem. Ann. Math. **141**, 443–551 (1995)
18. Zagier, D.: A one-sentence proof that every prime $p \equiv 1 \pmod 4$ is a sum of two squares. Am. Math. Monthly **97**, 144 (1990)

F. Jarvis, *Algebraic Number Theory*, Springer Undergraduate
Mathematics Series, DOI: 10.1007/978-3-319-07545-7,
© Springer International Publishing Switzerland 2014

Index

A

Above, 107
Adleman, Leonard
 Adleman columns, 247, 251, 252
Algebraic factorbase, 244, 250
Algebraic integer, 28, 29, 35
Algebraic number, 17, 20, 21, 23, 25, 28
Analytic class number formula, 207, 209,
 215–225
Archimedean norm, 229
Ascending chain condition, 86, 99
Associate, 66, 68, 75, 153, 196
Automorph, 129

B

Baker, Alan, 145
Bernoulli numbers, 205
Biquadratic field, 169, 184–188
Biquadratic fields, 58

C

Cantor, Georg, 18, 20
Catalan's conjecture, 15
Cauchy, Augustin-Louis
 Cauchy sequence, 230
Centrally symmetric region, 152
Characteristic of field, 26
Chinese remainder theorem, 103, 109, 110
Class group, 105–107, 120, 123, 144, 158,
 198, 201, 205, 218, 245
 narrow, 183
Class number, 106, 131–143, 158, 164, 198,
 201, 216
 explicit formulae, 224
Class number one problem, 145

Cohen-Lenstra heuristics, 184
Common factor, 3
Completion, 227, 230
Composite number, 2
Cone, 217
Conjugate, 40, 177
Conjugation, 184
Continued fraction, 170–180
 convergent, 172
 notation, 171
 period, 179
Convex region, 152
Coprime, 4, 6
Coset, 89
Cubic field, 169, 188–189
Cyclotomic field, 1, 191–206
Cyclotomic polynomial, 192, 195

D

Decryption key, 232
Dedekind, Richard, 72, 73, 214
 Dedekind zeta function, 214
 Euler product, 215
Degree of field extension, 22, 25
Diophantus, 14, 202
 Diophantine equation, 14
Dirichlet, Peter Gustav Lejeune, 207, 214,
 215
 Dirichlet L-function, 225
 Dirichlet character, 224
 Dirichlet's unit theorem, 149, 164, 167,
 168, 188, 189, 215, 216, 247
Discrete, 150
Discriminant, 39, 47–51, 53–57, 121, 156,
 193, 195, 196, 201, 216
Divides, 2

F. Jarvis, *Algebraic Number Theory*, Springer Undergraduate
Mathematics Series, DOI: 10.1007/978-3-319-07545-7,
© Springer International Publishing Switzerland 2014

Division algorithm, 4, 6, 21
Divisor, 2

E
Eisenstein, Gotthold
    Eisenstein's criterion, 193
Embedding, 40–44, 153, 157, 169
    complex, 51, 153
    conjugate, 153
    real, 50, 153
Encryption key, 231
Equivalent factorisation, 66, 70
Euclid
    Euclid's algorithm, 3, 4, 20, 22, 26, 74,
        82, 118, 170, 233, 253, 270
    Euclid's theorem, 2
    Euclidean domain, 82, 116, 119, 145
    Euclidean function, 82
Euler, Leonhard, 145, 200, 202, 207
    Euler product, 207, 209, 215, 225
    Euler's formula, 13, 259

F
Factor, 2
Factorbase, 234, 241, 244, 250, 251
Fermat, Pierre de, 1, 9, 202
    Fermat number, 254
    Fermat's last theorem, 1, 191, 202–206
        first case, 202, 204, 205
        second case, 202, 205
Field, 76
    Euclidean, 116
Finitely generated, 31
First Isomorphism Theorem, 89, 92
Fractional ideal, 99
    principal, 99
Free abelian group, 51, 84
Functional equation, 213, 231
Fundamental region, 51, 149
Fundamental theorem of arithmetic, 7, 8
Fundamental unit, 168, 183, 215, 219, 247

G
Gamma function, 208
Gauss, Carl Friedrich, 1, 8, 65, 70, 145
    Gauss sum, 198, 199
    Gauss's Lemma, 30, 250
    Gaussian integers, 9, 11, 15, 17, 28, 67,
        116
General linear group, 122
General number field sieve, 255

Greatest common divisor, 3, 103

H
Heath-Brown, Roger, 13
Hecke, Erich, 215, 231
Heegner, Kurt, 145
Hermite, Charles, 18
Highest common factor, 3, 83, 103

I
Ideal, 1, 73–81, 204, 214
    coprime, 79, 103
    coset of, 89
    discriminant, 156
    finitely generated, 78
    generating set, 76–79
    maximal, 87, 94–95
    norm, 104–105, 157
    prime, 76, 87, 96–98
    principal, 77, 82
Ideal class, 158
Ideal number, 71
Image of ring homomorphism, 88
Imaginary quadratic field, 164, 169
Index calculus, 234, 236, 238, 240, 241
Inert, 108
Inertia degree, 108, 109
Integral basis, 39, 51–53, 57, 195, 197
Integral domain, 75, 97
Irreducible element, 67
Irregular prime, 205

K
Kernel of ring homomorphism, 88
Kronecker, Leopold, 193
Kummer, Ernst Eduard, 1, 71, 202, 204–206,
    214

L
Lagrange, Joseph-Louis
    Lagrange's theorem, 3, 107
Large prime variant, 237, 251
Lattice, 149, 168
    complete, 149, 151
    volume, 151
Legendre, Adrien-Marie
    Legendre symbol, 13, 198, 200, 252
Length, 155
Lindemann, Ferdinand von, 18
Liouville, Joseph, 18, 19

M

Mihăilescu, Preda, 15
Minimal polynomial, 20, 21, 29
Minkowski, Hermann
    Minkowski bound, 162
    Minkowski's theorem, 152, 157, 161
Module, 31, 73
Monic, 20
Monogenic, 57, 111, 112, 197
Multiple, 2

N

Noether, Emmy, 84
    Noetherian ring, 78, 84–86, 99
Nonarchimedean norm, 229
Norm, 44–47, 67, 104–105, 227
    p-adic, 227
    equivalent, 229
    norm of a complex number, 9
    relative ideal norm, 200
    trivial norm, 227
Number field, 17, 25, 28
Number field sieve, 231, 255

O

Order, 35
Ostrowski, Alexander, 228
    Ostrowski's Lemma, 228

P

Pell's equation, 169–180
Pomerance Carl, 233
Power basis, 57
Prime
    inert, 108
    ramified, 108, 111
    splits, 108
    unramified, 111
Prime element, 67
Prime number, 2
Primitive element, theorem of the, 27
Principal ideal domain (PID), 81–84, 270

Q

Quadratic field, 25, 36, 67–71, 79–81, 108,
    113
    imaginary, 113–147
        Euclidean, 116–120
        units, 113–115
    prime, 112

real, 169–184
Quadratic form, 120–147, 201
    binary, 121
    class number, 131–143
    counting, 143–147
    discriminant, 121
    equivalence, 122
    positive definite, 121
    properly equivalence, 122
    reduced, 123, 201
    reduction, 123–130
Quadratic reciprocity, 13, 112, 198, 200, 272
Quadratic sieve, 231, 233, 237, 238, 254
Quotient map, 90
Quotient ring, 89

R

Ramification, 108
Ramification index, 108
Ramified, 108, 111
Real quadratic field, 164, 169, 180
Regular prime, 205
Regulator, 216
Relative ideal norm, 200
Relatively prime, 4, 6
Riemann, Bernhard, 207, 214
    Riemann zeta function, 205, 207–215
        Euler product, 207, 209
        functional equation, 208, 211–214
        meromorphic continuation, 208
Ring, 17
    Euclidean ring, 12
Ring homomorphism, 87
Ring isomorphism, 88
Ring of integers, 28, 35, 37
Root of unity, 191
    primitive, 191
RSA cryptosystem, 231, 232

S

Schur, Issai, 193
Sieve, 236
Simple field extension, 26
Smooth, 234, 239
Smoothness bound, 234, 236, 238, 241, 246,
    250, 253, 254
Special linear group, 122
Special number field sieve, 255
Splits, 108
Splitting field, 189
Splitting of primes, 107
Squarefree, 25

Stark, Harold, 145

T
Tate, John, 215, 231
Taylor, Richard, 1, 202
Trace, 44–47
Transcendental number, 17
Triangle inequality, 227

U
Ultrametric inequality, 229, 285
Unique factorisation, 1, 81, 119, 204
Unique factorisation domain (UFD), 81, 116
Unit, 11, 15, 66, 75, 76, 113, 153, 169, 247
Universal side divisor, 119
Unramified, 111

V
Valuation, 227
Vandermonde determinant, 49
Volume, 155

W
Wagon, Stan, 14
Wiles, Andrew, 1, 202
Wilson's theorem, 11

Z
Zagier, Don, 13
Zero divisor, 75
Zeta function, 205